HOLT

PHYSICAL SCIENCE

HOLT
PHYSICAL
SCIENCE

Mapi M. Cuevas
Professor, Department of Natural Sciences
Santa Fe Community College
Gainesville, Florida

William G. Lamb
Physics Teacher
Oregon Episcopal School
Portland, Oregon

SENIOR EDITORIAL ADVISOR
Curriculum and Multicultural Education

John E. Evans, Jr.
Science Education Specialist
Philadelphia, Pennsylvania

HOLT, RINEHART AND WINSTON

Austin • *New York* • *Orlando* • *Chicago* • *Atlanta*
San Francisco • *Boston* • *Dallas* • *Toronto* • *London*

ACKNOWLEDGMENTS

Content Advisors

Iva Brown, Ph.D.
University of Southern
    Mississippi
Department of Science Education
Hattiesburg, Mississippi

Robert Fronk, Ph.D.
Science Education Department
Florida Institute of Technology
Melbourne, Florida

Theophilus F. Leapheart, Ph.D.
Chemist
Midland, Michigan

Walt Tegge
Chief of Visitor Services
National Historic Oregon Trail
U.S. Department of the Interior
Bureau of Land Management
Baker City, Oregon

Curriculum Advisors

Barbara Durham
Physical Science Teacher
Roswell High School
Roswell, Georgia

Debbie Gozzard
Physical Science Teacher
Lynnhaven Middle School
Virginia Beach, Virginia

Charles Kish
Science Teacher
Saratoga Springs Junior High
    School
Saratoga Springs, New York

Gustavo Loret-de-Mola, Ed.D.
Science Supervisor, Dade County
Fort Lauderdale, Florida

Jim Pulley
Science Teacher
Oak Park High School
North Kansas City, Missouri

Rajee Thyagarajan
Physics Teacher
Health Careers High School
San Antonio, Texas

Reading/Literature Advisors

Philip E. Bishop, Ph.D.
Professor, Department of
    Humanities
Valencia Community College
Orlando, Florida

Patricia S. Bowers, Ph.D.
Associate Director
Center for Mathematics and
    Science Education
University of North Carolina
Chapel Hill, North Carolina

Cover Design: Didona Design Associates

Cover: Fireworks. Photo by Gabe Palmer/The Stock Market

Printed in the United States of America   ISBN 0-03-032517-X

4 5 6 7 041 97 96

For permission to reprint copyrighted material, grateful acknowledgment is made to the following sources:

George G. Blakey: From *The Diamond* by George G. Blakey. Copyright © 1977 by George G. Blakey.

Butler Institute of American Art: Illustration by Albert Bierstadt from the front and back cover of *The Oregon Trail* by Leonard Fisher.

Children's Television Workshop: From "CAT Scans: High-Tech Medicine Can Save Animals, Too" from *3-2-1 Contact*, May 1991. Copyright © 1991 by Children's Television Workshop. All rights reserved.

Cobblestone Publishing, Inc: From "The Trouble with CFCs" from *Odyssey*, vol. 11, no. 1, January 1989. Copyright © 1989 by Cobblestone Publishing, Inc., Peterborough NH 03458.

The Continuum Publishing Company: From "Elective Affinities" from *The Sufferings of Young Werther and Elective Affinities* by Johann Wolfgang von Goeth, edited by Victor Lange. Copyright © 1990 by the Continuum Publishing Company.

The Crown Publishing Group: From "Child Labor in the Mines" from *A Pictorial History of American Mining* by Howard N. Sloane and Lucille L. Sloane. Copyright © 1970 by Howard N. Sloane and Lucille L. Sloane.

Current Science®: "The Flashy Science of Fireworks" from *Current Science®*, vol. 76, no. 18, May 10, 1991. Copyright © 1991 by The Weekly Reader Corporation. "How Safe is Irradiated Food?" from *Current Science®*, vol. 77, no. 16, April 10, 1992. Copyright © 1992 by The Weekly Reader Corporation. From "Sneaker Science: How to Choose your Shoes" from *Current Science®*, vol. 77, no. 15, March 27, 1992. Copyright © 1992 by The Weekly Reader Corporation. "Steroids Build Muscles and Destroy Health" from *Current Science®*, vol. 74, no. 9, January 6, 1989. Copyright © 1989 by The Weekly Reader Corporation. "You Are There: Riding a Roller Coaster-Thrills, Chills, and Fun Physics" from *Current Science®*, vol. 76, no.18, May 10, 1991. Copyright © 1991 by The Weekly Reader Corporation.

J. M. Dent & Sons, Ltd: From "Communications: Messages by Wire" from *A History of Invention* by Egon Larsen, with illustrations by George Land. Copyright © 1961, 1969 by Egon Larsen. Published by Everyman's Library.

Doubleday, a division of Bantam Doubleday Dell Publishing Group, Inc.: From "Dreaming is a Private Thing" from *The Complete Stories*, Volume 1, by Isaac Asimov. Copyright © 1990 by Nightfall, Inc.

Down Beat Magazine: From "Wynton Marsalis: 1987" from *Down Beat Magazine*, vol. 54, no. 11, November 1987. Copyright © 1987 by Down Beat Magazine.

Holiday House: From *The Oregon Trail* by Leonard Everett Fisher. Copyright © 1990 by Leonard Everett Fisher.

Kids Discover: From "Bubbles" from *Kids Discover*, April 1992. Copyright © 1992 by Kids Discover.

Lansing Lamont: From *Day of Trinity* by Lansing Lamont. Copyright © 1965 by Lansing Lamont.

Los Alamos National Laboratory: Illustrations of sectional view of atomic bomb and six photographs of test explosion of 1st atomic bomb. Copyright © 1945 by Los Alamos National Laboratory.

Lothrop, Lee & Shepard Books, a division of William Morrow & Co., Inc.: Excerpt and 1 illustration from *Near The Sea: A Portfolio of Paintings* by Jim Arnosky. Copyright © 1990 by Jim Arnosky. From *To Space and Back* by Sally Ride and Susan Okie. Copyright © 1986 by Sally Ride and Susan Okie.

National Geographic Society: From "Chernobyl-One Year After" by Mike Edwards from *National Geographic*, vol. 171, no. 5, May 1987. Copyright © 1987 by National Geographic Society.

National Geographic WORLD: From "A New World At Your Fingertips" from *National Geographic WORLD*, no. 192, August 1991. Copyright © 1991 by National Geographic Society. Art by Dave Jonason. Copyright © 1991 by National Geographic WORLD.

Newsweek: From "The Longest Night" from *Newsweek*, November 22, 1965. Photograph by Bernard Gotfryd. Copyright © 1965 by Newsweek, Inc. All rights reserved. Reprinted by permission.

Noble, M. Lee: "Walk on a Winter's Day" by M. Lee Noble. Copyright © 1994 by M. Lee Noble.

Random House, Inc: From *The Wright Brothers: Pioneers of American Aviation* by Quentin Reynolds. Copyright © 1950 by Random House, Inc.; renewed copyright © 1978 by James J. Reynolds & Frederick H. Rohlfs, Esq.

Sierra Club Books: From *Ascent: The Mountaineering Experience in Word and Image* by Allen Steck and Steve Roper. Copyright © 1989 by Sierra Club Books. From "The Yosemite" from *The American Wilderness in the Words of John Muir* by the Editors of Country Beautiful. Copyright © 1973 by Country Beautiful Corporation.

CONTENTS

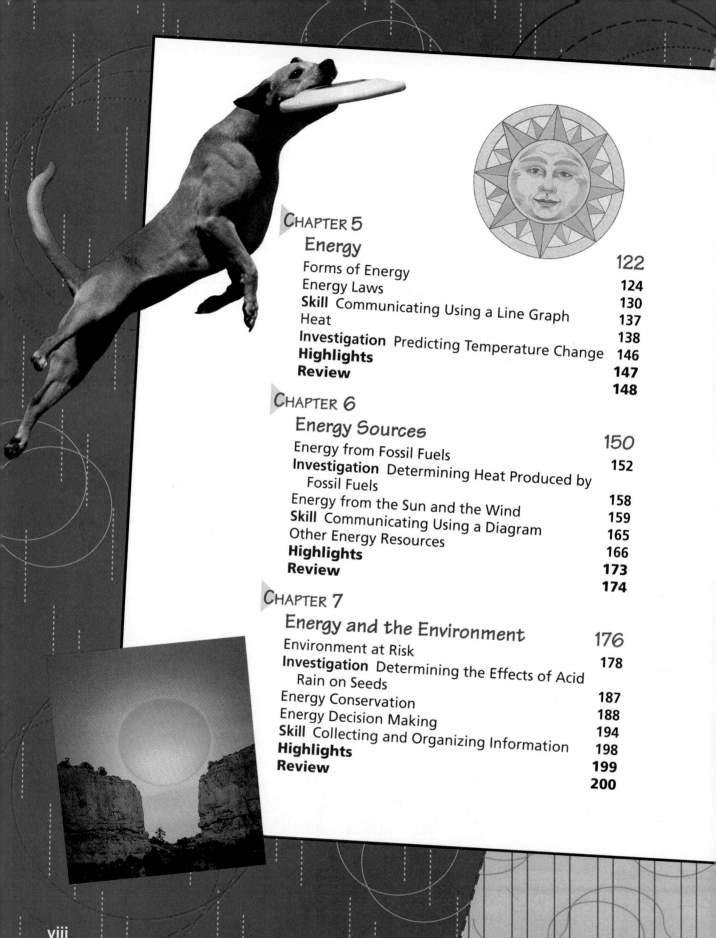

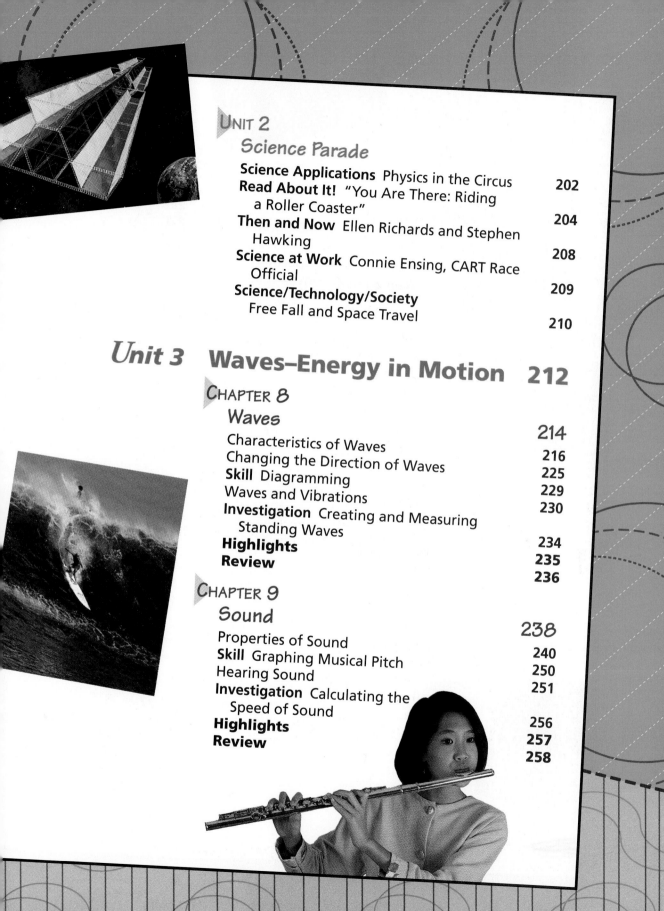

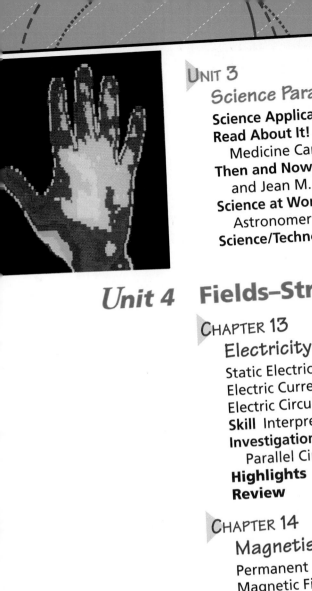

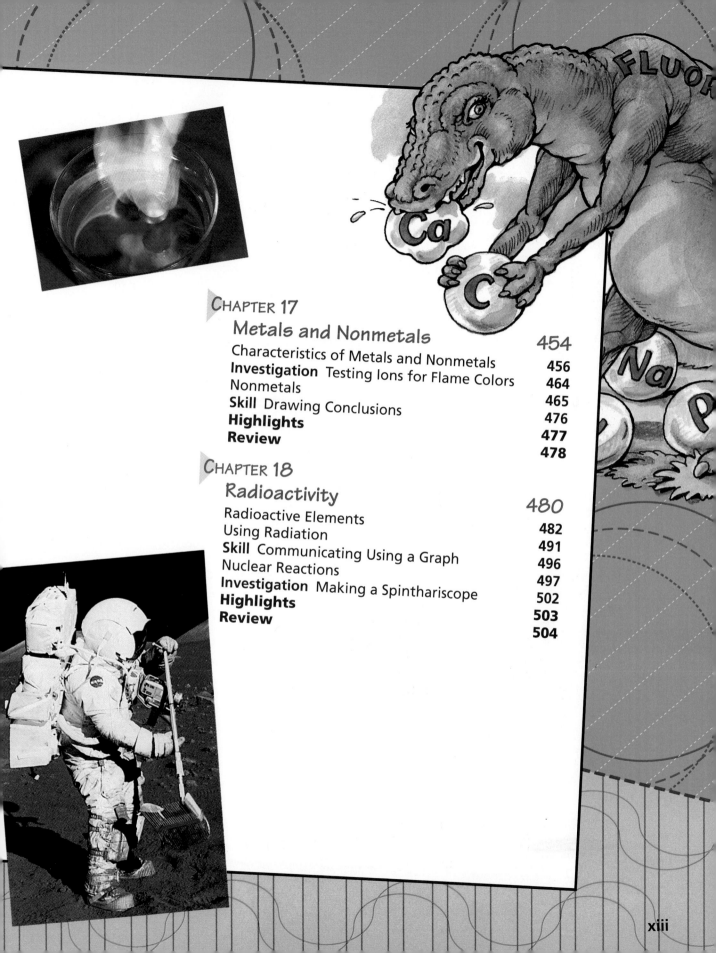

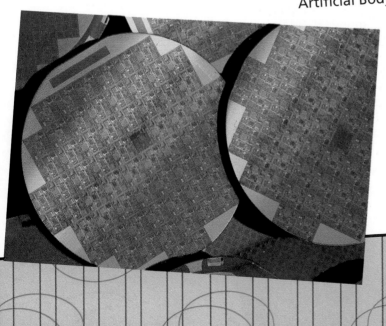

REFERENCE SECTION

SKILL

INVESTIGATION

ACTIVITY

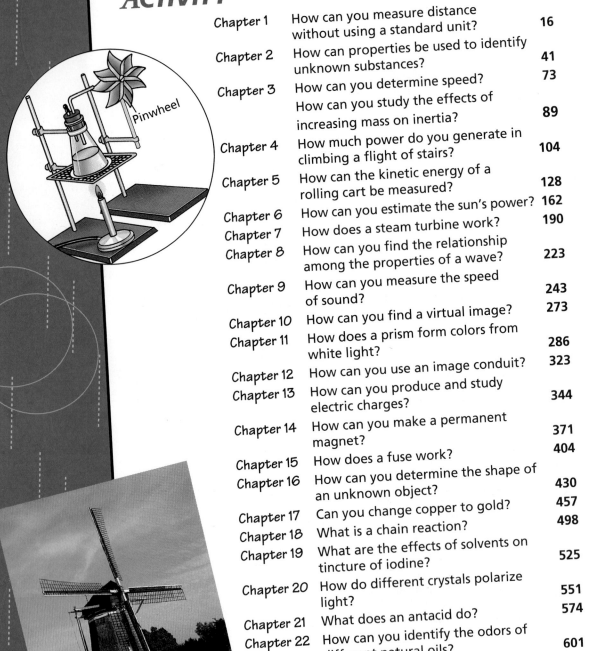

Pinwheel

DISCOVER BY

About Holt Physical Science

Share the Wonder

Why does ice melt? How much work do you do when you ride a bike? How does our use of energy affect the environment? These are just a few examples of questions asked by physical scientists over the years. The answers to these and many of your own questions about matter and energy can be found in HOLT PHYSICAL SCIENCE.

Explore the Nature of Science

Have you ever seen lightning flash in the sky? Have you changed a tire on a car and washed your hands to get them clean? If so, you have experienced the study of physical science. The study of matter and energy and how they react is what physical science is all about. Science is knowledge gained by observing, experimenting, and thinking. When you study physical science, you use your knowledge to understand how and why things happen the way they do. HOLT PHYSICAL SCIENCE will help you add to your knowledge through reading, discussion, and activities.

Work Like a Scientist

Much of science is based on observations of events that occur in your daily life. From these observations, you can form questions and develop experiments to expand on your observations. During your study of physical science, you will be working and thinking like a scientist. You will develop and improve certain skills, such as your observation skills. With new information, you will try to answer questions, to think about what you have observed, and to form conclusions. You will also learn how to communicate information to others, as scientists do.

Keep a Journal

One way to organize your ideas is to keep a journal and make frequent entries in it. Like a scientist, you will have a chance to write down your ideas and then, after doing activities and reading about scientific discoveries, go back and revise your journal entries. Remember, it's OK to change your mind after you have additional information. Scientists do this all the time!

Make Discoveries

Physical science is an ongoing process of discovery—asking and answering questions about the interaction of matter and energy. If you choose a career in the physical sciences, you may find answers to questions of students and scientists who lived before you. Through curiosity and imagination, physical scientists are able to create a never-ending list of questions about the world in which we live.

Prepare yourself for discovery. Through your studies, you will learn many interesting and exciting things. Allow yourself to explore and to discover the ideas, the information, the challenges, and the beauty within the pages of this book. HOLT PHYSICAL SCIENCE can help you understand the world around you. You will begin to ask new questions and to find new answers. You will discover that learning about physical science is important and fun.

1 INTRODUCTION TO PHYSICAL SCIENCE

"**W**ait, did you pack your toothbrush? Will they have toothpaste? What about your science project?" The questions from your father go on and on. Gee, you think to yourself, it's only a science field trip. People have been visiting and living in space for years.

Although the above paragraph sounds like a fantasy, humans may inhabit space in the near future. When they do, they will make and study new materials that have never been made on Earth.

CHAPTERS

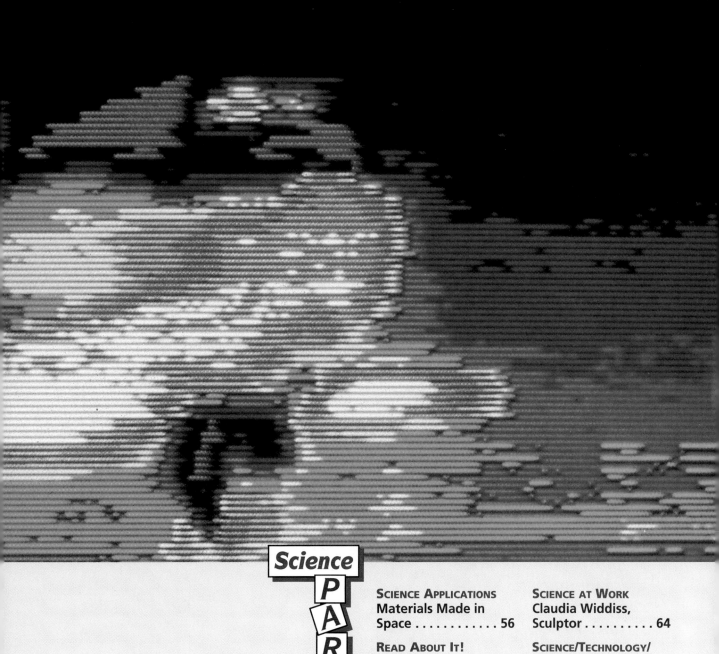

Science PARADE

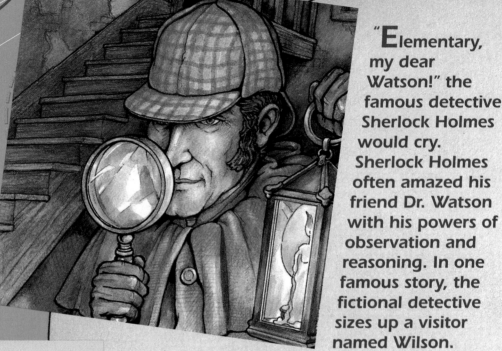

"**E**lementary, my dear Watson!" the famous detective Sherlock Holmes would cry. Sherlock Holmes often amazed his friend Dr. Watson with his powers of observation and reasoning. In one famous story, the fictional detective sizes up a visitor named Wilson.

"Beyond the obvious facts that he has at some time done manual labour, that he has been in China, and that he has done a considerable amount of writing lately, I can deduce nothing else."

. . . "How, in the name of good fortune, did you know all that, Mr. Holmes?" [Mr. Wilson] asked. "How did you know, for example, that I did manual labour?"

"Your hands, my dear sir. Your right hand is quite a size larger than your left. You have worked with it, and the muscles are more developed."

You may not realize it, but you solve problems every day. Usually, you solve these problems using a scientific method. Scientists use these methods to study nature, synthesize new chemicals, and create new technologies.

"But the writing?"

"What else can be indicated by that right cuff so very shiney for five inches, and the left one with the smooth patch near the elbow where you rest it upon the desk."

"Well, but China?"

"The fish which you have tatooed immediately above your right wrist could only have been done in China. . . . When, in addition, I see a Chinese coin hanging from your watch-chain, the matter becomes even more simple."

Mr. Jabez Wilson laughed heavily. "Well, I never!" said he. "I thought at first you had done something clever, but I see that there was nothing in it after all."

from *The Adventures of Sherlock Holmes*
by Sir Arthur Conan Doyle

Of course, there was something in Holmes' deductions, a scientific method. While Sherlock Holmes uses his methods to solve crimes, scientists use observation and reasoning to understand the universe.

For Your Journal

✏ What methods do you use when solving a problem?

✏ How does measurement help in solving problems?

✏ How do scientists solve problems and answer questions?

Asking Questions and Solving Problems

Objectives

Describe *the steps of a scientific method.*

Define *and* **distinguish** *among the terms* hypothesis, law, *and* theory.

Explain *how variables are used in scientific experiments.*

Have you ever had a problem that you knew you had to solve? Of course you have. Scientists are experts in solving problems. In fact, they have developed specific methods to solve problems. To understand how this works, let's set up a common problem that you or a friend might have.

Suppose you need to buy a new pair of athletic shoes. How do you choose the right kind for your feet and for the activities that you do?

Defining the Problem

You have a problem. Your shoes are worn out, and you need a new pair. Knowing you need new shoes is the first step in solving your problem.

When scientists work on a problem or question, they usually use a method that has several steps. These steps usually include defining the problem, stating a hypothesis, performing experiments, drawing conclusions, communicating results, and suggesting an answer. These steps are often called a *scientific method*. However, it is important to remember that the steps in a scientific method are not always used in the same order.

Usually, however, asking questions and solving problems cannot begin until a question or problem has been identified. This is called *defining the problem*. Your problem is to find the best pair of athletic shoes for your needs.

Figure 1–1. Choosing the right athletic shoe can be a problem.

ASK YOURSELF

Why is defining the problem an important part of any scientific method?

Suggesting an Answer

Sometimes scientists experiment in the laboratory or observe nature without any clear idea of what they are looking for. This method is most commonly used when scientists are interested in a subject about which almost nothing is known. More often, however, when scientific questions are being investigated, scientists have an idea of what they expect an experiment to show. Why do scientists have ideas about what to expect? The reason is because they research the problem and gather as much information as possible before beginning to experiment.

Scientists record what they think the answer, or solution, will be before they do an experiment. This educated guess about what might happen is called a **hypothesis**.

 ASK YOURSELF

What is your hypothesis for the question "Which shoes are the best for my feet and for the activities that I do?"

Experiments Tell You What You Need to Know

A hypothesis is usually tested in one of two ways—observation of nature or laboratory experimentation. Observations collected in a natural setting instead of in the laboratory are called *field studies*. Since you can't observe the action of athletic shoes in nature, you need to experiment to find the answer to your problem.

Whenever you conduct an experiment, you have to decide how to deal with variables. A *variable* is any part of an experiment that can be changed. There are often many variables in an experiment and in field studies. The trick is to test only one variable at a time. That way you will know whether that variable is the one that makes the difference.

Figure 1–2. These scientists are experimenting in a laboratory to gather new information.

Some of the variables related to athletic shoes are cushioning, control, and torsion. Of course, you should also consider how they fit! Read the following article to find out more about these variables.

Sneaker Science

How to Choose Your Shoes

from *Current Science*

Dr. Benno Nigg is a sneaker specialist at the University of Calgary in Canada. Here are some of the features he says any good athletic shoe should have.

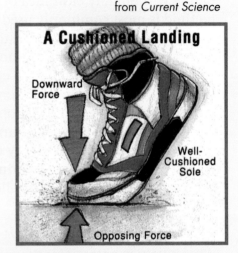

A Cushioned Landing

Downward Force

Well-Cushioned Sole

Opposing Force

- *Cushioning.* Most sports involve a lot of jumping, and it's important to remember that when you drop through the air, you carry a great deal of momentum. And the bigger you are, the greater is your momentum. When you land, the ground takes that momentum away from you by applying an opposing force to your feet.

  If the sole of a shoe is not well cushioned, the opposing force pushing against your feet can injure the delicate cartilage—tissues that lines your joints—the places where your bones meet. "Injured cartilage is the beginning of arthritis, a disease of the joints," says Dr. Nigg.

Out-of-Control Landing

Torque

Bulky Heel

- *Control.* Dr. Nigg says that the sole of any sneaker should be shaped to help you land naturally. If the heel is too bulky, you may land on the outside of your foot and experience a twisting type of force called *torque.* "If I twist your arm, it hurts because I've created torque. This type of force can break a bone," he says.

  "It's important that the shape of a shoe be as close as possible to the shape of the foot. Shoes with big bulky heels and shoes that look like machines can be dangerous because they don't let you land naturally," says Dr. Nigg.

- *Torsion.* Finally, Dr. Nigg says that good running shoes should always allow the front part of the foot to move independently of the rear of the foot. This kind of movement is called *torsion.* "Sometimes when your feet land on the ground, the front part of your foot will rotate. If your shoe has a rigid sole, it will force the rear of your foot to rotate too, and you could sprain your ankle," says Dr. Nigg.

  To test for torsion, Dr. Nigg suggests holding one hand on the front part of the shoe and the other hand on the heel. "If you can twist the shoe, then the sole is not rigid and the shoe will allow for torsion," he says.

  "These features are more important," says Dr. Nigg, "than whether a shoe costs $100 or has the latest gimmick."

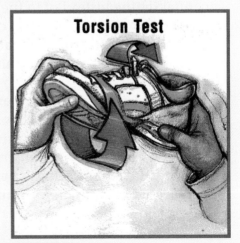

Torsion Test

Scientists must keep all variables the same except for the one being tested. This procedure is called *controlling variables*. This is the only way to make the experiment a fair test. After an experiment is completed, the results are carefully recorded.

Even though you can't observe athletic shoes in nature, there are problems that can be solved this way. Scientists often test a hypothesis by making observations of nature to see how certain variables seem to be related to the problem. Scientists measure all the variables thought to be important in the experiment.

ASK YOURSELF

Describe a situation in which it would be better to observe nature rather than experiment in the laboratory. Explain your choice.

Figure 1–3. These scientists are using collected data to create a map of the moon. They can then analyze the map to further their understanding of the moon's surface.

Performing Experiments

In all kinds of research, measurements of variables must be made and recorded. In science, information such as measurements and observations is called *data*.

What kind of data could you collect about athletic shoes, and how could you collect it? First you would have to set up a controlled experiment. Let's see. Suppose you decided to test five types of shoes that you thought would be good choices. For our purposes we will call them Brands A through E. The variables you want to test are cushioning, control, torsion, and—of course—fit. The first thing you do is make a table in which to record your data as you collect it. It might look something like this one.

	Brand A	**Brand B**	**Brand C**	**Brand D**	**Brand E**
Cushioning					
Control					
Torsion					
Fit					

Now, you need to decide how you are going to collect your data. One way would be to go to an athletic shoe store and try on one brand, testing for all the variables. Another way would be to go to the store and try on each shoe and test for one variable at a time. Each method has advantages and disadvantages.

 **DISCOVER BY** *Problem Solving* _____

List the advantages and disadvantages of different methods of data gathering. Choose the method you would use to test athletic shoes, and explain why you think it would be a fair test. ✐

Once data is collected, it is analyzed. Data can be analyzed in a number of ways. It can be used to calculate new quantities that describe the variable or to make charts and graphs. Usually data from laboratory experiments is much easier to analyze and interpret than data from field studies because the variables in an experiment are easier to control.

Scientists face similar problems when they are experimenting in the laboratory. There are many different methods for collecting data. Sometimes the method chosen will have an effect on the results, and sometimes it doesn't make any difference at all. Regardless, the choice has to be made and recorded. Then if people want to look at the research, they can repeat the experiment.

▼ **ASK YOURSELF**

Why is recording the method of data collection important?

Drawing Conclusions

Once you have collected your data, you have to analyze it and make a decision about which shoes you are going to buy. In scientific terms, you have to draw a conclusion. A **conclusion** is a judgment based on the analysis of data from experiments or field studies.

You need to draw a conclusion about your athletic shoes. The conclusion will prove or disprove your hypothesis. Remember, hypotheses are often incorrect. However, an incorrect hypothesis can lead to new discoveries and new information as readily as a correct hypothesis.

Which brand of shoes is best for you? Your conclusion should be based on the data you have collected. However, choosing shoes isn't a totally scientific matter. Your opinion about style and color will also affect your final choice. But, since you've approached the problem in a scientific manner, you won't be choosing your shoes based only on a brand name!

Figure 1–4. You draw a conclusion when you choose which shoes to buy.

Scientific conclusions may lead to the formulation of theories or laws. Probably the testing of athletic shoes won't lead to any new scientific ideas, but that doesn't mean your research is unimportant. Very little scientific research actually leads to new theories or laws. What is the difference between a scientific law and a scientific theory?

Scientific Laws   In science, a **law** is a summary of many experimental results and observations. A law only describes what happens, not why it happens. Often scientific laws are summarized as mathematical equations. One example is the laws of motion—you will learn more about them in Chapter 3.

Scientific Theories   An explanation of why things work the way they do is called a **theory**, or a model. A theory explains the results of many different kinds of experiments, observations, and occurrences. A theory also predicts the results of experiments that have not yet been done.

Figure 1–5. A detective who claims to have a "theory" to explain a crime, in a scientific sense has only a hypothesis.

Many people confuse the words *hypothesis, law,* and *theory.* Therefore, it may help to review their meanings. A hypothesis is an educated guess based on accurate and relevant information. A hypothesis states what a scientist thinks will happen in a particular experiment or field study. A law is a summary of experimental results and observations that tells what will happen but does not explain why. A theory is an explanation for a group of related results, observations, and occurrences that can be used to predict what will happen in new situations.

The meaning of *theory* in science is often misunderstood because many people use the term to mean an opinion. For example, you might hear a detective on television say something such as "I have a theory to explain Marsha's disappearance." The detective is making an educated guess. A scientist would call this guess a hypothesis.

To a scientist, a theory is much more than a guess. A theory ties together and explains many different kinds of observations. A theory also applies to situations that have not yet happened. If the detective had a scientific theory, he or she would be able to explain why all disappearances occur. The detective would also be able to predict disappearances that had not yet occurred.

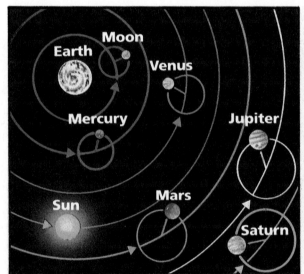

Figure 1–6. Scientific theories may change as new data is found. For example, it was once thought that Earth was the center of the solar system (left). Today we know that the sun is the center (right).

Under certain circumstances, scientific theories may be changed or replaced. Sometimes new observations are made that were not predicted and cannot be explained by a currently accepted scientific theory. Scientists often argue with one another over the meaning of data that does not fit an accepted theory. The scientists who report the new data are sometimes accused of doing poor laboratory work. Other scientists repeat the experiments to verify the new data. If they report the same data, the accepted theory is revised or replaced. Not only must the new theory explain the new data, but it must also explain everything the old theory did.

Scientific theories are never changed without discussion and rechecking of data and experiments. A scientific theory is never replaced unless the theory replacing it is clearly better. You can see that scientists take their theories very seriously. Whenever a scientist uses the word *theory,* he or she means a lot more than just an opinion.

 ASK YOURSELF

What are the differences among a hypothesis, a law, and a theory?

Communicating Results

Remember the data you collected and the conclusion you drew about your new athletic shoes? Once you have chosen your shoes, you need to communicate your choice. First, you will need to tell the salesperson at the store which shoes you prefer. Then your friends and classmates will see your new shoes. If you wish to, you can even tell people why you chose your shoes.

The final step in any problem-solving method is to communicate results and conclusions. For scientists, this communication occurs in several ways. Scientists may write articles that

Figure 1–7. Scientists usually share their findings and discoveries with other scientists.

are published in a scientific journal or magazine. They may write books that report the results of research. Or, scientists may give talks at a meeting of a scientific society. Often, they may do two or even all three of these things.

Scientists also talk to one another about their research. This informal communication network is often as important as articles, books, seminars, and meetings. Why do you think this might be so?

As an example of a scientist investigating a problem, consider American physicist Ephraim Fischbach (EE fruhm FIHSH bahk). In 1985 Fischbach first discovered evidence indicating that in a vacuum (a space from which all matter has been removed), lighter objects actually fall a little faster than heavier objects.

Figure 1–8. Ephraim Fischbach reported evidence that lighter objects fall slightly faster than heavier objects in a vacuum.

Fischbach and his colleagues reported their findings in a scientific journal. When other scientists read about Fischbach's work, there was a great deal of disagreement. Disagreement like this is very common in science. Only after many experiments and much discussion do scientists agree.

Some scientists didn't believe that Fischbach's evidence was strong enough to change the existing theory. These and other scientists conducted experiments to test Fischbach's conclusions. Between 1985 and 1992, literally hundreds of experiments were conducted all over the world. Because most of these experiments did not agree with Fischbach's reports, the theory of gravity was not changed.

You can use scientific methods to solve many problems that occur in your own life. Try the next activity to get a little practice.

DISCOVER BY Writing

Suppose that you are walking home from school and you see a burned log across the sidewalk. Your problem is to determine how the log got there. Write a hypothesis and design an experiment to test your hypothesis. Draw conclusions from your imaginary data. Communicate your results to your classmates. Remember that disagreement is common in the scientific community. ✐

Figure 1–9. Everyday mysteries can be solved using a scientific method.

 ASK YOURSELF

Why is disagreement in the scientific community an important part of a scientific method?

SECTION 1 REVIEW AND APPLICATION

Reading Critically
1. What are the steps in a scientific method?
2. Why must scientists control variables in their experiments?

Thinking Critically
3. Propose an experiment to answer the question "Which brand of battery lasts the longest?"
4. Define a problem from everyday life, and state a hypothesis that proposes an answer to it. Should this hypothesis be tested in the laboratory or in the field? Explain what difficulties you might encounter in testing the hypothesis.

Section 1 Asking Questions and Solving Problems   **13**

SKILL *Measuring Correctly*

▶ MATERIALS
- graduate, 100 mL ● water ● rubber stopper ● balance ● ruler

▼ PROCEDURE

1. Make a table like the one shown.

TABLE 1: MEASUREMENTS			
	Volume Readings	Mass Readings	Length Readings
Eye level			
Below or left			
Above or right			

2. Fill the graduate about half full of water. Notice that the surface of the water is curved. This curve is called the *meniscus*. To read the volume, look at the water with your eye (a) at the level of the meniscus, (b) below the meniscus, and (c) above the meniscus. Record your readings in the table.

3. Place the rubber stopper on the balance. Look at the pointer with your eye (a) directly lined up with the zero point on the balance, (b) either below or to the left of the zero point, and (c) either above or to the right of the zero point.

(The position depends on the type of balance your school has.) Record your readings in the table.

4. Measure and record the width of this textbook with a centimeter ruler. Read the ruler with your eye (a) directly over the mark, (b) to the left of the mark, and (c) to the right of the mark. Record your measurements.

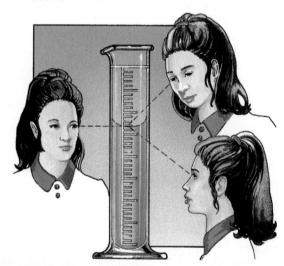

▶ APPLICATION

1. How did the readings vary as you moved your eye? Why do you think they changed?

2. What patterns do you notice in these variations? Explain why you think these patterns occur.

3. In making measurements, the change that you noticed is called the parallax effect. What can you do to avoid this effect?

4. What does this activity tell you about how you should make measurements?

✳ *Using What You Have Learned*
You are making a batch of muffins for your family. Your little sister is helping you. You see that she is measuring the milk while looking down at the measuring cup. How could this affect your muffins?

Measurements

Objectives

Identify *SI units for commonly measured quantities.*

Distinguish *between accuracy and precision.*

Calculate *density.*

Whentyou go to try on athletic shoes, what is the first thing the salesperson asks you? "What's your size?" If you don't know your size, the salesperson will measure your feet. Then, he or she will bring out several pairs of shoes in your size. What would happen if there were no standard sizes? Each shoe would have to be custom-made!

Shoe sizes are not the only things you measure. You measure your height, your weight, your clothes size, the temperature—it goes on and on. Without measurement, your life would be very different.

The Invention of Measurement

No one knows when measurement was first invented, although archaeologists (scientists who study ancient civilizations) can give us a hint about the age of measurement. They have discovered a bone in central Africa that is about 10 000 years old. This bone is important because it has a series of equally spaced marks on it. An analysis of the marks indicates that the bone could have been used as a calendar—to measure time between the phases of the moon. Since early humans were either hunters or farmers, we can hypothesize that they measured time to predict the best hunting or planting seasons.

It is likely that neighboring groups of early humans used the same units of measurement. When everyone agrees to use the same units, they are called **standard units**. It does not really matter what the standard is as long as everyone agrees to use it. Many different types of standard units have been used over the centuries. They have been as varied as the length of a barleycorn and the distance to a star. You can discover more about standard units by doing the next activity.

Figure 1–10. The Ishango bone, shown here, is approximately 10 000 years old. This bone is thought to have been used by early people to measure time.

ACTIVITY

MATERIALS

A hallway with beginning and ending points marked so that everyone can measure the same distance

PROCEDURE

1. Make your own measurement of the chosen distance. Measure the distance in "feet" by placing the heel of one foot against the wall on one side of the hall and then touching the heel of the second foot to the toe of the first foot. Do this along the distance, counting the number of steps you take.

2. Repeat step 1 two times, and average your results. Record your average number of "feet" in the chart your teacher has drawn on the chalkboard.
3. When all students have recorded their measurements on the chalkboard, use the results to answer the Application questions.

APPLICATION

1. What are two reasons that the distance measured by different students produced different numbers of "feet"?
2. Why did you measure three times and calculate the average?
3. How does this activity show the need for standard units?

▼ **ASK YOURSELF**

What is a standard unit?

The International System of Units—SI

For thousands of years, small groups of people used similar measures. However, people in different lands used different systems. As travel became more common and trade more important, people found that the different systems of measurement were a problem. This was as true in science as in any other area. For example, when a scientist in Egypt wanted to verify the experiments done by a scientist in China, all the measures had to be converted. Often, this resulted in delays and errors. Something had to be done to standardize measurement!

The problem was solved when a standard system was adopted by scientists all over the world. Today, the International System of Units, or SI, is the standard measuring system for all scientists. The system is called SI from the French name, *le Système International d'Unités.* Using the same standards of measurement makes it easier for scientists to communicate with one another.

SI has units for measuring all types of quantities. The unit is written after the number to show what the number refers to. All measurements must include both a number and a unit of measurement. For example, if someone wants to know your weight and you say 125, the person does not know whether your weight is 125 pounds, 125 newtons, or 125 medium-sized elephants! When recording a number in science, you must always include the unit; otherwise, your measurements and calculations will have no meaning. Common SI units and symbols appear in Table 1–1.

Table 1-1: **SI Units and Symbols**

Length	
kilometer	km
meter	m
centimeter	cm
millimeter	mm
micrometer	μm

Area	
square kilometer	km^2
hectare	ha
square meter	m^2
square centimeter	cm^2

Mass	
megagram	Mg
kilogram	kg
gram	g
milligram	mg
microgram	μg

Volume of Solids	
cubic meter	m^3
cubic centimeter	cm^3

Volume of Liquids and Gases	
kiloliter	kL
liter	L
milliliter	mL

Acceleration	
meter per second squared	m/s^2

Speed or Velocity	
meter per second	m/s
kilometer per hour	km/h

Frequency	
megahertz	MHz
kilohertz	kHz
hertz	Hz

Density	
kilogram per cubic meter	kg/m^3
gram per liter	g/L
gram per cubic centimeter	g/cm^3

Force	
kilonewton	kN
newton	N

Energy or Work	
joule	J
kilowatt hour	kWh

Power	
kilowatt	kW
watt	W

Temperature	
kelvin	K
degrees Celsius	°C

SI works by combining prefixes and base units. Each base unit can be used with different prefixes to define smaller or larger quantities. Each prefix stands for a number by which the base unit is multiplied. For example, the prefix *kilo-* stands for 1000. Therefore, one kilometer (km) is equal to 1000 meters (m). Likewise, the prefix *milli-* stands for one thousandth, or 0.001. Therefore, one milligram (mg) is 0.001 grams (g). Table 1–2 lists common SI prefixes.

Table 1-2: SI Prefixes

Multiplication Factor		Prefix	Symbol	Pronunciation	Term
1 000 000 000 000 000 000	$= 10^{18}$	exa	E	EHKS uh	one quintillion
1 000 000 000 000 000	$= 10^{15}$	peta	P	PEHT uh	one quadrillion
1 000 000 000 000	$= 10^{12}$	tera	T	TEHR uh	one trillion
1 000 000 000	$= 10^{9}$	giga	G	JIHG uh	one billion
1 000 000	$= 10^{6}$	mega	M	MEHG uh	one million
1 000	$= 10^{3}$	kilo	k	KIHL uh	one thousand
100	$= 10^{2}$	hecto	h	HEHK toh	one hundred
10	$= 10^{1}$	deka	da	DEHK uh	ten
0.1	$= 10^{-1}$	deci	d	DEHS uh	one tenth
0.01	$= 10^{-2}$	centi	c	CEHN tuh	one hundredth
0.001	$= 10^{-3}$	milli	m	MIHL uh	one thousandth
0.000 001	$= 10^{-6}$	micro	μ	MY kroh	one millionth
0.000 000 001	$= 10^{-9}$	nano	n	NAN oh	one billionth
0.000 000 000 001	$= 10^{-12}$	pico	p	PEEK oh	one trillionth
0.000 000 000 000 001	$= 10^{-15}$	femto	f	FEHM toh	one quadrillionth
0.000 000 000 000 000 001	$= 10^{-18}$	atto	a	AT oh	one quintillionth

DISCOVER BY *Writing*

Use Table 1–2 of SI prefixes to help you think of as many prefix puns as you can in five minutes. Here are two examples:

● 1 000 000 000 lings is a form of laughing (*giga*ling or giggling)

● 0.01 rella is a fairy tale heroine (*centi*rella or Cinderella)

Write your ideas in your journal. Share your favorites with your classmates. ✐

ASK YOURSELF

What are the advantages of all scientists' using SI?

Measuring Quantities in SI

Any quantity you wish to measure can be measured in SI units. Distance, volume, mass, weight, and temperature are some of the quantities you will be measuring using SI units.

Length　Have you ever bought a new pair of jeans that were too long? What did you do about it? You probably measured (or had someone else measure) the excess length, cut a little of the material off, and rehemmed them. You probably used a measuring tape to make sure they were the same length all the way around.

　　The most common method for measuring length is to use a meter stick, a ruler, or a measuring tape. You simply put one end of the measuring device at the beginning of what is being measured and read where the end of what is being measured crosses the measuring device.

Figure 1–11. This frame represents a volume of one cubic meter (1 m³).

　　Length and distance in SI are measured in meters. One meter is about the same as the length of a baseball bat or the height of a doorknob from the floor. Units are created by attaching SI prefixes to the word *meter*. A kilometer, or 1000 m, is about the same distance as 10 soccer fields or 10 football fields. A centimeter, or 0.01 m, is about the same distance as the thickness of your little finger.

Figure 1–12. In a graduate, the measurement should be read from the middle of the meniscus.

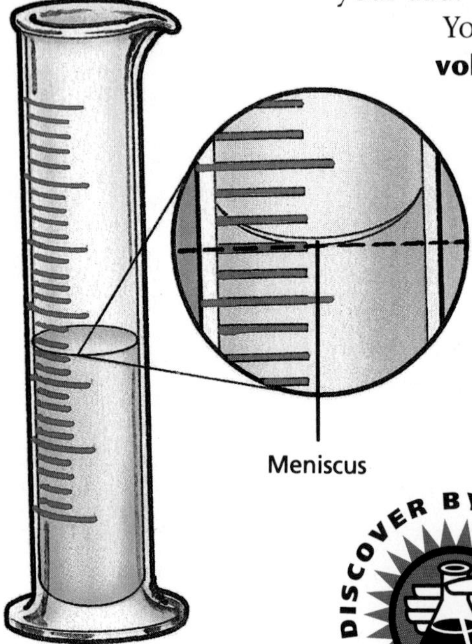

Meniscus

Volume

Quick! Your dad is making breakfast on Saturday morning and is calling to you to measure a cup of milk for the pancake batter. What do you do? You get a measuring cup, pour milk into the cup up to the 1-cup line, and give the milk to your dad.

You measure the volume of liquids all the time. The **volume** of an object is how much space it takes up. A common instrument for measuring liquid volume is the graduate, or graduated cylinder. Like a measuring cup, a graduate has marks on the side that indicate volume. When you read liquid volume, you should follow the same steps you did in the Skill activity on page 14. You can also measure the volume of solids in a graduate. Try it for yourself.

DISCOVER BY Doing

Put some water into a graduate, and record the volume. Place a small rock or marble into the graduate, and record the new volume. Subtract the first volume from the second volume. This is the volume of the solid.

The method you just used in the activity is called *displacement* because the solid displaces, or moves, some of the water. The volume of the water that the solid displaced, equals the volume of the solid.

In SI the standard unit of volume is the cubic meter (m^3), which is very large. For many everyday and scientific activities, the cubic centimeter (cm^3) is more useful. A cubic centimeter is about the same volume as about 20 drops of a liquid. A soft-drink can holds about 300 cm^3. However, many types of soft drinks are sold in liter (L) bottles. A liter equals 1000 cm^3. For small volumes, you may use a unit called the milliliter (mL), or 0.001 L. A milliliter and a cubic centimeter are identical measures of volume.

Mass

Suppose you are on the wrestling team. You have an important match today, and you must make sure you have "made weight." That means that you need to check and see if you are within the limits for your weight class. You step on the scale and find that you are right on target! Neither your weight nor your mass has changed.

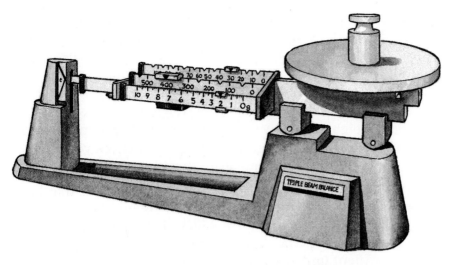

Figure 1–13. A balance, such as this triple-beam balance, is used to measure mass.

Another quantity you measure is mass. **Mass** is the measure of the amount of matter in an object. To measure mass in the laboratory, you use an instrument called a *balance*. Instructions for using a balance may be found on page 630 in the Reference Section.

Wait! Weren't you trying to "make weight" for your wrestling match? Nobody said you needed to "make mass." People often confuse mass with weight, but the quantities are not the same. The mass of an object is the amount of matter in it; mass is a constant value. That means it stays the same no matter where you are. *Weight,* however, is a measure of the force of gravity pulling on the mass of an object. The weight of an object changes depending on the amount of gravitational force acting on it. This is why objects weigh less on the moon than they do on Earth. Because the moon has less gravity, there is less force pulling on the object, making the object lighter. Even though the weight changes, the amount of matter in the object stays the same. Whether on Earth or on the moon, the mass of an object remains constant.

Figure 1–14. All of these objects have approximately the same mass even though they are different sizes.

Well, since you obviously won't be wrestling on the moon, you could consider your mass and your weight the same. For everyday purposes, we consider mass and weight to be the same, although this is only really true at sea level. As you move away from the center of the earth, the pull of gravity decreases. The farther away you are, the less you weigh. How does this fact affect your weight from day to day? If you live in the mountains and you have to wrestle at sea level, you better check your weight carefully. It could go up, even though your mass doesn't change.

The standard SI unit for mass is the kilogram (kg). A 1-L bottle full of water has a mass just a little larger than 1 kg. For many everyday and scientific activities, however, the gram is a more convenient unit to use for mass. A medium-sized paper clip has a mass of about 1 g.

Temperature
A quantity that you must often measure in the science laboratory is temperature. One of the instruments most commonly used to measure temperature is the thermometer.

The kelvin (K) is the SI unit for measuring temperature. Water freezes at 273 K and boils at 373 K. Average body temperature is 310 K.

Kelvin thermometers are very large and cumbersome and are not practical for most common or laboratory uses. Therefore, degrees Celsius (°C) is often used for temperature measurement. To change kelvins to degrees Celsius, subtract 273 from the temperature in kelvins. Water freezes at 0°C and boils at 100°C. Average body temperature is 37°C.

Figure 1–15. A thermometer is used to measure temperature. What is the temperature shown on this thermometer?

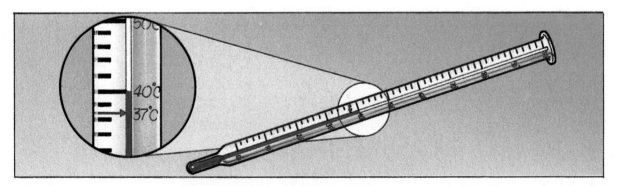

ASK YOURSELF

What unit would you use to measure the height of your desk? What instrument would you use to measure it?

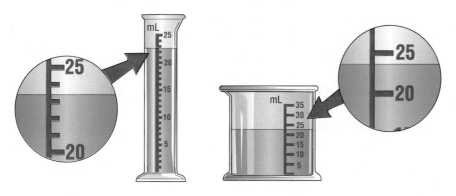

Figure 1–16. These two graduates contain the same volume of liquid. The graduate on the left provides a more precise measurement.

Accuracy and Precision in Measurement

When you measure anything, there are two things you must consider. The first of these is *accuracy*. That is, whether the measured value is the same as the real value. For example, suppose the known mass of an object is 1 kg. If you measure the object's mass and get 1 kg, your measurement is accurate. If you measure its mass and get 2 kg, your measurement is inaccurate.

A second important factor in measuring is *precision,* or the exactness of a measurement. Precision is limited by the smallest division on the scale of the measuring tool. You always estimate to one decimal point beyond the measure marked on the instrument. For example, suppose you used the fat graduate shown in Figure 1–16 to measure volume. Each line on the graduate represents five milliliters (mL). So your measurement could only be precise to 0.5 mL (0.0005 L). The thinner graduate measures to the nearest mL, and it would be precise to 0.1 mL (0.0001 L). The thinner graduate, therefore, gives the more precise measurement.

So, accuracy tells you how good a job you did measuring the liquid. Precision tells you how exact the measuring tool is at showing you the measurement. Accuracy has more to do with the person who is measuring. Precision has more to do with the tool itself.

Both volumes in Figure 1–16 have the same accuracy because they give about the same number. However, the measurement from the thinner graduate is more precise.

Figure 1–17. This person's height is being measured with two different measuring tapes. Which tape gives the more precise reading? Why?

 ASK YOURSELF
What is the difference between accuracy and precision?

Derived Quantities: Area and Density

Sometimes you need to know a quantity that cannot be measured directly. These quantities must be calculated from two or more measurements. A quantity that is calculated from two or more measurements is called a *derived quantity.*

One derived quantity you may already know about is area. *Area* is the measure of how much surface an object has. To find the area of a square or rectangle, for instance, you would measure the length of two adjacent sides and multiply them. Because area is calculated from two directly measured quantities—the length and the width—area is a derived quantity.

Another property that is derived from measurements is density. **Density** is the ratio of the mass of an object to its volume. Density is often used to help determine what an object is made of. The formula for calculating density is written as follows:

$$\text{density} = \frac{\text{mass}}{\text{volume}} \qquad \text{OR} \qquad d = \frac{m}{V}$$

Suppose your teacher has given you two objects and has asked you to determine which is more dense without using any equipment. What do you do? Test your ideas in the next activity.

Figure 1–18. By using a balance, you can compare the densities of substances. Since the smaller volume has a greater mass, it must have a higher density.

DISCOVER BY Problem Solving

Your teacher will give you two objects. Determine which is more dense without using any special equipment. Write your method and your results in your journal. ✐

Now you know that you can estimate differences in density. You can also use more precise densities to help identify certain substances.

DISCOVER BY Calculating

Imagine that you are an explorer who has discovered a block of something that looks like gold. You measure the volume as 1000 cm³. You use a balance and find that the block has a mass of 19 320 g. Use the formula to find the density of the block. Write your calculations in your journal. ✐

You look up the density of pure gold in a reference book and find that gold has a density of 19.32 g/cm³. Is your block gold? While chances are good that it is, you can't base your answer on density alone. In order to be sure, you would have to do some other tests. You will learn about some of these tests in other chapters.

 ASK YOURSELF

How is a derived quantity different from a measured quantity?

SECTION 2   REVIEW AND APPLICATION

Reading Critically

1. Why are standard units of measurement important?

2. Name an instrument used for measuring each of the following quantities: length, mass, liquid volume, and temperature.

Thinking Critically

3. Aluminum has a density of 2.70 g/cm³. What is the mass of a piece of aluminum that has a volume of 100.0 cm³?

4. The width of a table, measured with a ruler that is marked in centimeters, is 65.4 cm (0.654 m). The width of the same table, measured with a ruler marked in millimeters is 652.7 mm (0.6527 m). Which measurement is more precise? Why? Which measurement is more accurate? Why?

INVESTIGATION

Finding Mass, Volume, and Density

▶ MATERIALS
- pennies (10) ● balance ● graduate, 100-mL

▼ PROCEDURE

PART A

1. Make a table like the one shown.

TABLE 1: MASS OF PENNIES	
Penny Number	**Mass of Penny in Grams**
1	
2	

2. Put the pennies on a piece of paper. Write a number on the paper under each penny, and keep them in this order throughout Part A of this investigation.

3. Use a balance to find the mass of each penny in grams as precisely as possible. Round off the values to the nearest 0.1 g. Record the mass of each penny in your table.

4. Your teacher will have a table on the chalkboard that is marked in 0.1 g units. Make a mark in the correct column of the table for each penny your group measures.

5. Compare the masses of the pennies for each group. Are there any noticeable patterns based on mass?

PART B

6. Make two piles of pennies based on the class data. Put each pile on a sheet of paper, and mark the piles "Heavier" and "Lighter." If you cannot decide into which pile a particular penny goes, leave it out.

7. Measure and record the mass of each pile. Write the mass on the paper you are using to identify the pile.

8. Fill a graduate half full of water, and determine the volume as precisely as possible. Record the volume of the water.

9. Carefully put one pile of pennies into the graduate. Measure and record the volume again.

10. Determine the volume of the pennies by displacement. Record the volume.

11. Calculate and record the density of the pile by dividing the volume of the pennies into the total mass of the pennies.

12. Record the density of the pennies on the piece of paper that identifies the pile.

13. Repeat steps 8 through 12 for the other pile of pennies.

14. Record the density of each pile of pennies in the table on the chalkboard.

▶ ANALYSES AND CONCLUSIONS

1. When the pennies were massed, two groups were identified. What was the average mass of each group?

2. Do the measured densities reported by all laboratory groups show two groups of pennies? Explain.

3. Are there any clues that would allow you to separate the pennies into the same two groups without massing them? Explain.

4. How is it possible for the pennies to have different masses?

▶ APPLICATION

You are given 50 coins, some of which are gold. The others are brass. Since the coins look alike, how would you separate them?

 Discover More

Once you have identified the difference between the heavier and lighter pennies, find out when and why the change occurred.

The Big Idea

Science is using methods that involve observing, thinking, and formulating and reformulating ideas about the things around us. Science is using special instruments, solving problems, and trying out new methods of operations. Science is the constant search for information. Scientists must respect evidence that supports new ideas and think carefully about those ideas. They must design fair tests to see whether their ideas are correct.

Measurement is an important part of science. The use of standard units allows scientists to communicate their data to one another easily. Units of measure in SI enable scientists to describe things that are microscopically small and astronomically large.

Look back at the solution to your problem that you entered into your journal at the beginning of the chapter. Reevaluate how your solution was like a scientific method. How could you have made your problem solving more scientific?

Connecting Ideas

This flowchart shows how scientists might use a scientific method. Remember that measurement is an important part of collecting data. Copy the chart into your journal. Use it to show how you would solve a problem in your everyday life. As information is collected, the problem may have to be redefined, new experiments designed, or further observations made. Scientific methods are not one-way streets.

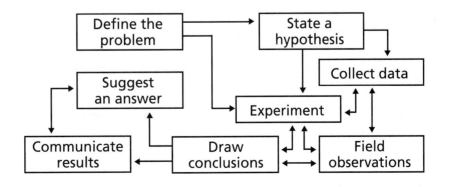

REVIEW

Understanding Vocabulary

Match each term with the correct definition.

1. explanation for a group of related results, observations, and occurrences
2. ratio of the mass of an object to its volume
3. educated guess about what might happen
4. summary of experimental results that can be used to predict what will happen and why they happen the way they do
5. amount of matter in an object
6. judgment based on the analysis of data

a. conclusion (8)
b. density (24)
c. hypothesis (5)
d. law (9)
e. mass (21)
f. theory (9)

Understanding Concepts

MULTIPLE CHOICE

7. In an experiment, controlled variables are
   a) allowed to change.
   b) limited in number.
   c) kept the same.
   d) difficult to change.

8. A person says, "I think that a steel ball bearing and a marble will hit the ground at the same time if I drop them from the same height at the same time." That person is stating a
   a) hypothesis.   b) law.
   c) theory.   d) model.

9. A rock has a volume of 5.43 cm^3 and a mass of 17.86 g. Its density is
   a) 0.30 cm^3/g.   b) 0.30 g/cm^3
   c) 3.29 g/cm^3.   d) 3.29 cm^3/g.

10. Which would be the most precise unit of measurement to use to measure the length of a car?
    a) cm^3 b) mm
    c) cm   d) km

SHORT ANSWER

11. Why are degrees Celsius commonly used in the laboratory rather than kelvin temperatures?

12. What is the difference between an object's mass and its weight?

13. Explain the differences among a hypothesis, a conclusion, a theory, and a law.

Interpreting Graphics

14. The same volume of water was added to each graduated cylinder. What is the volume of the marble in the second cylinder?

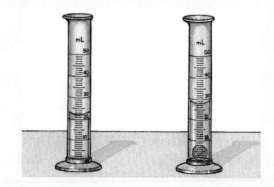

Reviewing Themes

15. *Systems and Structures*
How can two cubes of equal volume have different masses?

16. *Systems and Structures*
How can a measuring device be made more precise?

Thinking Critically

17. Two blocks of wood have the same mass, but the volume of one of the blocks is three times greater than the other. Compare the densities of the two blocks.

18. Jack wishes to determine the effect of darkness on the leaf color of plants. Design an experiment for Jack that controls the necessary variables. Which variable is being tested?

19. How might your life be different if there were no standard system of measurement?

20. How could you increase the density of a snowball without adding more snow to it?

21. Why is it important for scientists to communicate the results of their research to other scientists?

22. Look at the three timepieces shown below. Which is the most accurate? Which is the most precise? Explain your answers.

23. Which of the timepieces is the most practical? Explain why you think so. Include accuracy and precision in your answer.

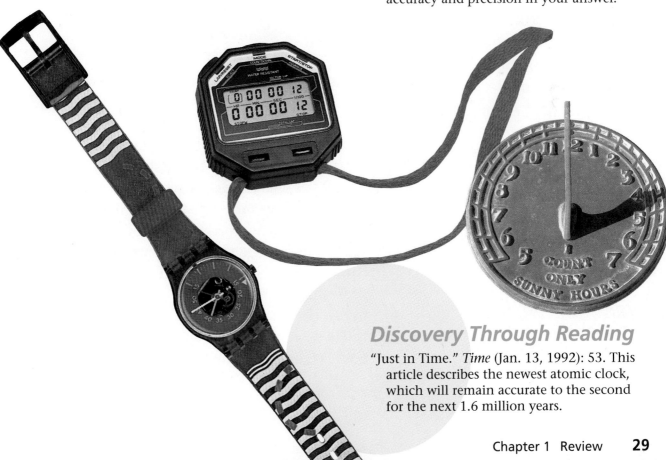

Discovery Through Reading

"Just in Time." *Time* (Jan. 13, 1992): 53. This article describes the newest atomic clock, which will remain accurate to the second for the next 1.6 million years.

CHAPTER 2

FUNDAMENTAL PROPERTIES OF MATTER

Water has unique properties that make it both useful and dangerous. Not even scientists study water's properties as carefully as a climber picking a path through the debris of last week's avalanche.

"We started traversing the glacier sometime after midnight, before the sun could soften the snow. But somehow we have underestimated the distance to the col, or the weight of the packs, or the sheer effort of moving at 18,000 feet. All this slows us down, and now the sun moves across the sky toward noon faster than we are heading toward the col. I prefer traveling across the glacier at night: the headlamp illuminates just those next ten feet of space in front of you. Daylight illuminates too much: the

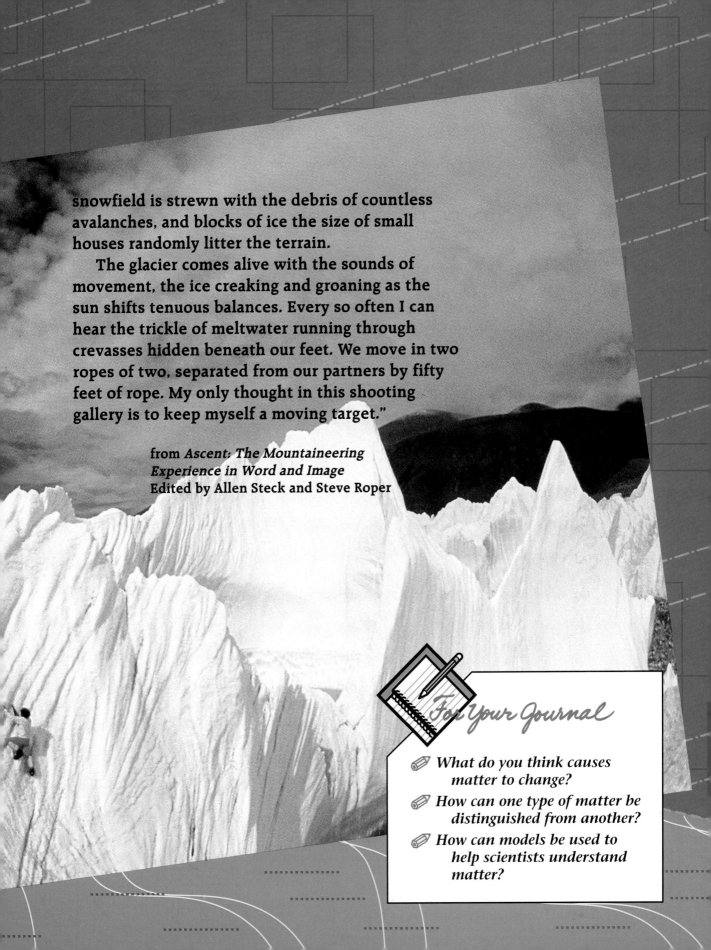

snowfield is strewn with the debris of countless avalanches, and blocks of ice the size of small houses randomly litter the terrain.

The glacier comes alive with the sounds of movement, the ice creaking and groaning as the sun shifts tenuous balances. Every so often I can hear the trickle of meltwater running through crevasses hidden beneath our feet. We move in two ropes of two, separated from our partners by fifty feet of rope. My only thought in this shooting gallery is to keep myself a moving target."

from *Ascent: The Mountaineering Experience in Word and Image* Edited by Allen Steck and Steve Roper

For Your Journal

- What do you think causes matter to change?
- How can one type of matter be distinguished from another?
- How can models be used to help scientists understand matter?

Matter

Describe *the ways matter can be classified.*

Distinguish *between elements and compounds.*

Propose *a method for separating the components of a mixture.*

Melting icicles drip onto the frozen ground. Grass and leaves glisten with dew on a cool morning. In a junk-yard, automobiles silently rust. Overhead, the night sky is illuminated by exploding fireworks. These are just a few examples of matter changing.

You see many more changes in matter every day. The study of the ways in which matter changes is an important part of science. Learning to predict how matter will change has helped scientists produce useful products, such as nylon, plastics, and medicines. Changes in matter will help produce new products in the future.

Figure 2–1. This wulfenite crystal is one example of a solid.

States of Matter

Matter can be classified by its state, purity, and properties. The glacier and the climbers you just read about are all made of matter. Even the air you breathe is made of matter. Matter is anything that has mass and takes up space. It is divided into four states, or phases: *solid, liquid, gas,* and *plasma.*

A solid is anything that has both a definite volume and a definite shape. A rock is a solid, as is a baseball bat and a crystal of salt. In most cases, solid matter can be picked up and carried around without having to place it in a special container.

A liquid is anything that has a definite volume but no definite shape. A key property of liquids is that they flow and can be poured. Liquids take the shape of any container into which they are poured. Look at the photograph, and you will see that if you pour a liquid into a short, wide container, such as a beaker, it will form a short, wide shape. If you pour the same amount of liquid into a tall, narrow container, it will have a tall, narrow shape.

A gas has no definite volume and no definite shape. A gas always takes both the volume and the shape of any container into which it is placed. Try the next activity to find this out for yourself.

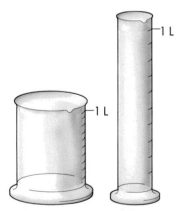

Figure 2–2. The liter of liquid shown here has a different shape depending on the graduate in which it is placed.

Your teacher is going to place an open container of perfume in the classroom. In your journal, write down when you first smell the perfume. Does the odor of perfume grow stronger or weaker compared with when you first began to smell it? What does this tell you about gases? ✎

On the earth, a sample of gas would spread throughout the atmosphere. In fact, the atmosphere itself is a mixture of gases that is held near the earth by gravity.

Throughout your day you come into contact with solids, liquids, and gases. This probably isn't much of a surprise to you. However, did you ever stop to consider how often you eat or drink different states of matter?

DISCOVER BY *Doing* _____

Make a list of all the foods and beverages served in the school cafeteria at lunch today. Then, make a table classifying these foods as solids, liquids, gases, or combinations of these states of matter. Explain how you determined the category for each food. Compare your results with a classmate's. How do your results differ? How are they the same? ✎

Figure 2–3. Plasmas are composed of electrically charged particles. A fire is an example of a plasma. Also, a fusion reaction, such as that of the sun, takes place in a plasma.

The fourth state of matter, plasma, has no definite volume or shape and is composed of electrically charged particles. These charged particles are generally associated with large amounts of energy. A flame is one example of a plasma. The sun is also a plasma.

 **ASK YOURSELF**

How is a plasma like a gas?

Pure Substances

The second way matter is classified is by its purity. If matter is a **pure substance**, it contains only one kind of molecule. Molecules are small particles that make up matter. They are made up of even smaller particles, called *atoms*.

It's Elementary Aluminum, copper, and charcoal are each made of only one kind of atom. A pure substance that is made of only one kind of atom is called an **element.** Copper wire is made up of only copper atoms, and charcoal is made up of only carbon atoms.

If you looked for pure elements in nature, you would be lucky to find more than five or six. Why is this? It is not because there are few elements to look for. There are more than 100 elements, and 92 of them can be found in nature. So, you see, there are plenty of elements to find.

Figure 2–4. Iodine (left) and sulfur (right) are examples of elements. The drawings show their chemical makeup and structure. Each element is made up of only one type of atom.

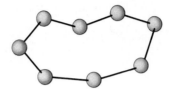

Elements—Compounded Daily

The reason it is hard to find pure elements in nature is that they usually react to form compounds. **Compounds** are pure substances whose molecules are made of two or more different kinds of elements joined together. You can see examples of elements reacting to form compounds by trying the next activity.

DISCOVER BY *Doing*

You will need three pieces of steel wool, three cups, an egg, vinegar, and water. Label the cups "Egg," "Vinegar," and "Water." Put a piece of steel wool in each cup. Then add the egg, vinegar, or water. Wait until the next day, and remove the steel wool from the cups. Rinse off the steel wool. Then record what you observe. 🖉

In the activity, you found that elements react when exposed to some chemicals. Molecules of elements and compounds are often represented by colored balls. Each color represents a specific kind of atom. When different kinds of atoms (colored balls) are combined, they make compounds.

Water is a common example of a compound. Water is a transparent, colorless liquid at room temperature. However, water is made of the elements hydrogen and oxygen, both of which are gases at room temperature. As you can see, compounds do not necessarily have the same characteristics as the elements from which they are made.

Oxygen

Hydrogen ◯

Carbon ●

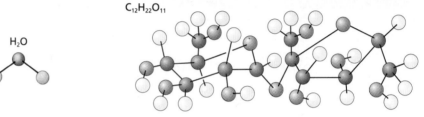

$C_{12}H_{22}O_{11}$

H_2O

Figure 2–5. Water (left) and sucrose (right) are examples of compounds. Water is a compound made of hydrogen and oxygen atoms. Sucrose is a compound made of hydrogen, oxygen, and carbon atoms. In this photograph, coloring has been added.

One important characteristic of compounds is that they cannot be separated into the atoms they are made of by physical methods. The atoms in a compound are joined chemically. Water, for example, can be frozen or boiled, but the ice or water vapor will still be water. Freezing or boiling cannot separate water into hydrogen and oxygen atoms or molecules. Other physical methods such as straining, filtering, passing the water through a magnetic field, or dissolving it in alcohol cannot separate water into hydrogen and oxygen atoms or molecules.

▶ **ASK YOURSELF**

What are the two categories of pure substances?

Mixtures

Most matter is not pure but is made of a combination of two or more types of substances. These combinations of matter are called mixtures. A **mixture** is a combination of two or more different kinds of molecules. The key characteristic of a mixture is that each kind of molecule in the mixture keeps its own identity. This means that if you take a mixture apart, the molecules of the substances will be the same as they were before you put them together.

Many mixtures occur in nature. For example, air is a mixture of gases made up mostly of nitrogen and oxygen with small amounts of other gases. The human body is a complex group of mixtures. Your body is mostly water mixed with proteins, fats, minerals, and carbohydrates. A pizza can also be thought of as a mixture. This mixture includes dough, sauce, cheese, and toppings.

In some mixtures, however, it is difficult to see the different substances, or parts. For example, it is difficult to see the sugar in a mixture of sugar and salt because both solids are white crystals.

One way to identify a mixture is to separate it into its different substances. Unlike compounds, mixtures can be separated into their parts using physical methods. For example, you could use a fork to separate the parts of a pizza. The method does not change the characteristics of the parts. Not all mixtures are as easy to work with as the pizza. Some physical methods of separating parts of a mixture are shown in the table below. These are just a few ideas, and some methods work better than others, depending on the type of mixture.

Table 2-1: Some Physical Methods of Separating Parts of a Mixture
Boil away (evaporate) water leaving solids
Collect metallic parts with magnet
Pick out solid parts by hand or with tweezers
Separate solids from liquids by filtering or straining
Dissolve in water or other liquid

Try the next activity. You'll have a chance to discover more about mixtures, substances, compounds, and elements.

DISCOVER BY Researching

Look at the labels on items that can be purchased at a grocery store or a drugstore. Use the information to find two examples of mixtures and two examples of pure substances. Use resources in your school library to find out whether the pure substances are compounds or elements. Record your results. ✐

Figure 2–6. This reverse osmosis plant uses high pressure and filters to separate salt from sea water.

▼ **ASK YOURSELF**

List three examples of mixtures. How do you know they are mixtures?

SECTION 1 *REVIEW AND APPLICATION*

Reading Critically

1. What are two properties common to all matter?

2. Name the four states of matter, and describe how they differ from one another.

Thinking Critically

3. Pictured here are models of two molecules of different substances. Which one is an element? Which one is a compound? Explain how you can tell the difference.

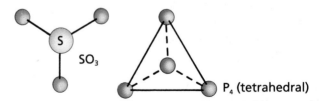

SO_3

P_4 (tetrahedral)

4. The ability to separate mixtures is important for modern society. Explain this statement, and give at least two examples.

SKILL Hypothesizing

▶ **MATERIALS**
materials will vary according to hypotheses tested

A *hypothesis* is a statement that describes what you think has happened or will happen. A hypothesis usually begins with the word *if* or *when* and often has the word *then* in the middle.

Here is an example of the use of a hypothesis. Sarah accidentally mixed salt and pepper as she was filling the shakers. She thought she could separate the mixture by filtering the salt out of the pepper. This was her hypothesis: "If I use a strainer to filter the mixture, then I can separate salt from pepper." To test her hypothesis, Sarah got a strainer and slowly poured the mixture into it.

▼ **PROCEDURE**

1. Here is a situation for which you can write a hypothesis. Suppose you have a mixture of iron filings, salt, and sand. How can you separate them? Write a hypothesis that explains your method. Be sure to include which substance you will separate and what method you will use. (The table on page 37 may be helpful.)
2. Now test your hypothesis. Make a list of the equipment you will need, and give it to your teacher. After you have set up your equipment, try your method. Keep careful records of what you observe.

3. Is your hypothesis correct? One thing to remember is that it is okay if your hypothesis is incorrect. In fact, most scientists make many more incorrect hypotheses than they do correct ones. And they can learn as much from incorrect hypotheses as they can from correct ones. Why do you think this is?

▶ **APPLICATION**
What did you learn from this activity about separating mixtures?

✳ **Using What You Have Learned**
Suppose you were going to teach someone else to write a hypothesis. What would your instructions be?

Characteristics of Matter

Suppose you're watching a game between the Chicago Bulls and the Portland Trailblazers. How can you tell the two teams apart? How can you tell in what city they are playing? You can tell by their uniforms, of course! Substances can be identified in a similar way.

Properties of Matter

A trait that identifies a substance is called a *property*. The properties of a substance stay the same, even if its shape or volume changes. Color, boiling point, odor, density, and chemical composition are all examples of properties.

You can use properties to identify many things. For example, when the Chicago Bulls and the Portland Trailblazers are playing basketball, you can use the property of the color of the uniforms to distinguish between the two teams and to determine which team is playing at home. Most teams wear white at home; so if the Bulls are in red uniforms, that means they are playing in Portland.

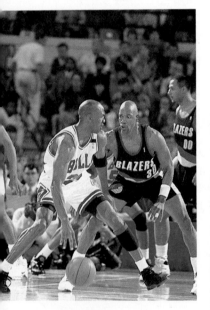

Figure 2–7. Each player's uniform helps you identify the players during a game, just as properties of matter allow you to identify different substances.

The color of the uniform is a property that can change. However, the name of the team also appears on the uniform. The name of the team is a property that you can use to identify the team, regardless of the color of the uniform. A property that always stays the same is called a **characteristic property** because it is characteristic of a particular kind of matter. How can you use properties to identify unknown substances? Try the next activity to find out.

ACTIVITY

How can properties be used to identify unknown substances?

MATERIALS
bottles with tops (4), unknown liquids (4), paper cups (4)

PROCEDURE
1. Make a chart like the one shown.

TABLE 1: IDENTIFYING LIQUIDS			
Liquid	Prediction One	Prediction Two	Final Prediction
A			
B			
C			
D			

2. Without removing the tops of the bottles, examine the liquids and predict what they are. Record the predictions in your chart.

3. CAUTION: Never put your nose directly over a container. The safe way to smell anything in the laboratory is to open the container and wave the fumes toward you. Take the top off each bottle, and smell the contents. Record what you now think each liquid is.

4. Each liquid is safe to taste. Taste each liquid by pouring a small amount of liquid into a paper cup. Put the liquid

on your tongue, and swish it around in your mouth. Spit the liquid back into the cup, and rinse out your mouth with water. Record your final predictions in your chart.

5. Repeat step 4 for each liquid, using a new cup each time.

6. Your teacher will identify each liquid for you. Compare this information with your predictions.

APPLICATION
Were your predictions accurate? Explain why or why not.

Figure 2–8. In addition to his uniform, this player can be identified by his individual playing style. This trait can be considered a characteristic property.

Every substance has two kinds of properties—*physical properties* and *chemical properties*. A chemical property is a trait that relates to a change in the composition of a substance.

Basketball players also have their own characteristic properties. Each name is a characteristic property of that player. Players may also be identified by their special style of play. Either of these is a characteristic property of that player. These properties do not change, regardless of the uniform the player is wearing.

A **physical property** of matter is one that can be observed or measured without changing the composition of the substance. For example, a physical property of water is that it freezes at 0°C. Freezing water does not change its chemical composition—it is still water. Physical properties are often used to identify substances. There are hundreds of physical properties; some of the most common ones are listed in Table 2–2.

Table 2-2 **Some Physical Properties of Matter**

Property	Definition
Boiling point	the temperature at which a substance char.ges from a liquid to a gas
Condensation point	the temperature at which a substance changes from a gas to a liquid; same temperature as boiling point
Density	the mass of a specific volume of a substance
Freezing point	the temperature at which a substance changes from a liquid to a solid; same temperature as melting point
Melting point	the temperature at which a substance changes from a solid to a liquid
Resistance	the opposition a substance has to the flow of electrical current
Solubility	the degree to which a substance will dissolve in a given amount of another substance, such as water

In many cases, physical properties depend not only on the substance itself but also on the environment of the substance. For example, the temperature at which a substance boils depends upon the atmospheric pressure. Because the atmospheric pressure is lower at high altitudes, water boils at a lower temperature on mountaintops than it does at sea level. You'll learn more about physical properties in Section 3.

 ASK YOURSELF

Think of a favorite athlete or musician. What characteristic properties can you use to describe that person?

It's a Matter of Change

When you stretch a rubber band or crush a cube of ice, you are causing physical changes. Spreading margarine on toast, bending the tab on a pop-top can, and squeezing a tube of toothpaste are other examples of physical changes. A **physical change** does not produce a new kind of molecule. In each of the previous examples, the molecules remained the same. Only certain characteristics of the matter, such as shape and volume, were changed. Try the next activity to learn more about physical changes.

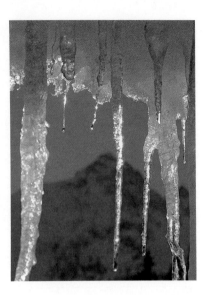

Figure 2–9. When a substance such as water evaporates, freezes, and condenses, it undergoes phase changes. Even though the phase of the water changes, its composition remains the same.

DISCOVER BY *Doing*

Measure a small amount of water, and place it in a sealable plastic bag. Put the bag in a freezer, and leave it there until the water freezes. Then remove the ice from the bag, and place it in a beaker. Cover the beaker, and let the ice melt. Then remeasure the liquid. How does your second measurement compare with your first? Explain your findings. ✐

Freezing, Melting, Boiling—Make Up Your Mind!

Freezing is a common physical change. Melting and boiling are also physical changes. These changes all involve a change in phase, so they are called *phase changes.*

When a liquid freezes, it changes from a liquid to a solid. The temperature at which this occurs is called the *freezing point.* Water, for example, turns to ice at its freezing point, 0°C. An increase in heat causes a change in phase. For example, if ice is heated, it changes back into a liquid. The temperature at which a substance changes from a solid to a liquid is called its *melting point.* For water, the melting point is 0°C. If the water is then boiled, it changes from a liquid phase to a gaseous phase. The temperature at which this happens is called its *boiling point.* Water's boiling point is 100°C.

More Changes of Phase
Additional phase changes include condensation, evaporation, and sublimation. Condensation is the phase change from a gas to a liquid. This change is the opposite of boiling. You have probably seen condensation

44   Chapter 2 Fundamental Properties of Matter

many times in the form of water droplets on the outside of a glass holding a cold drink. The cold surface of the glass causes water vapor in the air to condense into water drops on the glass.

Evaporation is the opposite of condensation. In evaporation, a liquid changes to a gas and moves into the surrounding atmosphere. A wet sponge gradually dries out, and a fogged window clears by evaporation.

Sublimation is a phase change in which a solid changes directly into a gas. Dry ice is an example of a substance that sublimes. When it does so, the solid dry ice changes to a smoke-like gas without ever melting into a liquid. For this reason, dry ice is often used to simulate fog in television and theatrical productions.

Figure 2–10. Dry ice is an example of a substance that sublimes at room temperature.

Phase changes are related to temperature. Hydrogen, for example, freezes at −259°C and boils at −252°C. It is the only substance that freezes and boils at these particular temperatures. If you have a colorless liquid that boils at 100°C and freezes at 0°C at sea level, you can be reasonably certain the substance is water. The temperature at which a substance changes phase is a characteristic property.

These phase changes are examples of physical changes because whether the water is solid, liquid, or gas, it is still water. Its physical properties, such as volume, density, hardness, visibility, have changed, but it is chemically the same—water.

 ASK YOURSELF

Describe the physical changes that occur when you make a cup of hot chocolate.

SECTION 2   *REVIEW AND APPLICATION*

Reading Critically

1. Define the term *property*, and give three examples of physical properties.
2. Explain why the temperature at which a substance changes phase is considered a characteristic property.

Thinking Critically

3. Water boils at a lower temperature on mountaintops than it does at sea level. How would this property of water affect your cooking time if you were trying to make hard-boiled eggs on the top of Mt. Everest?
4. After heavy snowfalls, the snow is often removed from roads and put in piles. As time passes, these piles become smaller, even when the temperature stays below freezing. What do you think causes the snow piles to become smaller?

A Model of Matter

Objectives

Explain *three features of a scientific model.*

Apply *the kinetic molecular model of matter to explain thermal expansion.*

Analyze *phase changes of matter using the kinetic molecular model.*

Figure 2–11. Models are also useful to scientists.

Y ou have probably seen or used many kinds of models. Perhaps you have built model ships or planes, or you have seen a model train. Models are an enjoyable hobby for people of all ages.

In science a model is used to represent an unfamiliar idea or a thing that is either invisible or too large or too small to see. One type of model is a three-dimensional model, such as a model airplane or boat. A second type of model is a diagram or a description. The purpose of a model is to provide a description of an unfamiliar idea or thing using as many familiar ideas as possible. This makes it easier for people to understand.

Models in Science

Figure 2–12. Models are often used to represent ideas or phenomena that are either invisible or too large or small to see. This wave machine is used as a model to represent the movement of sound waves.

One example of a scientific model is the wave model of sound. This model compares sound waves to the waves in water. As you know, sound is invisible, which makes it difficult to study. The wave model allows scientists to learn more about sound by studying the visible waves produced in water.

> *A scientific model has three features.*
>
> *1. It explains and ties together different phenomena.*
> *2. It is simple to understand and use.*
> *3. It can be used to predict natural occurrences and the results of experiments that have not yet been conducted.*

The history of science shows that models, like theories, evolve. Even good models are subject to change. Progress in science constantly produces observations that were not predicted by current models and theories. As this happens, the current models and theories must be changed to reflect new information.

 ASK YOURSELF

How can models help scientists understand new ideas?

Moving Particles as a Model of Matter

Many models have been proposed to describe how matter behaves under various circumstances. The currently accepted model is called the *kinetic molecular model of matter.* This model is based on the idea that all matter is made up of particles and that these particles are in constant, disordered motion.

The idea that matter might be made up of particles is not new. It was first mentioned in the writings of Leucippus (loo SIHP puhs), a Greek philosopher born about 490 B.C. The idea was more fully developed by one of his students, Democritus. The idea that matter is made of particles continued with the work of the British scientist John Dalton (1766–1844).

Another British scientist, Robert Brown (1773–1858), extended the idea of matter and molecules when he used a microscope to observe the movement of particles suspended in drops of water. He proposed that these particles were being hit by even smaller moving particles that we now call molecules. The zigzag motion of small particles suspended in a liquid or gas was named *Brownian motion* after Brown. You can see Brownian motion by observing the dust specks in a beam of light.

Scientists found that many other phenomena could be explained by assuming that matter was made up of tiny particles in random motion. These observations led to the development of the kinetic molecular model of matter.

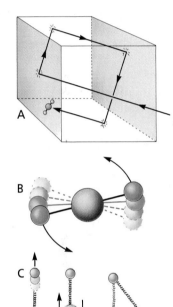

Figure 2–13. Diagram A shows how a gas molecule moves in a closed container. Diagram B shows how the molecule rotates as it moves. Diagram C shows how the bonds holding the atoms of the molecule bend and stretch.

DISCOVER BY *Researching* _____

Go to your school library or to the public library, and read about the ideas of Democritus that were very similar to modern scientific ideas. Find out why his ideas were not accepted during his lifetime. Record your findings in your journal. ✎

▼ ASK YOURSELF

On what two characteristics of molecules is the kinetic molecular model of matter based?

Figure 2–14. The molecules of a gas are not attached to one another. When placed in a closed container, the gas molecules move until they fill the volume of the container.

Molecules and Physical Properties

Several physical properties can be explained by the kinetic molecular model. For instance, density is the mass of a substance in a given volume. Density tells us how close molecules are to each other. If the molecules of a substance are close together, density is high because there are more molecules in a certain amount of space (volume).

The phases of matter can also be explained by the motion and attachments of molecules. In the gas phase of a substance, molecules are not attached to each other in any way. Each molecule is free and moves about very quickly. The speed of a typical gas molecule is several hundred kilometers per hour. The individual molecules keep moving, often colliding with one another, until they bump into the container walls. This is why they fill any container in which they are put. The distances between the molecules are very large compared with the sizes of the molecules.

The liquid phase is much more dense than the gas phase and the distances between molecules are not large compared to the size of the molecules. The molecules bump into one another much more frequently and the attractive forces between them play a much larger role.

Figure 2–15. Although there are forces attracting liquid molecules to one another, the molecules can slip past each other, as if playing musical chairs.

Liquids take the shape of any container into which they are poured. This is because the molecules are free to move past each other. The molecules are close enough together that attractive forces between them keep liquids from changing volume as they are poured from container to container.

In the solid phase, the molecules are also close together. But they have a very limited motion with respect to one another because they are bonded together in a three-dimensional pattern. Therefore, the solid keeps its shape as well as its volume. In solids, the most common molecular motions are the bending and stretching of bonds.

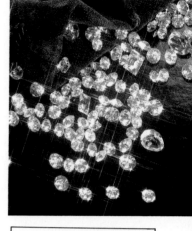

 **DISCOVER BY Writing**

In your journal, diagram the molecular arrangements of a solid, a liquid, and a gas. Label the parts of the diagram, and note the relative motions of the molecules.

▶ **ASK YOURSELF**

How does the movement of molecules in a solid compare with the movement of molecules in a gas?

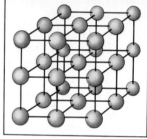

Figure 2–16. All of the molecules of solids are arranged in a regular crystalline pattern and have limited motion. The molecules in these gems, for example, do not move as much as the molecules of liquids and gases.

Molecules and Physical Changes

One commonly observed physical change occurs when the volume of a substance increases as the temperature increases. This physical change is called **thermal expansion.** Most solids, liquids, and gases exhibit thermal expansion.

Hot and Cold   The kinetic molecular model can help you see what causes thermal expansion. Increasing the temperature of a substance increases the motion of its molecules. In solids, in which stretching and bending are the most common forms of motion, faster stretching and bending pushes the molecules farther apart. This occurs even though the molecules remain firmly attached to one another.

DISCOVER BY Problem Solving

Fill a balloon with air. Then measure its circumference. Put the balloon in a freezer for 15 minutes. Remeasure the balloon. Are the two measurements different? If so, how do you explain this?

Figure 2–17. Understanding thermal expansion of solids enables scientists and engineers to plan solutions to the problems this phenomenon might cause.

The expansion joint in the bridge shown here provides space so the bridge material can expand on hot days without buckling. The joint also allows the bridge to contract on cold days. The space between the joints is very important. In order for the joints to prevent the walkway from buckling, they must be far enough apart to allow for expansion. Knowing how much the pavement will expand enables engineers to construct joints with the proper separation.

Reaching a Boiling Point   Have you ever filled a pot with soup and put it on the stove to heat? If you filled it too full, you may have seen it spill over. You might say it "boiled over." Careful observation shows, however, that the soup spills before it starts to boil. This is because the soup expands as its temperature is increased.

When liquids are heated, the individual molecules begin to move farther apart. The distance between molecules increases because they are moving faster and farther with respect to each other.

When gases are heated, the gas molecules move faster and farther, causing the gas to expand. In a closed container, the molecules strike the walls of their container with greater force. For this reason, tire manufacturers recommend that tire pressure be checked only when the tires are cool. If you checked tire pressure in tires when they were hot, the pressure would appear too high and you would let out some air. Then when the tires cooled, they would be underinflated.

 ASK YOURSELF

Use the kinetic molecular model of matter to describe why you should check tire pressure when tires are cool.

A Change of Phase

Phase changes can also be explained by the kinetic molecular model. When a solid gets hot enough, its molecules stretch and bend so much that groups of molecules break away from the solid. When the entire solid has separated into free-moving molecules,

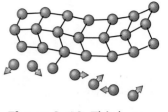

Figure 2–18. This lump of gallium easily melts from a solid to a liquid. During this phase change, the connected molecules separate into individual molecules in close association.

the material is no longer a solid—it is a liquid. The substance has melted.

Freezing is the reverse of melting. If the molecules in a liquid slow down enough, they are able to attach to each other and hold on. Once all the molecules are attached to each other, the substance is no longer a liquid—it is a solid.

During boiling, individual molecules separate from the liquid and move off independently. When this happens, the liquid becomes a gas. When gases condense into liquids, the individual molecules slow down enough that the weak attractive forces are once again able to stick them to each other. Once this happens, the substance is a liquid with molecules bonded together rather than a gas with independent molecules.

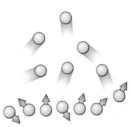

Figure 2–19. Liquid oxygen evaporates very quickly at room temperature. As this substance boils, its molecules gain enough energy to move off as individual gas molecules.

 ASK YOURSELF

What happens to the molecules when a solid changes into a liquid?

SECTION 3 REVIEW AND APPLICATION

Reading Critically

1. Name the three features of a scientific model.
2. Use the kinetic molecular model to explain the differences among solids, liquids, and gases.
3. Explain how the density of substances can be related to the kinetic molecular model of matter.

Thinking Critically

4. Dry ice is the solid phase of carbon dioxide. As you read in Section 2, dry ice does not melt. Instead, it sublimes. Use what you have learned from the kinetic molecular model to explain how this happens.
5. Why do musicians "warm up" their instruments before they tune them?

INVESTIGATION

Measuring Freezing and Melting Points

▶ MATERIALS

- safety goggles ● laboratory apron ● ring stand ● buret clamps (2)
- 500-mL beaker ● hot plate ● large test tube ● moth flakes
- thermometers in notched corks (2) ● stirring rod ● stopwatch ●
graph paper

▼ PROCEDURE

CAUTION: Put on safety goggles and a laboratory apron.

1. Work in a group, and set up the apparatus as shown.
2. Fill the beaker about 2/3 full of water, and place it on the hot plate. Put the moth flakes in the test tube and place it in the water.
3. Heat the water until the moth flakes have melted.
4. When the moth flakes have melted, place a thermometer in the test tube. Place another thermometer in the water, as shown in the picture. **CAUTION: Do not at-**

tach the clamp directly to the thermometer. Stir the water in the beaker with a stirring rod.

5. Make a table like the one shown. One group member should read the stopwatch. A second member should read the thermometers. A third member should record the information. If there are four members, two of them can read the thermometers.
6. When everyone is ready, the timer says "start" and turns off the hot plate. The recorder should write down the starting time. After the

start, the timer will say "reading" every 30 seconds for 20 minutes. The readers look at their thermometers; the recorder asks for the water temperature and then the moth flake temperature.

7. Watch the moth flakes closely. Note the time when solid flakes first appear and when no more flakes are forming from the liquid.
8. Record your data on a graph with temperature on the vertical axis and time on the horizontal axis.

TABLE 1: TEMPERATURE READINGS		
Time	Water Temperature	Moth Flake Temperature

▶ ANALYSES AND CONCLUSIONS

1. Using your data, determine whether the water and the moth flakes cooled at the same rates or at different rates. Explain.
2. Is there a section on your graph that shows a period when the temperature of the moth flakes stayed the same? What was the temperature during this flat section (plateau)?

▶ APPLICATION

Why might knowing the freezing points or melting points of foods be helpful to people who prepare meals?

✳ Discover More

Test the melting points of different frozen fruit bars to see whether different juices affect the melting points.

The Big Idea

All matter is made up of atoms, has mass, and takes up space. Matter can be classified according to its phase, its purity, and its properties. Scientists use models to show their ideas about the structure of matter and the interaction of its particles. As scientists gain new information from experiments, they may revise their models. Scientists today use the kinetic molecular model of matter to explain how matter behaves.

For Your Journal

Look back at the ideas you wrote in your journal at the beginning of the chapter. How have your ideas changed? Revise your journal entry to show what you have learned about matter.

Connecting Ideas

Read the following poem. Which phases of water are mentioned? Explain how the information in the poem does or does not match what you know about physical changes. Record your answers in your journal.

Walk on a Winter's Day
by Lee Noble

A walk by the lake
on a cold winter's day
the tip of my nose
chilled by biting frosted wind.
Each breath comes as puffs of clouds
that linger for a moment and quickly disappear.
Once shimmering lake no longer rippling
a mass of quiet stillness waiting for the touch of the sun.
Hanging from trees once green with leaves
glittering icy ornaments poised to quietly slip away.
A flurry of soft and gentle flakes
hurriedly drifting and twirling
cover me briefly in a blanket of down
which dissolves and leaves no trace.
The chill of the air, the bite of the wind
the elements on my face
like old friends that come for a visit
and then must go away.

REVIEW

Understanding Vocabulary

1. Make a list of the ways you could distinguish between a mixture (36) and a pure substance (34).

2. For each pair of terms, use the definitions of *element* (34) and *compound* (35) to help you determine which is an element and which is a compound.
   a) iron, iron oxide
   b) chlorine, sodium chloride
   c) copper nitrate, copper

Understanding Concepts

MULTIPLE CHOICE

3. When a substance freezes, it changes from the liquid phase to the solid phase because
   a) molecules stick together.
   b) groups of molecules break up into individual molecules.
   c) individual molecules stick together and then break apart.
   d) molecules slide past each other in disordered motion.

4. Thermal expansion
   a) is caused by atoms and molecules moving farther apart.
   b) is only observed in liquids and gases.
   c) is caused by atoms and molecules getting larger.
   d) always happens to a gas when it is cooled.

5. Which of the following is *not* a characteristic property of a substance?
   a) phase at room temperature
   b) color at room temperature
   c) volume at room temperature
   d) density at room temperature

SHORT ANSWER

6. How does the kinetic molecular model of matter illustrate the three features of a scientific model?

7. Describe and distinguish between the behavior of molecules in each of the three phases: solid, liquid, and gas.

8. Explain the difference between elements and compounds.

Interpreting Graphics

9. What physical change is taking place when the ethylene glycol is at 198°C? How can you tell?

10. Describe what is happening to the ethylene glycol molecules at 198°C.

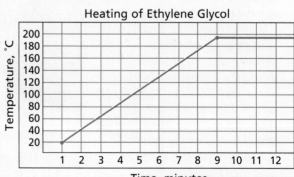

Heating of Ethylene Glycol

Reviewing Themes

11. *Systems and Structures*
A mercury thermometer measures the thermal expansion of liquid mercury as it moves through a very thin tube. Explain how this happens, using the kinetic molecular model of matter.

12. *Systems and Structures*
A pure substance contains only one kind of molecule. Check the labels on foods at home. List some of the pure substances you find.

Thinking Critically

13. Propose a method for separating a mixture containing salt, sand, water, methyl alcohol, and wood chips.

14. The air in the earth's atmosphere is a mixture. Research the gaseous components of the earth's atmosphere. Which is the most abundant gas in the mixture?

15. In the early history of the United States, people would search out sandy stream beds in which small particles of gold were mixed with the sand. The separation process used was called "panning." What characteristic property of the two substances made panning possible?

16. The idea that all matter can be classified in terms of basic elements has been accepted for a long time. In one ancient theory, there were four "elements": earth, air, fire, and water. How does fire differ from the other three "elements"?

17. One characteristic property of a substance is density. Would this be a good property to choose when identifying a gas? Explain.

18. The pictures here show the outside of a flight simulator and what the screen inside looks like. Flight simulators like this one are used to train pilots before they actually fly a plane. Using what you learned in this chapter, explain why a simulator could be considered a scientific model.

Discovery Through Reading

Newton, David. *Consumer Chemistry Projects for Young Scientists.* Franklin Watts, 1991. This illustrated book offers a variety of at-home projects, including several that focus on the properties of matter.

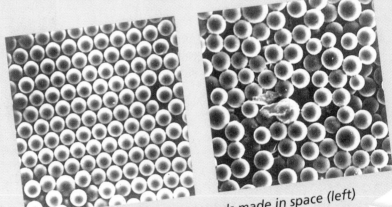
Science PARADE

MATERIALS MADE IN SPACE

"MADE IN THE USA." You've seen that manufacturing tag many times but not given much thought to it. However, you'd probably look twice at a tag that read "MADE IN OUTER SPACE." You might grin and think that this sounds like a not-very-clever idea from a science fiction book. You can stop grinning. The tag may not exist yet, but new materials are being produced in outer space.

IT CAME FROM OUTER SPACE

Materials with extraordinary—and valuable—properties can be produced in space. For example, polystyrene (tough plastic material) spheres of very uniform size can be manufactured in space. Scientists use these perfectly uniform plastic spheres to calibrate scientific instruments back on Earth. The space balls help measure everything from dust to face powder. The first batch of space-made polystyrene spheres was valued at about $23,000 per gram!

Comparison of polystyrene beads made in space (left) with those made on Earth (right)

HOW DOES YOUR CRYSTAL GROW?

The microgravity (near weightlessness) of space is an ideal environment for growing crystals. When crystals grown on Earth reach a certain size, their own weight causes them to break. In microgravity, scientists can grow larger and more complex crystals. On the space shuttle, astronauts baked crystals in individual furnaces that reach temperatures of more than 500°C.

The materials for space crystals range from metals like cadmium to proteins. Some vital space research focuses on growing crystals of the protein interferon, a drug used to treat AIDS, and other AIDS-related proteins. Such space-grown crystals are returned to pharmaceutical labs on Earth, where scientists study them further. Space-made medicines may someday become part of the battle against AIDS.

One space-grown protein may make computers remember more. Scientists extracted a light-sensitive protein from a type of bacterium growing in salt marshes. On the space shuttle, astronaut scientists injected the protein into a polymer carrier.

Protein grown in free fall

The space-manufactured protein may someday be installed into computer memories. Computer engineers think it might make computer memory capacity 10 to 30 times greater. And think of it: from a salt marsh to the space shuttle to a computer memory chip. Quite a journey for a humble protein! ◆

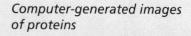

Computer-generated images of proteins

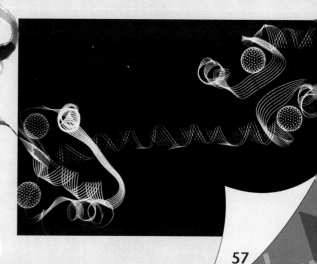

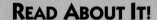

Bubbles

from *Kids Discover*

Bubbles Everywhere

Bubbles are everywhere. There are bubbles in your bathtub. Bubbles in your soft drink. Bubbles in your shirt. Wait a minute. Bubbles in your shirt? It's true. If you are wearing a cotton shirt made by a certain manufacturer, there may be air bubbles in your shirt—put there to reduce shrinkage when you dry the shirt in a clothes dryer.

What exactly is a bubble? In the purest sense, a bubble is a ball of air or gas in a liquid or with a thin skin of liquid around it. Soap bubbles and carbonated beverages fit this definition. But in recent years, people have given the name of bubble to many things that don't fit this classic definition. The expanded definition of bubble is something like "anything that is shaped like a sphere and hollow inside and that looks and acts like a bubble."

Kids of all ages love to have fun with bubbles. But there is a serious side to bubbles. Scientists and inventors are always studying bubbles, looking for new ideas that might help people. Among other things, they have tried nicotine bubbles to help people stop smoking.

Professor Bubbles, Richard Faverty, makes a gigantic soap bubble that lasts about 30 seconds.

Bubbles with Gas

Whether you call it pop, soda, tonic, soft drink, or carbonated beverage, there's one thing it's got to have—bubbles. Whenever you reach for a carbonated beverage, the sparkle will come from the same source—carbon dioxide. Carbon dioxide (CO_2) is a colorless, odorless, tasteless gas. When it's trapped with water in a closed container (or underground), the carbon dioxide dissolves in the water. Opening the container reduces the pressure on the water. When the pressure is reduced, the gas expands, and out come those tiny bubbles.

By around 1820, bottled water became popular. Soon flavorings—ginger and lemon—and mineral salts were added.

How do they get the tiny bubbles in the pop? Syrup or concentrate and water are put into a pressurized vessel where the atmosphere is carbon dioxide. This gas is absorbed by the liquid. Later, as the liquid is passing through a pipe (approximately three inches around), very fine bubbles of carbon dioxide are injected into it through a sparger. The sparger's job is to break up large bubbles into tiny bubbles so they can be more easily absorbed into the liquid.

By the mid-1800s, druggists were dispensing carbonated water from soda fountains in their drug stores. Fizzy water was slowly being converted from a health tonic to a pleasure drink. The drug-store soda fountain was fast becoming an American institution. But it didn't help quench the thirst of people who lived in rural areas. So by the late 1800s, local bottlers were producing local brands of soft drinks. Eventually the best, or those with the best advertising, became national brands, while the others died out.

In 1772, Joseph Priestley, nicknamed "the father of the soft drinks industry," suggested that with the help of a pump, water would be impregnated with fixed air, thus making carbonated water.

For centuries fashionable people "took the waters" at health spas in Europe and the United States. Many of the spas were located at places where water (containing dissolved carbon dioxide) mysteriously bubbled out of the ground. This water was believed to cure various diseases and generally to invigorate those who bathed in it and drank it. Before the spa resort was in operation at Saratoga Springs in upstate New York, George Washington visited Saratoga as a guest of the Schyler family. He tried the water and was so impressed that he tried to buy land near the springs, but it had already been purchased.

Saratoga Springs spa park today

In the 17th century, the first marketed soft drinks appeared. They were a mixture of water and lemon juice sweetened with honey. Spa water was first bottled in the late 1700s. So people who couldn't travel to a spa could still benefit from the waters.

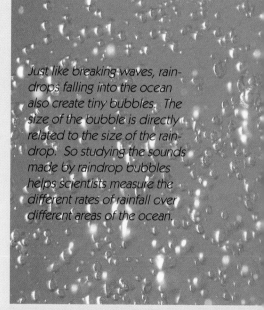

Nitrogen bubbles in the blood cause a dangerous, even deadly, condition called the bends, which results when divers dive too deep, stay down too long, or return too quickly from a deep dive.

Just like breaking waves, raindrops falling into the ocean also create tiny bubbles. The size of the bubble is directly related to the size of the raindrop. So studying the sounds made by raindrop bubbles helps scientists measure the different rates of rainfall over different areas of the ocean.

Bubble Gum Fun

Most kids love it. Most parents hate it. Most doctors say it can't hurt you. But some people have dislocated their jaws chewing it. It's bubble gum, of course. And it is over 65 years old.

Actually, bubble gum is one of the newer things that people have tried chewing. The ancient Greeks chewed the gum of the mastic tree, and over a thousand years ago the Mayans of Central America chewed chicle, a white gummy substance from the sapodilla tree. In the mid-1800s, a Mexican general brought some chicle to the United States and gave some of it to Thomas Adams, an inventor and businessman. When he and his father added sugar and flavoring for taste, the chewing gum industry began. But it wasn't until 1928 that Walter Diemar, working at the Fleer chewing gum company, came up with the right ingredients to make a gum that was elastic, moist, and not too sticky to peel off faces once a bubble had popped.

In 1928, when Walter Diemar put together the right ingredients for bubble gum, he had only pink food coloring at hand, and that's why bubble gum has been mainly pink ever since.

Those who have to clean up discarded bubble gum complain bitterly about it. But others have found constructive uses for it. Some dentists and speech therapists recommend chewing bubble gum to strengthen jaws. A Columbia University study showed that chewing gum makes people more relaxed. One scientist even claims that liquid bubble gum can be used as an insecticide. When plant-eating bugs eat it, their jaws stick together and they can no longer chew plant leaves.

It's been estimated that the gum chewed by Americans in the years since 1928 would make a stick 113 million miles long. That's long enough to reach to the moon and back more than 200 times.

Some card companies no longer include bubble gum with their sports cards because collectors were complaining that the bubble gum was staining the cards!

The new generation of bubble gum includes such exotic flavors as bananaberry split and watermelon. There's sugar-free bubble gum and even bubble gum in the middle of a lollipop!

Bubbles in the Laboratory

Some bubbles are just for fun. But scientists are finding that certain bubbles may help answer some age-old questions about our universe while they raise some new ones. Air bubbles trapped in amber may hold the key to such ancient mysteries as why the dinosaurs became extinct. Bubbles locked in Antarctic ice may help us understand the world we live in by showing the relationship between carbon dioxide in the atmosphere and temperature. And bubbles in outer space raise some questions about the structure of the universe. ◆

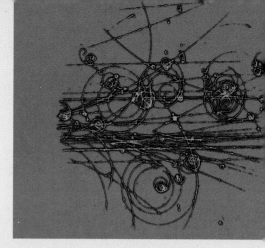

There are countless theories about why the dinosaurs became extinct. Here's one of them. Air bubbles trapped in amber that fossilized 80 million years ago suggest that the air at that time contained 50 percent more oxygen than today's air. If that is true, the dinosaurs may have slowly suffocated as the oxygen in the air was gradually reduced.

A bubble chamber helps in the study of the tiny particles that are the building blocks of matter and the forces that bind them together. In a bubble chamber, liquid is heated high above its boiling point and kept under high pressure to prevent boiling. When the pressure is suddenly reduced, charged particles speeding through the liquid create strings of bubbles. High-speed photographs of the bubbles make possible precise measurements of the particles and their activities.

Eiffel Plasterer, a bubbleologist, sealed many bubbles so they could last longer. How long did his longest bubble last? (340 days!)

When a wave, driven by the wind, breaks at sea, it traps a large amount of air in the water. The trapped air quickly becomes millions of tiny bubbles, which radiate sound as they move in the water. Scientists use sound-detecting equipment to track the bubbles and learn more about ocean currents and waves as they move across the ocean.

Francis Bacon (1561-1626)

Englishman Francis Bacon was a politician and lawyer who lived during the time of Queen Elizabeth I and King James I. Francis Bacon was also a philosopher who wrote about learning from observation and experimentation. The methods of experimental science had been formulated earlier by scientists such as Galileo. Bacon, however, was the first person to write about those methods in a way that convinced others that observation and experimentation were useful ways to learn things.

Bacon's main interest was in discovering practical applications of science that would make life better for the people of the world. In fact, he worked with many skilled artisans from all over Europe to improve methods of metal working and mining. He encouraged artisans to use his methods of problem solving to develop new techniques for use in industry.

Bacon published his ideas, and his work soon attracted widespread interest. In one sense, Francis Bacon's writings helped launch the scientific age in which we now live. ◆

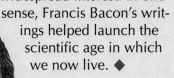

Shirley Jackson (1946-)

Shirley Jackson has always been interested in physics. After graduating as valedictorian of her high-school class, she was admitted to the Massachusetts Institute of Technology (MIT). Four years later she received her bachelor's degree in physics. Jackson decided to continue her education at MIT and became the first African American woman to receive a Ph.D. from that institution. She continues to be involved with MIT as a member of its educational council.

Jackson's first area of research was in the field of high-energy physics. During the early years of her career, she worked in this field at the National Accelerator Laboratory in Batavia, Illinois. Later she was a scientific associate for the European Organization for Nuclear Research in Geneva, Switzerland, and a lecturer at the NATO Advanced Study Institute in Antwerp, Belgium. Now employed at AT&T Bell Laboratories in Murray Hill, New Jersey, Jackson's current research is in low-energy physics. She has written numerous articles on such topics as nutrino reactions and channeling in metals and semiconductors. ◆

CLAUDIA WIDDISS
Sculptor

Claudia Widdiss knows about the properties of matter. She uses the properties of stone daily to choose what she will create. Widdiss is a sculptor.

Widdiss begins this limestone sculpture from a block of stone that was discarded by another artist. The block weighs about 181 kg. Widdiss makes a few scale models based on her roughed out block of stone. Scale models are a way of sketching in three dimensions.

Widdiss carves the stone with a hammer and a pointed chisel. Her hammer and chisel are designed much like those used in prehistoric times, except her tools are made of steel instead of stone. It takes three days for her to rough out the basic shape of the sculpture. The basic shape must be enhanced by the natural properties and characteristics of the stone.

After working by hand with hammers and chisels, Widdiss begins more detailed work on the stone. She uses an air hammer, which is activated by compressed air. The air hammer presses a chisel against the stone and carves it into the exact shape she wants. Widdiss carves all around the stone and works on the sculpture from many different positions and angles. The tools that she uses can be very dangerous. That is why she wears protective clothing, such as a mask, gloves, and ear coverings.

Here are two views of Widdiss's finished sculpture. After she secures the sculpture to its base, it will be ready for all to see.◆

● Discover More

● For more information about
● careers in sculpting, write to the
● Sculptors' Guild
110 Greene Street
● New York, NY 10012

Biodegradable Plastics

Scientists and manufacturers once thought plastics were the perfect material. Plastics made from petroleum could be shaped into virtually any form—from garbage bags to automobile bodies. Even better, synthetic plastics lasted forever. Unfortunately, plastic's durability has turned out to be a problem.

The Plastic Problem

Synthetic plastics consist of long, tangled carbon polymers. Polymers are giant molecules made by bonding many small molecules together. They resist being broken down by natural forces. Plastic bags, diapers, and bottles are clogging landfills. In response, scientists are developing the technology to produce biodegradable plastics. These natural plastics dissolve under contact with organisms in the earth and water.

Food for Bacteria

To achieve biodegradability, scientists are trying to develop plastics with a special set of properties. First, the material must be an inviting food for naturally occurring bacteria and enzymes. Most biodegradable plastics begin with starch molecules made of carbon, hydrogen, and oxygen. Second, the material must hold together well enough to serve as a plastic. This means that the starch molecules must form long, twisted polymers, like the carbon polymers in synthetic plastic. Finally, natural plastics must degrade into substances harmless to human life and the environment.

It looks and feels like polystyrene, but it isn't! ECO-FOAM is a packaging material based on cornstarch.

Research labs are working on several ways to manufacture biodegradable plastics. In one method, scientists begin with powdery starch from corn or potatoes. As scientists heat the starch to 177°C, the starch molecules form twisted polymers much like synthetic plastics. In another method, scientists grow bacteria that manufacture their own plastic polymer, a polyester similar to what is used in fabrics. The bacteria make and store the polyester for future consumption, much as the human body stores fat. Scientists kill the bacteria and use the naturally produced polymer to make plastics.

A Dissolving Act

Some biodegradable plastics have already reached consumers. Physicians can stitch up internal organs with plastic sutures that eventually dissolve on their own. Some mail-order companies already cushion their shipments with a starch-based packaging material. The packaging is made of 95 percent starch and 5 percent polyvinyl alcohol. Within seconds after getting wet, the plastic packaging dissolves. Another biodegradable plastic used in hospitals holds dirty laundry. Workers simply toss the bags into the washing machine, and the bags dissolve as the clothes are laundered. After this plastic has dissolved, the alcohol in it attracts bacteria in the water. The bacteria attack the alcohol and use the starch as food.

Not So Fast!

Scientists have a harder time making biodegradable plastics that dissolve slowly. No one wants a disposable diaper that dissolves quickly on contact with water!

Some farmers now cover their fields with synthetic plastic to prevent weed growth. However, they must pay to remove the oil-based plastic, which ends up in a landfill.

Chemists hope to manufacture a starch-based plastic sheeting to serve as a crop mulch. With a slow-dissolving biodegradable plastic mulch, farmers could let the plastic dissolve harmlessly in their fields.

The Search Continues

Biodegradable plastics are not an easy answer to the problem of plastic waste. The first "biodegradable" plastic bags were a mixture of starch and synthetic plastics that did not fully degrade. Scientists warned that such products might degrade into toxic compounds. Also, biodegradable materials must come in contact with soil and water. Science and technology continue the search for a plastic that will serve consumers' needs without harming the environment. ◆

Even starch-based plastics will not degrade if crammed into a landfill.

"**L**adies and gentlemen and children of all ages . . . Please direct your attention above the center ring . . ."

Just about everyone loves a circus. You may wonder how the high-wire performer balances on that thin wire, how the acrobats land from great heights without getting hurt, or how a trapeze artist does somersaults in midair. All of these things depend on forces, motion, and energy.

Science PARADE

CHAPTER 3

ACCELERATION, FORCE, and MOTION

"You just wait," his mother said. "We'll build a better sled than Ed Sines has. Now get me a pencil and a piece of paper."

"You goin' to build a sled out of paper?" Orville asked in amazement.

"Just wait," she repeated. . . .The outline of a sled began to appear on the paper. As she drew it she talked. "You see, Ed's sled is about four feet long. I've seen it often enough. We'll make this one five feet long. Now, Ed's sled is about a foot off the ground, isn't it?"

Orville nodded, his eyes never leaving the drawing that was taking shape. It was beginning to look like a sled now, but not like the sleds the other boys had.

Scientists and engineers pursue the goals of faster, farther, and higher, in everything from automobiles to aeronautics. With wind tunnels and computer simulations, they carefully calculate the factors of acceleration, force, and motion.

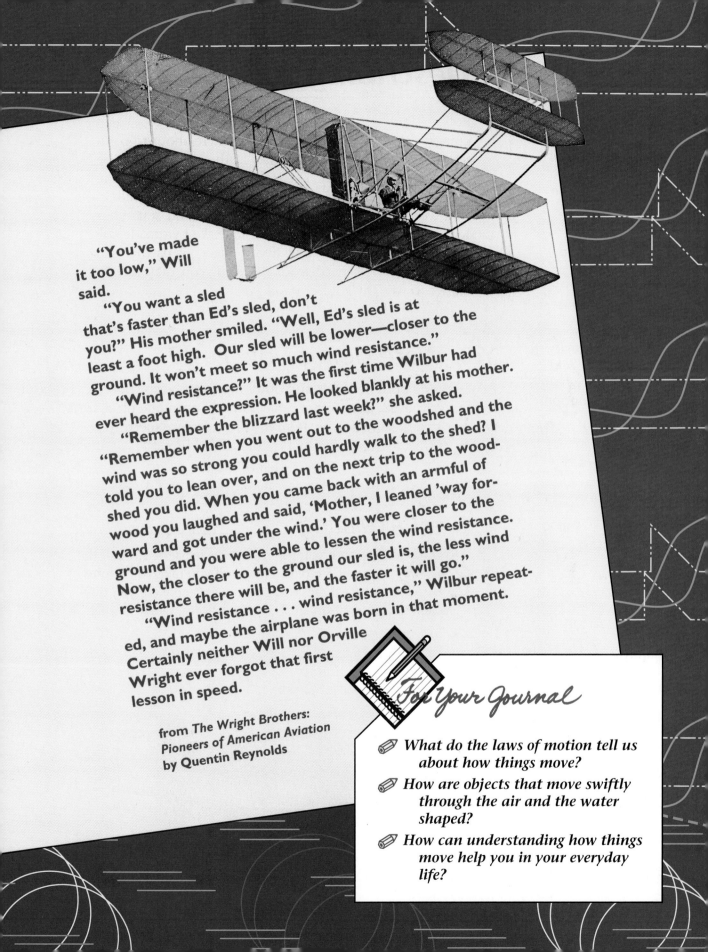

"You've made it too low," Will said.

"You want a sled that's faster than Ed's sled, don't you?" His mother smiled. "Well, Ed's sled is at least a foot high. Our sled will be lower—closer to the ground. It won't meet so much wind resistance."

"Wind resistance?" It was the first time Wilbur had ever heard the expression. He looked blankly at his mother.

"Remember the blizzard last week?" she asked. "Remember when you went out to the woodshed and the wind was so strong you could hardly walk to the shed? I told you to lean over, and on the next trip to the woodshed you did. When you came back with an armful of wood you laughed and said, 'Mother, I leaned 'way forward and got under the wind.' You were closer to the ground and you were able to lessen the wind resistance. Now, the closer to the ground our sled is, the less wind resistance there will be, and the faster it will go."

"Wind resistance . . . wind resistance," Wilbur repeated, and maybe the airplane was born in that moment. Certainly neither Will nor Orville Wright ever forgot that first lesson in speed.

from *The Wright Brothers: Pioneers of American Aviation* by Quentin Reynolds

For Your Journal

- What do the laws of motion tell us about how things move?
- How are objects that move swiftly through the air and the water shaped?
- How can understanding how things move help you in your everyday life?

Describing Motion

Imagine that you are an Olympic athlete about to compete in a bobsled event. You gaze at the icy, steep course and take a deep breath. The starting bell rings. Grasping the sled, you and your partner push and run alongside it. After a running start, you both jump on and start your first run down the treacherous course. You've made the fastest start of the day! Now, if you and your partner can coax every bit of speed from the sled, you may have a chance to win the gold!

You Were Going How Fast?

In describing both bobsleds and other moving objects, the term *speed* is used. How fast an object moves from one place to another in a specific time is its **speed.** To calculate speed, you divide the distance traveled by the total travel time.

$$\text{average speed} = \frac{\text{total distance}}{\text{total time}} \quad \text{OR} \quad v = \frac{d}{t}$$

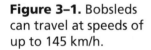

₃ᶜᵒᵛᵉᴿ ᴮʸ Calculating

Suppose you ride your bicycle to the store. If the store is 5 blocks away, and you take 10 minutes to get there, what average speed were you traveling? Explain your calculations in your journal. ✎

Figure 3–1. Bobsleds can travel at speeds of up to 145 km/h.

Of course, you know from your bicycling experiences that speed changes when you pedal fast to avoid a snarling dog or when you stop to fix a flat tire. Speed calculated by dividing distance by time is really your average speed. In the next activity you'll have a chance to calculate average speed.

DISTANCE ÷ TIME = SPEED

Figure 3–2. Speed is calculated by dividing distance by time taken to travel that distance.

ACTIVITY

How can you determine speed?

MATERIALS
meter stick, toy car, stopwatch, masking tape

PROCEDURE
1. Lab partners should divide up the following jobs: driver, starter, timer, flagger, and recorder.
2. Set up the materials as shown. Tape the meter stick to the floor or table in an area where the car can move freely. You will use the meter stick to measure the distance the car travels.
3. The driver should place the car so that its nose aligns with the front end of the meter stick. On a signal from the starter, the driver should release

the car so that it travels next to the meter stick. The timer should start the stopwatch at the signal, too.
4. The flagger should signal the instant the nose of the car passes the end of the meter stick. The timer should stop the stopwatch on the flagger's signal.
5. The recorder should record the trial number (first, second, third, and so on), the distance traveled in meters, and the elapsed time in seconds.
6. Repeat steps 3 through 5 to make at least three trials.
7. Calculate the car's average speed for each trial. First average the total distance traveled and the total time elapsed for all trials. Then, divide the distance by the time. Record the average speed on the chalkboard.

APPLICATION
1. How might the distance traveled by the toy car be measured with greater precision?
2. Why is speed considered a derived measure rather than a direct measure?

ASK YOURSELF

Suppose that the speedometer on your bike doesn't compute average speed. How could you find your average speed?

Figure 3–3. The cars on this stretch of highway are constantly changing velocity, even when their speed remains constant. Explain why.

It's Not the Speed . . .

Speed tells us how fast an object is moving. Another word used to describe motion is *velocity*. **Velocity** is the speed of an object in a particular direction. It is easy to confuse speed and velocity because they are both based on the same measurements—distance traveled in a certain amount of time. The difference between speed and velocity is that speed can be motion in any direction, but velocity identifies a certain direction.

Imagine that you are at the roller rink, skating at a steady pace, smoothly following the curve of the rink. Putting one foot in front of the other, you're pushing just hard enough to keep going. Suddenly a friend skates up to you and yells, "Your velocity is changing! Your velocity is changing!" For a moment, you think your friend has lost his mind. You are puzzled, though, because you know you are covering the same distance each minute. Then you remember science class; you're studying speed and velocity. You recall that even though you are traveling at a constant speed, your velocity is changing because you are constantly changing direction to follow the curve of the rink.

Here is another example. If you were to travel in a car at a constant 50 km/h in a straight line, your speed and velocity would both be the same. If you were to speed up from 50 km/h to 70 km/h while going in a straight line, both your speed and velocity would change by the same amount. If you were to travel at a constant 50 km/h in a big circle, your speed would stay the same but your velocity would continually change. In order to have constant velocity, both speed and direction must stay the same.

ASK YOURSELF

A velodrome is a curved, steeply banked track designed for bicycle racing. Explain how the speed and velocity of a cyclist on this track would differ from that of a cyclist on a flat, straight stretch of road. Now, consider the cyclists in a mountain bike race. Their race course consists of rugged terrain that includes mountains and streams. How would their speed and velocity compare to cyclists riding on a velodrome and cyclists riding on a flat, straight road?

Figure 3–4. The parachute on the space shuttle (left) helps bring it to a stop. The dragster's parachute (bottom) also opens to help bring it to a stop.

Slowing Down and Speeding Up

Picture yourself sitting in a dragster, revving the engine, and waiting for the starting light. The track is short and straight; the race will be over in a few seconds. The green starting light flashes, you stomp on the accelerator, and the noise of the squealing tires drowns out the cheers of your crew. As you speed up, your body is pushed back into the car seat. You are feeling acceleration. Both you and the car keep accelerating as the car crosses the finish line. Then you stop the car by using the brakes and releasing a parachute that flares out the back. As the car slows down, you stop accelerating, right? Wrong!

Any time your velocity changes, whether by speeding up, slowing down, or changing directions, you are accelerating. The rate at which velocity changes is called **acceleration.** So when your dragster slows down, its velocity changes and, thus, it accelerates. Your body tends to keep moving forward as the dragster slows. Your seat belt keeps you from traveling forward.

Figure 3–5. These cars are maintaining a fairly constant speed. However, these cars are also accelerating because their direction is changing.

Now imagine that you are trying to qualify for the Indianapolis 500. Only those drivers with the fastest qualifying times get into the race. You drive around the oval track as fast as the car, and your nerves, allows. Remember, because acceleration also applies to changes in direction, you are accelerating even when you come into and get out of a curve. In fact, you are accelerating even if your speed is constant. Because acceleration depends on velocity, the only time you are not accelerating is when both your speed and direction are constant. In the next activity you'll have the opportunity to discover how often you accelerate throughout the day!

DISCOVER BY *Doing*

Today as you are walking from class to class or from class to lunch, notice how you accelerate by speeding up, by slowing down, or by changing direction. Record the results in your journal. Share your results with a classmate. How do they compare? 🖋

▼ ASK YOURSELF

How does acceleration affect velocity?

Momentum

Figure 3–6. The momentum of this football player moves him forward even though he is being tackled around the legs.

Think about all the speeds, velocities, and accelerations that occur during a football game. The players are moving all the time. What affects their motion?

Imagine that you are watching a football game on television. One sports announcer exclaims, "Unbelievable! Number 36 was hit at the three-yard line, but he was just too heavy to stop. His momentum carried him into the end zone for a touchdown!" The other sports announcer says, "It's a good thing Number 28 wasn't carrying the ball. He weighs 40 pounds less than Number 36 and wouldn't have had the momentum to score."

Both of the sports announcers are using a physical science term here. In physical science, **momentum** is an object's mass multiplied by its velocity. Momentum, like velocity and acceleration, has definite direction. Unlike acceleration and velocity, the momentum is related to the mass of the moving object.

Figure 3–7. The bicycle and the truck are traveling at the same velocity. Which has more momentum?

 **BY** *Problem Solving*

Suppose you are playing football on the defensive team. Which sort of ball carrier would you rather face? (a) a small, slow runner (b) a small, fast runner (c) a large, slow runner or (d) a large, fast runner. In your journal, write down your answer. Then explain your choice. ✐

If objects have the same velocity, it will be harder to make the more massive object change velocity. If objects have the same mass, it is harder to change the faster object's velocity.

This applies to people, cars, bobsleds, bicycles, or any other moving object. For example, a small car can corner more easily than a large car (assuming the tires, brakes, and so forth are the same). Because the small car has a lot less mass, it has less momentum at the same velocity. It is easier to accelerate the object with less momentum. Since cornering requires acceleration, the car with less momentum corners better.

▼ **ASK YOURSELF**

Think of a sport other than football, and then describe how momentum affects the game.

SECTION 1   *REVIEW AND APPLICATION*

Reading Critically

1. How are speed and velocity alike? How are they different?

2. How does stopping a car make it accelerate?

Thinking Critically

3. When engineers design cars, they are very concerned with the overall weight of the car. Why is this?

4. Why do you think cars have speedometers instead of velocity meters?

Forces and Motion

Define the terms force, friction, *and* gravity.

Explain *how forces can cause changes in motion and how mass and distance affect gravity.*

Compare *friction, drag, and air resistance.*

If you want to do more than describe motion, you must understand what affects motion. What starts it? What stops it? What must you do to make the motion of an object like a bobsled change? The answer is simple. It just takes a push or a pull!

Pushes, Pulls, and Motion

A push or a pull is required to change motion—even in bobsledding. When the bell rings, the bobsled racers *push* the sled from the starting line and jump on as it increases velocity. After that, gravity *pulls* the bobsled downhill, causing the bobsled and its passengers to accelerate. The bobsled accelerates until the brakes and the opposing motion between the sled's runners and the ice *push back* enough against the forward motion to stop the acceleration.

Figure 3–8. This soapbox derby racer is accelerating down the slope as a result of the force of gravity.

Throughout the day you experience pushes and pulls. When you walk down the street, your feet push on the street, moving you forward. When you write a homework assignment, you push on the pencil, making it move across the paper. Every time there is a change in the motion of anything, a force has made that change in motion happen. A **force** is a push or a pull.

In SI, force is measured in newtons (N), named after the British scientist Sir Isaac Newton. One **newton** is the force required to cause a 1-kg mass to accelerate 1 meter per second each second. If you applied a force of 2 N to a 1-kg brick, it would accelerate by 2 m/s each second. If you only applied 1/2 N to the 1-kg brick, it would accelerate by 1/2 m/s each second.

Newtons = kilograms × meter/second/second

Figure 3–9. Because of the different masses of these two objects, different amounts of force are required to move them.

How Much Acceleration?

The amount of acceleration depends on both the amount of the force used and the mass of the object being pushed or pulled. If you apply a small force to a massive object like a railroad car, you will cause a tiny acceleration. If you apply a large force to a small object like a golf ball or a pencil, you will cause a large acceleration. If you apply a tiny force to a tiny object or a large force to a large object, you will cause a medium acceleration.

Whenever a force is applied to an object, the object's motion is changed. When you kick a soccer ball, you change its motion by applying a force. The kick is a push. When you push a bobsled, you change its motion from no motion at all to moving forward. When you drop a fork in the school cafeteria, the earth pulls the fork down. The earth's pull, gravity, is a force that changes the fork's motion downward. The fork also pulls at the earth, but the earth's acceleration is much smaller.

Balancing Act   One force can cancel the effects of another force. For example, you can hold a fork without dropping it. When you do, you are pushing up with exactly the same force as gravity is pulling down. If a force does not change the motion of an object to which it is applied, the force must have been balanced by another force, because there is no movement when opposing forces are balanced. Only when forces are not balanced does the object accelerate.

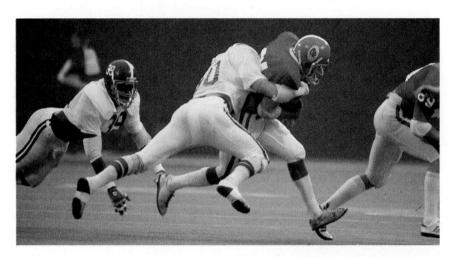

Figure 3–10. When a football player is successfully tackled, the player's forward motion is stopped by a force produced by the tackler.

Think about some situations in which objects accelerate—the drag racer, for example. In order for a drag racer to stop quickly, a large, unbalanced force must be applied in the direction opposite the dragster's motion. To apply such a force, a parachute is released from the back of the dragster. The force of the air against the parachute quickly opposes the forward motion of the dragster. The force of the air against the parachute is unbalanced, and the dragster slows down.

Figure 3–11. The result of forces on an object

Forces on a stationary object	Result	Forces on a moving object	Result
Balanced	Object remains stationary	Balanced	Object remains at constant velocity
Unbalanced	Object accelerates	Unbalanced	Object accelerates

ASK YOURSELF
What happens to an object that has a force applied to it?

Friction

Imagine that you are pushing a box loaded with books across the floor. As you push, the floor seems to oppose the motion of the box. What's wrong? Nothing!

You are experiencing **friction,** the force that opposes motion between two surfaces that are touching. In order for you to slide the box of books across the floor, you have to overcome friction.

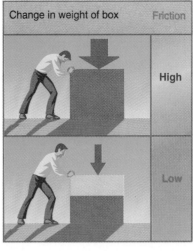
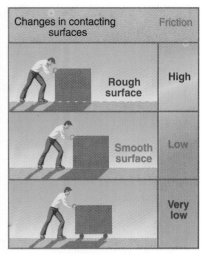

Figure 3–12. The person in this illustration must overcome friction to move this box.

The amount of friction between two surfaces depends on how hard the surfaces are forced together, the materials that make up the surfaces, and the smoothness of the surfaces. The more books you load into the box, the harder it is to push across the floor. If the floor is covered with rough carpet instead of smooth wood, the box is harder to push. The two surfaces "grab" at each other more. Try the next activity to find this out for yourself.

DISCOVER BY Doing

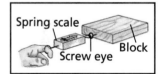

Attach a screw eye and a spring scale to one end of a wooden block as shown. Use this device to experiment with friction. First, drag the block along a sidewalk, and note the amount of pull on the spring scale. Then, place a heavy weight on top of the block, and pull it across the same surface. Finally, run water on the sidewalk as you pull the block and weight. Also try pulling the block along a smooth surface like a table top or a waxed floor. What conclusions about friction can you draw from your experiment? ✐

If you placed the box of books on wheels, you could push it across the room much more easily because there would be only a small amount of friction between the wheels and the floor and the wheels and their axles. Friction is not completely eliminated, however. The rolling box would begin to slow down as soon as you stopped pushing it.

Figure 3–13. As these boats speed across the water, they encounter both drag and air resistance.

What a Drag!   Friction exists with liquids and gases as well as solids. Imagine that you are piloting a motor boat through the water. Wind and the spray of water blow past you. At top speed, you are skimming through the water at 50 km/h.

As the boat moves, its lower surface passes through the water and its upper surface passes through air. There is no direct contact between solid surfaces. However, the boat still encounters friction. Even though the boat's motor continues to apply a force, the boat stops accelerating when it reaches a speed of 50 km/h. And when the force of the motor stops pushing, the boat quickly comes to a stop.

Both the water and the air offer an opposing force against the moving boat. Friction between a solid surface and a liquid or a gas is sometimes called *drag*. When the gas is air, however, the friction is usually called *air resistance*.

Figure 3–14. In order to achieve high speeds, automobiles must encounter little air resistance. Automobile designers use wind tunnels like the one shown here to check the air resistance of their designs.

How Do They Go So Fast? The top velocity of speed skiers is 224 km/h. They can go this fast because they minimize air resistance. The friction between the skis and snow is very small. The main drag on speed is the air in front of the skiers. Skiers can cut air resistance by decreasing the amount of body exposed to the air. That's why they crouch while skiing. In what other ways can skiers decrease drag to increase speed?

Poles held
close to body

Aerodynamically
designed helmet

Tight-fitting,
smooth-surface
clothing

Buckles on boots
align with surface

Figure 3–15. A downhill skier's clothing and equipment minimize air resistance.

ASK YOURSELF

What factors determine the amount of friction between two solid surfaces?

Gravity

As you read at the beginning of this section, gravity is the force that pulls bobsledders and skiers down a mountain. In the seventeenth century, Sir Isaac Newton discovered that all objects with mass attract each other. In fact, he determined that gravity is a fundamental force of nature.

When you hold an object and then let it go, the object accelerates toward the earth because of the force of gravity pulling it. When you throw a baseball, it eventually hits the ground because the force of gravity pulls it down. Houses must have support beams to oppose the force of gravity that would otherwise pull the roof down. The moon circles the earth, and the earth circles the sun instead of heading off in another direction because the force of gravity keeps them in orbital paths. Gravity is everywhere in the universe because objects that have mass are present throughout the universe.

Falling Down   The force we call *gravity* is described by the universal law of gravitation, or the law of gravity. The law of gravity says that the strength of an object's gravitational force depends on its mass. The more mass an object has, the more gravitational force it exerts on other objects.

When you drop a fork in the kitchen, the fork falls straight down because the earth has a very large mass and a very large gravity to match. The refrigerator is also pulling on the fork, but you do not see the fork fall toward the refrigerator because, compared to the earth, the refrigerator has a very small mass and thus a very small gravitational pull. Sensitive instruments can measure the gravity pull between the refrigerator and the fork, however.

So the force of gravity between two objects depends on both of their masses—the bigger the mass, the bigger the force. The law of gravity also states that the farther apart two objects are, the weaker the gravitational force that pulls them together.

Figure 3–16.
Gravitational force, represented by the arrows, varies with the mass of the objects and their distance from one another.

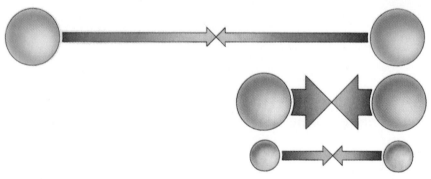

DISCOVER BY *Doing*

Use a slide to observe the acceleration of objects toward the earth caused by gravity. Go to the top of the slide, and release, at exactly the same time, a large glass marble and a steel ball bearing of the same size. Have someone at the bottom of the slide note when each ball reaches the bottom. What conclusions can you draw about their acceleration? ✎

How Heavy Is It? Any discussion of gravity requires a discussion of the meaning of the words *mass* and *weight.* Many people confuse those two words because they are so closely related. Mass is the amount of matter in an object. You may recall that SI units of mass include the kilogram (kg) and the gram (g). **Weight** is the force of gravity pulling one object toward the center of another object. Usually you think about gravity pulling an object toward the center of the earth. Because weight is a force, weight is measured in newtons (N).

You have probably heard people say they are trying to "lose weight." What they really mean is that they are trying to lose mass, because they want their bodies to have less matter. Of course, as long as they stay on Earth, they will lose weight when they lose matter. What if you were a space traveler? Since weight identifies the gravitational pull of objects, someone who weighs 600 N on Earth would weigh only 100 N on the moon. The moon has only one-sixth the gravitational pull of the earth and so attracts with only one-sixth the force. Although you would weigh less on the moon, the amount of matter would still be the same in both locations. To lose mass, moving to another planet isn't the answer!

Figure 3–17. Astronaut Edwin (Buzz) Aldrin weighed less while he was on the moon than when he was on Earth. His mass, however, was unchanged. Why?

 ASK YOURSELF

What is the difference between mass and weight?

SECTION 2 *REVIEW AND APPLICATION*

Reading Critically

1. What is the relationship between force and acceleration?

2. Explain why friction is a force.

3. What is gravity?

4. Give two examples of objects that are accelerating but are not speeding up. Explain your answer.

Thinking Critically

5. Two identical cars are traveling at identical speeds on identical roads. The only difference is that the second car is traveling in a rainstorm. Which car requires the greater force to move it forward? Explain your answer.

6. Imagine that the sun suddenly collapses into a black hole, with all of its original mass compressed into the volume of a marble or a bead. If the center of the black hole is in the same place as the current center of the sun, how will that affect the earth's orbit around the sun?

SKILL Interpreting Diagrams

▶ **MATERIALS**
 • paper • pencil

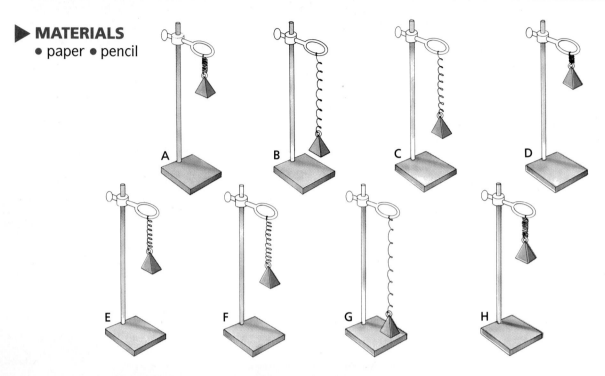

▼ PROCEDURE

1. Sketch the diagrams shown. If they are not in the correct sequence, reorder them as you do this step.
2. Show the direction in which the weight is moving with an arrow. If the weight is not moving, do not add an arrow.
3. Draw an arrow to show the force exerted by the spring on the weight. Make the length of your arrow represent the size of the spring's force relative to the size of the force of gravity. For example, if the spring's force is much greater than the force of gravity, the arrow you draw for it should be longer than the gravity-force arrow in the diagram.
4. Determine the direction in which the unbalanced force is acting, and show it in your diagram.
5. Indicate the direction of acceleration.

▶ **APPLICATION**

Does the weight on the spring ever move without any acceleration or unbalanced force acting on it? Explain why.

✳ ***Using What You Have Learned***
1. When are the acceleration and the direction of motion the same? When are they different?

2. Explain how it is possible to have acceleration in one direction and motion in the other.
3. Is acceleration always in the same direction as unbalanced force? Why or why not?
4. Is it wrong to say that the brakes of a car cause an acceleration opposite to the direction in which the car is moving?

Predicting and Explaining Motion

An Olympic bobsled race is an exciting place to observe motion, but you also observe objects in motion every day. In the late 1600s, Sir Isaac Newton formulated three laws that describe motion. These three laws identify some universal characteristics of objects, both in motion and at rest. The laws of motion can help you understand why objects move the way they do.

The First Law of Motion

Have you ever been in a car that had to stop suddenly? If so, you probably remember that your body kept traveling forward, even after the car had stopped. Your seat belt provided a balancing force that stopped you from continuing forward. You experienced one aspect of the first law of motion.

This law states that every object maintains constant velocity unless acted on by an unbalanced force. Unless an object receives an unbalanced push or pull, it will just keep doing what it is already doing. In general terms, this law says that objects at rest stay at rest and objects in motion with a constant speed stay in motion unless they receive an unbalanced push or pull, to change the objects' velocity.

Figure 3–18. In a car crash, the driver and passengers keep moving forward until they are stopped by another force. If seat belts are not worn, this stopping force may be the car's windshield.

Move It   Suppose you put two balls, a bowling ball and a soccer ball, in the middle of a level gym floor. Neither ball will begin to roll on its own. If you apply an unbalanced force to the balls, they will start to roll. However, because the soccer ball has less mass, less force is required to start it rolling. Much more force is required to start the bowling ball rolling.

Now suppose that a friend of yours has started both balls rolling toward you so that they are both traveling at the same velocity. Which ball will be harder for you to stop? The bowling ball, of course! The bowling ball is harder to start and harder to stop because of its greater mass.

The first law of motion also states that objects with greater mass require more force to change their velocity. The resistance any object has to a change in velocity is called **inertia** (in UHR shuh). This is why it is much harder to start and stop the bowling ball than the soccer ball. Because the bowling ball has greater mass, more force is required to start or stop it. It has more inertia than the soccer ball. This property makes small cars easier to handle than big ones and small football players easier to stop than big ones.

Figure 3–19. Because of the inertia of the objects on this table, the table setting remains intact momentarily as the table is quickly slid from underneath it.

Weight Loss Recall that inertia depends on the mass, not the weight of an object. Objects have inertia regardless of where they are. Inertia even applies to objects in space. In space, massive objects such as support beams will seem to have no weight even though they have large mass. However, once a beam is put in motion, slowing it down or stopping it will require a very large force. The inertia of the beam will be proportional to its mass, not its apparent weight. Try the next activity to learn more about inertia.

Figure 3–20. Astronauts rescuing satellite

ACTIVITY

How can you study the effects of increasing mass on inertia?

MATERIALS
dynamics cart with string tied to both ends, springs (2), string, 1-kg masses (6), stopwatch

PROCEDURE
1. Set up the dynamics cart as shown.
2. Make a table like the one shown.

TABLE 1: MASS AND INERTIA		
Trial	Mass of the Cart	Added Mass (kg)

3. When the cart is not moving between the two supports, it is at its equilibrium position. Displace it from its equilibrium position by pulling or pushing it toward either support.

4. After moving the cart from its equilibrium position, let it go and measure how many seconds it takes to repeat its back-and-forth motion 10 times. Each back-and-forth motion is called a period. Divide the time for 10 periods by 10 to get the average period, or time, for one back-and-forth motion. Do three trials, and average the results. Record the results in your table.
5. Add a 1-kg mass to the cart, and repeat steps 3 and 4.
6. Repeat the activity with 2 and then 3, 1-kg masses added to the cart.

APPLICATION
1. Construct a line graph with the period versus the added mass, putting the total added mass on the horizontal axis and the period on the vertical axis.
2. Use the first law of motion to explain the relationship shown in your graph.

 ASK YOURSELF

You have a research paper due next week but haven't started it. Your English teacher told you that you needed to overcome your inertia. Explain why he used this term.

The Second Law of Motion

The best way to understand the second law of motion is to imagine that you are rolling snow into a ball. When the ball is small, you can roll it with little force. You have little trouble accelerating the ball from a standstill. However, as the ball gathers more snow and increases in mass, more force is needed to roll it.

Figure 3–21. Rolling snowballs illustrates the second law of motion.

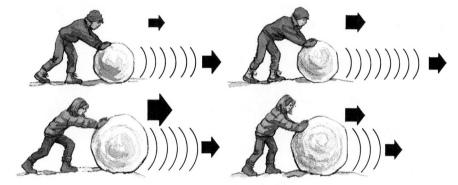

You have to apply more force to the large snowball than to the small snowball to produce the same acceleration. If you applied the same force to each of the snowballs, the small one would accelerate more quickly than the large one. When unbalanced force increases, acceleration will increase. When mass increases, acceleration is inversely proportional to its mass.

The second law of motion describes the relationship between acceleration, force, and mass. One part of this law states that any time an unbalanced force is applied to an object, the object accelerates. The law further states that the amount of acceleration depends both on the mass of the object and the amount of force applied to it.

When subjected to the same unbalanced force, a smaller mass receives a bigger acceleration than a larger mass. This point is illustrated in the photographs of the Olympic athletes.

Figure 3–22. Because of the different masses of these two objects, a similar amount of force imparts very different accelerations.

The second law of motion applies to many common situations. For example, the mass of a car determines how much force its brakes must apply to make it stop. Accidents sometimes occur when people overload their car and then assume the car will stop as easily as it did when empty. An empty car has less mass than the loaded car, so the braking force produces a greater stopping acceleration. A loaded car has more mass. If braking force and velocity are the same, the stopping acceleration of the loaded car will be much smaller than that of the empty car and it will take more time to stop it. This can be expressed as a mathematical equation.

$$\text{acceleration} = \frac{\text{force}}{\text{mass}} \quad \text{OR} \quad a = \frac{F}{m}$$

In the equation, a is the acceleration of an object, F is the unbalanced force pushing or pulling on the object being accelerated, and m is the mass of the object being accelerated.

 ASK YOURSELF

Why is it impossible to roll a bowling ball and a baseball at the same acceleration by using the same force?

The Third Law of Motion

Imagine that you and a friend are warming up for a hockey game. Suddenly, your friend skates into your path, so you give him a push. He rolls away as you intended, but to your surprise, you find yourself rolling off in the other direction. What happened? You have just experienced the third law of motion!

The third law of motion states that for every force there is an equal and opposite force. The movement of you and your friend shows that it is impossible for you to push something (action) without receiving an equal push in the opposite direction (reaction). These pushes, or forces, are often called *action forces* and *reaction forces*.

The original force is the action force. In this example, the push of your hand on your friend's back is the action force. The resulting opposite and equal force is the reaction force. In this case, the push of his back on your hand is the reaction force, and it is this reaction force that causes you to travel backward, away from the spot where you pushed him.

Figure 3–23. These skaters demonstrate the third law of motion. When one skater pushes the other, they both move in opposite directions.

The third law of motion applies to all other forces. You move when you walk because the force that your feet apply to the earth is matched by a reaction force that the earth applies to your feet. Remember that you have a lot less mass than the earth. Because of that, your motion when the earth pushes on you is a lot greater than the earth's motion when you push on it. In fact, the earth's motion when you push it is so small that it cannot even be measured. It is the reaction force that moves you forward. To test the third law, try the next activity.

DISCOVER BY Doing

Use two spring scales linked together to suspend a 1-kg mass. What does each scale read? This phenomenon relates directly to the third law of motion. Explain how. ✎

► ASK YOURSELF

Explain how action and reaction forces affect an object.

Conservation of Momentum

This is it—the last frame of the game. If you make this strike, you will win the bowling tournament of champions. You pick up the heavy ball, wipe it off, and grip it firmly. You line up at the starting point, hold the ball slightly above waist level, and face the pins. Moving forward, you slide gracefully close to the foul line and release the ball. The crowd watches. No one even whispers. The ball rolls toward the pocket between the head pin and the number three pin. Wham! The ball hits the pins, and they go spinning in all directions. You've made the strike!

The total momentum of the bowling ball and pins remains the same whether the bowler knocks down all ten pins or just a few. The momentum of the ball just before it hits the pins is equal to the combined momentum of the pins and ball just after impact. This example illustrates the **law of conservation of momentum.** This law states that momentum can be transferred from one object to another but cannot change in total amount.

Figure 3–24. The momentum of the ball before impact is equal to the total momentum of the pins and the ball after impact.

Testing the Law The law of conservation of momentum is fairly easy to see on a pool table. If one pool ball is shot directly at a second, identical pool ball, the first one will stop or "stick" when the two collide and the second one will shoot forward at the velocity the first one had just before the collision. Because they are identical, their masses are the same. Since the second ball has a velocity after collision equal to the velocity of the first one before collision, the momentum of the first ball has been transferred to the second ball during the collision.

 by Writing ——————————————

Imagine you and a friend are skating. Instead of skating at a slow speed, you are both skating fast. How would the reaction between gently pushing someone while skating at a slow speed to forcefully pushing someone while skating at a fast speed compare? Write your comparison in your journal.

Saving Up Rockets work in outer space because momentum is conserved. When fuel is forced out the back of the rocket, the change in momentum of the fuel is offset by the rocket's momentum change in the opposite direction. The total momentum of the rocket and the fuel remains the same.

ASK YOURSELF

How is momentum conserved when a batter hits a baseball?

Figure 3–25. In a simple water rocket, momentum from moving water is transferred to the rocket to make it rise. In this process, momentum is conserved.

SECTION 3   *REVIEW AND APPLICATION*

Reading Critically

1. How do the acceleration of a small rock and a large rock compare if they are thrown with the same force?

2. You kick a soccer ball, and the action force sends the ball flying through the air. What is the reaction force?

3. How does inertia relate to the safety of passengers in a car as it comes to a stop?

Thinking Critically

4. How is it possible for an object with a large mass acted on by a large gravitational force to have the same acceleration as an object with a small mass acted on by a small gravitational force?

5. A space station construction worker found herself floating free 100 m from the space station because her safety line somehow became unhooked. Attached to her spacesuit were her unhooked safety line, her tool belt and tools, and her oxygen tank. How could she get back to the space station without calling someone for help? In answering, explain the physics of your proposed method.

INVESTIGATION

Calculating Force, Mass, and Acceleration

▶ MATERIALS
- pulley with table clamp • dynamics cart • masking tape
- string • can • sand • standard masses • blocks of about the
same mass as the dynamics cart (4) • meter stick • stopwatch

▼ PROCEDURE

1. Set up the apparatus and make a chart like the one shown.

2. Put enough sand in the can so that when released the dynamics cart will roll at a constant velocity.

3. Remove the middle strip of masking tape to make one 0.50-m interval.

4. Place the cart in front of the first strip of tape and add a 5-g mass to the can.

5. On the timer's signal, release the cart and measure the time taken to travel the 0.50-m distance. Repeat three times, and average the values. Record the average time.

6. Repeat steps 4 and 5 until 25 g have been added to the can.

7. Add one block to the cart, and repeat step 5.

8. Add 25 g to the can, and repeat steps 5–7. Record the results in your chart.

9. Repeat step 5, adding one

TABLE 1: CALCULATING FORCE, MASS, AND ACCELERATION				
Mass Added to Can (g) (m/s/s)	Blocks Added to Cart	Average Time to Go 0.50 m	Calculated Quantities Force (N)	Acceleration

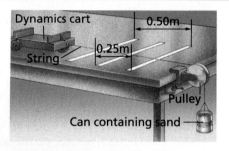

Dynamics cart · 0.50m · 0.25m · String · Pulley · Can containing sand

block each time until all 4 blocks have been added to the cart.

10. To calculate the unbalanced force, assume 0.01 N for every gram of mass added to the can in addition to the sand added to offset friction.

11. To calculate acceleration, use the following formula

$$a = \frac{d}{t^2}$$

12. Graph the results for steps 4–6 by plotting acceleration on the Y axis and force on the X axis.

13. Graph the results for steps 7–9 by plotting acceleration on the Y axis and cart mass on the X axis.

▶ ANALYSES AND CONCLUSIONS
In steps 5–7 you did not change the mass. What is the relationship between the acceleration and the force used to move the cart?

▶ APPLICATION
A delivery truck with a mass of 5 tonnes, which is traveling at a speed of 15 m/s, requires

20 m to stop. What is the stopping distance of a 10-tonne truck traveling at the same speed?

✳ Discover More
Repeat the Investigation, but use plastic jars of water that weigh about as much as the blocks. How do the results compare?

HIGHLIGHTS

The Big Idea

Motion can be described, explained, and predicted using a few interconnected ideas. If you can measure the distance an object travels in a certain time, its direction, and how fast its speed or direction is changing, you can describe its motion completely. If you want to know how hard it is to make the object change its motion, you must also measure its mass. To change an object's motion, you must apply a force to it by somehow pushing or pulling it. Common forces include friction and gravity. The three laws of motion explain why things move and why they come to a stop.

Look back at the ideas you wrote in your journal at the beginning of the chapter. How have your ideas changed? Revise your journal entry to show what you have learned about the laws of motion and how they allow us to predict how things will move.

Connecting Ideas

By looking at the illustration and writing a short essay, you can show how the big ideas of this chapter are related. In your journal, write an essay that describes what is happening or what is about to happen in the pictures.

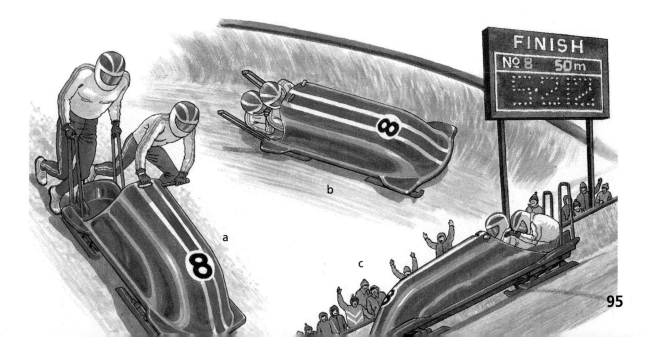

REVIEW

Understanding Vocabulary

1. Define the terms in each set in a way that explains the relationship of each term to the other.

a) speed (72), velocity (74)

b) acceleration (75), momentum (77)

c) force (78), friction (81), gravity (83)

d) action force (91), reaction force (91)

Understanding Concepts

MULTIPLE CHOICE

2. If you jog for one hour and travel 10 kilometers, 10 kilometers per hour describes your

a) momentum.

b) average speed.

c) velocity.

d) speed and velocity.

3. Which of the following objects is *not* accelerating?

a) a ball being juggled

b) a horse racing on an oval track

c) a bowling ball returned to you on a straight, level track

d) a braking bicycle

4. It would be easier to push a box of fruit up a board and into a truck if you

a) took some fruit out of the box.

b) covered the board with rough carpeting.

c) packed the fruit carefully.

d) There is no way to know.

5. A person's weight is less on the moon because

a) the moon has less mass than the earth.

b) the person's mass is less on the moon.

c) there is no gravity on the moon.

d) the moon circles the earth.

6. A loaded delivery truck and a compact car are both traveling 88 kilometers per hour. The truck will

a) have less inertia than the car.

b) travel the same distance more quickly than the car.

c) require less force to travel 100 kilometers per hour.

d) take longer to stop for a red light.

SHORT ANSWER

7. When could a small object have the same momentum as a large object?

8. Why does an arrow shot from a bow soon start to curve downward?

Interpreting Graphics

9. Look at the figure below. Which ball will travel faster? Explain why.

5N

10. Refer to the figure again. Which ball will roll back further in the opposite direction once it hits the wall? Explain your answer.

Reviewing Themes

11. *Systems and Structures*
A baseball and bat can be thought of as a system. Explain how they interact in accord with the law of conservation of momentum.

12. *Energy*
Energy may do many things; it may cause change or motion. How is energy related to the act of pushing a book off a table top?

Thinking Critically

13. How does the tread on tires help a car to stop quickly? Use your understanding of friction in your explanation.

14. Imagine that the earth is now twice as far from the sun. How would the earth's mass change? How would the gravitational attraction between the earth and sun change?

15. A marble and a basketball dropped at the exact same moment from the exact same height will accelerate at the same rate and reach the ground together. Using what you know about the force of gravity and the second law of motion, explain why this is true.

16. Imagine that a bathtub filled with water is rolling on frictionless wheels in a straight line along a level road. If the stopper pops out and the water drains, the momentum of the tub will lessen even though its velocity stays the same. Why is this true?

17. Why will a boat use more fuel (require more force) to travel 32 kilometers per hour against the wind in a rainstorm than it will on a sunny day with no wind?

18. Exercises such as running, swimming, calisthenics, and aerobics are designed to make you healthier and stronger. When you do these exercises, you are actually moving all or parts of your body in directions that oppose some invisible forces. What are the forces that oppose the following exercises: push-ups, running in place, running around a track, and swimming the length of a pool?

Discovery Through Reading

Macaulay, David. *The Way Things Work*. Houghton Mifflin, 1988. This award-winning book explains how a variety of machines depend on physical forces so they may operate.

WORK

*E*migrants bound for Oregon in the 1840s would have envied the motorized machines we have today for hauling materials. The hardships of the trek to the West gave them life-and-death lessons in the science of work.

Albert Bierstadt, "The OregonTrail" 1869, 31" x 49", oil on canvas.
The Butler Institue of American Art.

These days a moving van easily hauls a family's possessions from Missouri to Oregon in just a few days. Fresh produce speeds across half a continent in a few hours, before the peaches ripen and the melons spoil. A century-and-a-half ago, the same journey required immense physical labor.

The two-hundred-yard crossing was made difficult by treacherous currents and varying depths of fast-moving water. Groups of wagons had to be chained together, one behind the other, while extra teams of oxen alternately swam and walked as they pulled the heavily loaded wagons through the water. The wagons were guided across the river by horsemen who kept the oxen heading in the right direction. The wagons themselves were made as watertight as possible by having their cracks stuffed with clothing or cloth sheets soaked in tar or pitch. Most of the supplies and family possessions inside the wagon were either piled high off the wagon floor to keep them dry or unloaded altogether and put on a raft. At times the wagon wheels were well above the water. And just as suddenly they would disappear into deep water, flooding the wagon's interior. One emigrant wrote that they had to

" . . . ferry the wagons over. Had to take them apart and float the box (i.e. the wagon without its wheels) . . . got some groceries wet, some coffee, sugar dissolved . . ."

from *The Oregon Trail*
by Leonard Everett Fisher

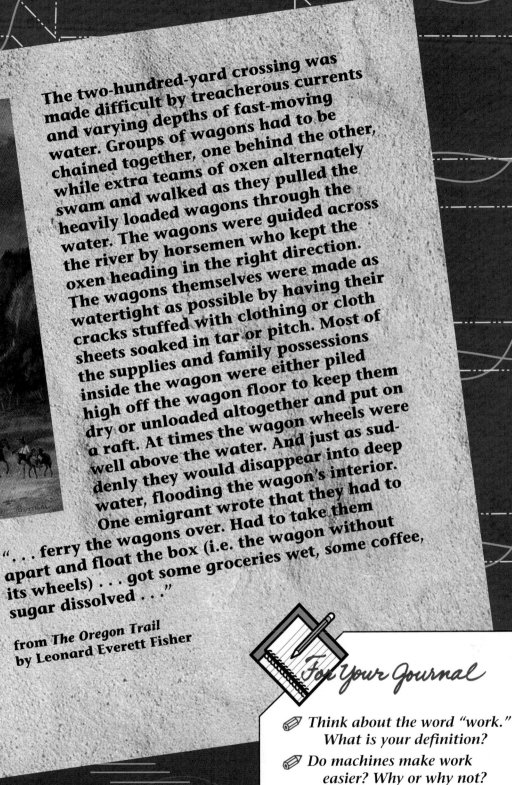

For Your Journal

- *Think about the word "work." What is your definition?*
- *Do machines make work easier? Why or why not?*
- *How are machines useful in your everyday life?*

Work and Power

Define work *and* power.

Explain how machines can multiply force without multiplying work.

Use the concept of mechanical advantage to explain how machines make work easier.

Imagine that it is 1843. You and your family have been traveling by foot, horseback, and wagon for six months. The 3218 kilometer journey along the Oregon Trail was grueling; many people and animals died along the way. But at last you have arrived in the Oregon Territory and are surrounded by vast evergreen forests. Here you will build a new home, a new life.

First Things First

By the time you reach the Oregon Territory, it is already late fall, too late to plant crops or build a home. Therefore, you and your family stay with others who have homes and plenty of food. During the winter months, you listen to their stories about life in the beautiful, but harsh, Willamette Valley.

At last spring arrives! Clearing the land is the first task. Crops need to be planted to provide the food supply for next winter. The need to clear the land is so great that you will put together a temporary shelter and live in that until the crops are planted. The trees that you clear from the land will be used later to make a log cabin and barn.

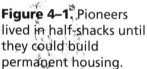

Figure 4–1. Pioneers lived in half-shacks until they could build permanent housing.

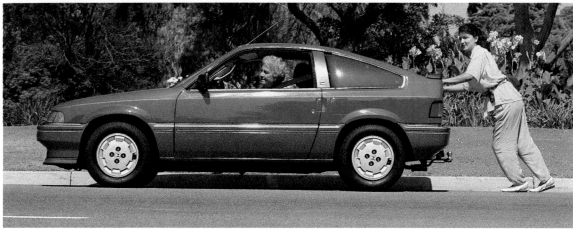

Figure 4–2. The work this person is doing while pushing the car is equal to the force she applies to the car multiplied by the distance the car moves.

You take a look at one of the smaller trees and figure you could probably push it over without too much effort. You push and push, but the tree doesn't move. Your hands and arms begin to ache, and you start sweating. Hard work, right? Wrong! You haven't done any work at all. To have done work, your push would have had to move the tree.

In science, *work* does not mean just making an effort. Work means accomplishing something. More precisely, **work** means not only applying a force but also using that force to make something move. In order to do work on the tree, you have to make it move. The work you do on the tree is the force you apply to it multiplied by the distance you make it move. The equation for work is

$$\text{work} = \text{force} \times \text{distance} \quad \text{OR} \quad W = F \times d$$

Pushing on a wagon or a barrel might allow you to do work on those objects, but you probably won't do much work pushing on a tree. To do work on a tree, you must exert a large force. Some of the other pioneers would recommend that you use a machine—an ax. A **machine** is a device that helps you do work by changing the size or direction of a force. Why do you think an ax is considered a machine?

Although the pioneers did not have modern-day farm machines to clear the land, they did have machines such as the ax and plow. They either brought these machines on the journey or made them once they arrived at their destination. In the next section, you will learn about different types of machines.

 ASK YOURSELF

Why is no work being done when you push on the side of a building?

Work, Work, Work

The land has been cleared, and the crops have been planted. Already the days have grown shorter; there is less daylight in which to get chores done. Now it is time to build a log cabin and after that, a barn. To do the work quickly, you need lots of power.

Power, like work, is a word that means one thing in ordinary language and another in the language of science. In everyday language, power is often used to mean "strength." Power has even been used as a synonym for force. In science, however, **power** is the amount of work done in a certain period of time. Power is a rate, like speed. As you read in the last chapter, speed is the distance traveled in a certain period of time. Since power can tell us the time it takes to do a certain amount of work, motors and engines are given power ratings to identify how fast they can do work. In mathematical terms, power is represented as

$$\text{power} = \frac{\text{work}}{\text{time to do work}} \quad \text{OR} \quad P = \frac{W}{t}$$

Figure 4–3. Although these runners are doing the same amount of work, they do not all have the same power. The runner in front is using more power than the others. Why?

Figure 4–4. In most situations, the force applied in doing work is not constant. In these situations, average force is used to calculate work.

DISCOVER BY *Writing*

Look up the words *work* and *power* in a thesaurus, and write the commonly accepted synonyms for each of them in your journal. Use a dictionary to look up the meanings of the synonyms with which you are unfamiliar. Tell whether each synonym is also a synonym for *work* and *power* as these words are used in science. ✎

More Power to You   To help you build your log cabin, you could ask your neighbors for assistance. Many people working together supply more power than just one person. They can do the same work at a greater speed, or they can do more work in the same amount of time. Try the next activity to find out the relationship between power and speed.

DISCOVER BY *Calculating*

Connect a spring scale to an object that you can pull across a table. Then slowly pull the object across the table while you take a reading on the spring scale. Repeat this procedure, only this time pull the object quickly. Finally, measure the distance across the table. Use this measurement and the spring scale readings to calculate the work and the power developed when pulling the object. What is the relationship of the speed with which the object is pulled to the amount of power developed? ✎

Measuring Force and Work   In SI, force is measured in *newtons* (N). Recall that a newton is the force required to cause a 1-kg mass to accelerate 1 m per second each second. For instance, if you held a 1 kg mass in your hand, the force pulling down on your arm would be about 9.8 N.

In SI, work is measured in **joules** (J) (JOOLZ). If you lift a weight of one newton one meter from the ground, you do one joule of work. In the next chapter, you will learn that energy is also measured in joules.

The name of the metric unit that measures 1 J of work done each second is the *watt* (W). One watt is a very small amount of power. Therefore, the kilowatt (kW), equal to 1000 W, is more commonly used to measure large amounts of power. You have probably heard the term *kilowatt* used in association with electric power. However, it is important to note that the watt is the standard unit of power, regardless of how the power is produced. In the next activity, you'll have a chance to measure how much power you can generate.

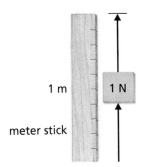

Figure 4–5. One newton lifted 1 m requires 1 J of work.

ACTIVITY

How much power do you generate in climbing a flight of stairs?

MATERIALS
stopwatch, meter stick

PROCEDURE

1. Determine your weight in newtons. If your school has a metric scale that provides mass in kilograms, multiply kilograms by 9.8 m/s/s to get your weight in newtons. If your school has a scale that weighs in pounds, multiply your weight by 4.45 N/lb to get your weight in newtons.

2. One student should use the stopwatch to time how long it takes each student to walk quickly up the stairs. Another student should record the times.

3. Measure the height of one step in meters. Multiply the number of steps by the height of one step to get the total height of the stairway.

4. Multiply your SI weight by the height of the stairs (in meters) to get the work you did in joules. The equation for work is
work = force × distance OR $W = F \times d$

5. To get your power in watts, divide the work you did by the time (in seconds) that it took you to climb the stairs.

6. Record your name and your power rating on the chalkboard.

APPLICATION

1. Who in your class was the most powerful? Was it the person you expected it to be?

2. Do the power ratings of the students provide any indication of endurance? Explain.

3. How powerful is someone who can lift a 100-kg mass a distance of 1 m in 1 s?

An elephant and a mouse raced each other up the stairs. The mouse beat the elephant by a full second but the elephant claimed, "I am more powerful than you are, and this race proves it." How can that possibly be true when the mouse won the race?

May the Force Be With You

An ax helps you cut down trees by changing the size or direction of a force. When a machine is used to change the size of a force, it does so by either multiplying force or multiplying distance. No machine can do both.

The worker in the photograph is using a crowbar to open a crate. The end of the crowbar pushes the top of the crate up with great force. However, the end of the crowbar that is pushing up on the crate moves only a short distance. The other end of the crowbar receives a much smaller force but moves a larger distance.

In this example, not only is the force multiplied, but the direction of the force is also changed. When you push down on the crowbar, the top of the crate is forced up. In many cases, a change in the direction of the force makes work easier.

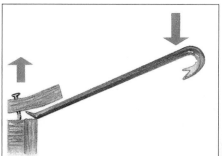

Figure 4–6. This crowbar converts a small force applied over a large distance into a large force applied over a small distance.

Figure 4–7. Efficiency is increased when friction is reduced. For example, skiers use wax to reduce friction between the skis and the snow.

A machine does not multiply work. Because of friction, which you read about in the last chapter, less work is always received from a machine than is put into it. When distance is increased, force is reduced. Perhaps you've used a screwdriver to open a can of paint. As you move one end of the screwdriver, the other end moves in the opposite direction. Your hand moves much farther than the paint-can lid. Increasing the distance you move decreases the force necessary to open the can. In either case, the work put into the machine is always greater than the work you get out of it.

ASK YOURSELF

Explain why a machine does not multiply work.

The Advantage of Machines

Pioneers used machines because machines multiply force or distance to make work easier. People today use machines for the same reason. The number of times a machine multiplies force is its **mechanical advantage (MA)**. Mechanical advantage is the ratio of the force that comes out of a machine to the force that is put into the same machine. This is expressed as

$$\text{mechanical advantage} = \frac{\text{force output}}{\text{force input}}$$

Some machines, such as a hammer, multiply distance instead of force. As force is increased, distance is reduced. When you use a hammer to pound in a nail, your arm moves a small distance while the hammerhead moves a much larger distance. You have increased the output distance of the hammerhead by reducing the output force. However, because the hammerhead has a large mass, its momentum increases as its speed is increased. An increase in the hammer's momentum increases the amount of force it can apply to the nail.

Figure 4–8. Without machines, it would probably take the collective force of hundreds of people to move this large stone.

Machines make work easier by changing the amount of force, distance, direction, or speed. However, machines never reduce the amount of work. They actually make more work by adding friction. Because of friction, no machine is 100 percent efficient. You can never get as much work out of a machine as you put in because some of the work must always go to overcome friction. Lubricants can reduce friction and increase efficiency.

 ASK YOURSELF

What is mechanical advantage?

SECTION 1   *REVIEW AND APPLICATION*

Reading Critically

1. In terms of science, how do work and power differ?

2. Explain why a machine can't be 100 percent efficient.

Thinking Critically

3. How might your life be different if there were no machines?

4. It requires just as much work to stop a pole-vaulter landing in a foam-filled pit as it does to stop a pole-vaulter landing on asphalt. However, a pole-vaulter landing on asphalt is subjected to much more power. Explain why this is true.

SKILL Comparing Force and Work Using a Spring Scale

▶ **MATERIALS**
- books (4) ● ruler ● wooden block with an eye hook
- spring scale ● 20-cm × 30-cm piece of heavy cardboard

▼ **PROCEDURE**

1. Make a table like the one shown to record your data.

TABLE 1: COMPARING FORCE AND WORK
Force × Height =
Force × Length =

2. Stack the books on the floor or on a table. Measure how high the stack is in meters.

3. Attach the block of wood to the spring scale. Lift the spring scale and block to the top of the stack. Record the amount of force needed in newtons.

4. Calculate the work you did by multiplying the force by the height of the books. Record your calculations.

5. Now place the cardboard against the books to make a ramp. Attach the block to the spring scale again. Place the block at the bottom of the ramp.

6. Pull the block up the ramp. Record the amount of force needed.

7. Calculate the work you did this time by multiplying the force by the length of the ramp. Record your calculations.

▶ **APPLICATION**

Look at your data. Which takes more force—lifting a block or pulling it up the ramp? Which takes more work? Explain.

✳ ***Using What You Have Learned***

Suppose you were going to teach someone to compare force with work using a spring scale. What directions would you give?

Types of Machines

Objectives

Name and **describe** the six types of simple machines.

Evaluate the mechanical advantage of simple machines.

Design and **construct** a compound machine.

The logs for your new home have been prepared. You used an ax to chop notches close to the ends. The notches will hold the logs to each other when they are fitted together to form the sides of the cabin. The sides of the log cabin will be about 2.4 meters high. You cannot lift the heavy logs without help, so you've gathered your neighbors together for a house-raising. However, neither the force you can apply with your muscles nor that of your neighbors will be enough to accomplish your goals. What is the solution?

One of your neighbors who has assisted at many house-raisings takes charge. He and your other neighbors use long pieces of wood to roll a log into place at the bottom of the cabin. Then they use poles and rope to roll the log up to the top of the cabin. Construction takes place before your eyes! What did your neighbors know that you didn't? They knew how to use the resources around them as machines. The machines they used are **simple machines,** machines that have only one or two parts. There are six simple machines: the lever, the wheel and axle, the pulley, the inclined plane, the wedge, and the screw. Which of these machines might be used at a house-raising?

Levers

Remember the long pieces of wood the workers put under the log to move it to the bottom of the cabin? They were using the wood as a lever. A **lever** is a bar used for prying or dislodging something. Many common tools, such as crowbars, rakes, and wheelbarrows, are examples of levers.

The part of the lever that you push on is called the *effort arm.* The part that pushes on the object you want to move is called the *resistance arm.* The point on which the lever pivots is called the *fulcrum.* There are three ways to arrange the parts of a lever. The arrangement of the parts determines whether the levers are first class, second class, or third class. The three classes of levers are illustrated in Figure 4–9.

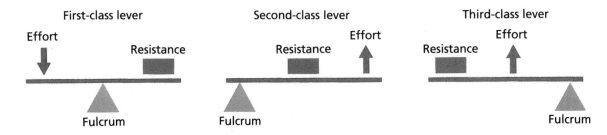

First-class lever   Second-class lever   Third-class lever

Effort   Resistance   Resistance   Effort   Resistance   Effort

Fulcrum   Fulcrum   Fulcrum

Figure 4–9. There are three classes of levers, as shown here.

One kind of lever you are probably familiar with is the seesaw in a playground. The support for the seesaw is the fulcrum, which allows the seesaw to pivot. The part of the seesaw resting on the fulcrum is the only part that will not go up or down.

Anyone who has been on a seesaw knows that changing how far you sit from the fulcrum of the seesaw changes how hard you must push down to make your partner go up. The farther you are from the fulcrum, the more your force is multiplied. Work becomes easier as you increase the length of the effort arm.

The mechanical advantage of a lever can be calculated by dividing the length of its effort arm by the length of its resistance arm. If you wanted to lift someone on the seesaw who weighed twice as much as you do, you would really need to sit a little farther than twice as far away because the friction requires you to put in a little extra force.

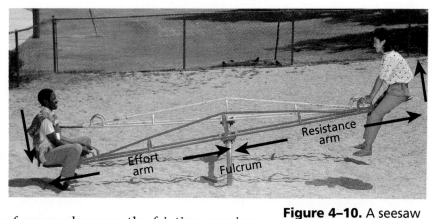

Figure 4–10. A seesaw is actually a first-class lever. The seesaw is supported at its center by a fulcrum.

Shovels, oars on a rowboat, balance scales, scissors, and pliers are all examples of levers. Identify the effort arm, the resistance arm, and the fulcrum of all these levers. What other tools can you think of that are levers? Try the next activity to find out more about the mechanical advantage of levers.

DISCOVER BY Doing

Use a meter stick, a pencil, and two 1-kg masses to observe the mechanical advantage of a lever. The meter stick will act as the lever, the pencil as the fulcrum, and the masses as the effort and resistance forces. Arrange the items as shown. Experiment with the lever by repositioning the fulcrum and one of the masses. What conclusions can you draw about the lever's mechanical advantage? ✐

ASK YOURSELF

Why does a long crowbar have a bigger mechanical advantage than a short crowbar?

Wheel and Axle

Have you ever made ice cream? There are electric ice-cream makers, and there are also the ice-cream makers that you crank by hand. When you make ice cream "by hand," it seems to take forever until the finished product is ready. It may appear that you are not using a machine at all, but you are. As a matter of fact, the ice-cream maker is an example of a **wheel and axle,** a simple machine consisting of a lever connected to a shaft.

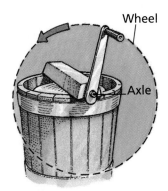

Figure 4–11. The crank on this ice-cream maker is a wheel and axle.

For something to be considered a wheel and axle, the wheel cannot spin on the axle. The pioneers used a wheel and axle to winch heavy loads of ore up from mines and occasionally to pull ferries across rivers.

Every time you turn a doorknob or a knob on a radio, you are using a wheel and axle. The knobs are the wheels, and the shafts are the axles. When you turn the knob, you are multiplying your force in turning the shaft.

Another example of a wheel and axle is the steering wheel of a car. A small force applied to the large steering wheel is multiplied as the smaller axle turns to create a very large force. In this application of a wheel and axle, the effort arm is the radius of the wheel, and the resistance arm is the radius of the axle. When a car is steered around a curve, the driver must move the steering wheel a large distance to move the axle a short distance. However, the small force on the steering wheel becomes a large force to turn the wheels of the car.

A wheel and axle also has mechanical advantage. Think about the common screwdriver. The handle of the screwdriver is the wheel, and the blade of the screwdriver is the axle. The radius of the wheel is the effort arm, and the radius of the axle is the resistance arm. This explains why it is easier to insert screws with a screwdriver that has a wide handle. The mechanical advantage is greater because the effort arm—the radius of the handle—is bigger than the resistance arm—the radius of the blade.

Figure 4–12. This person is using a wheel and axle in the form of a screwdriver to apply a twisting force to a screw.

 ASK YOURSELF

How does a wheel and axle make work easier?

The Pulley

Now both your new home and barn are complete, and the crops have been harvested. It is time to store the hand-bundled hay in the upper floor of the barn. What is the most efficient way to get the bundles up there? You could use a pulley. A **pulley** is a simple machine that consists of a wheel that is free to spin on its axle. This is different from the wheel and axle, which must move together. Most pulleys have a groove around the circumference to hold a rope or a cable. Force is applied to the rope that travels in the groove of the pulley. The rope is attached to the object to be moved, in this case, the bundle of hay.

Pulleys can be either fixed or movable. If they are fixed, they are attached to something that does not move, like a building or a tree. Fixed pulleys only spin; they do not move up and

MA=2 MA=3 MA=4 MA=?

down with the weight being lifted. Therefore, the mechanical advantage is one. It multiplies whatever force you put in by one. Remember, though, that because of friction it will really be less than one and you will get less force out than you put in.

In most cases, fixed pulleys are used to change the direction of force. They cannot be used by themselves to multiply force. By using a fixed pulley, you can pull down on the rope to cause an upward force on the hay bundles and get them to the upper floor of the barn.

Figure 4–13. The mechanical advantage of a system of pulleys can be determined by counting the number of ropes supporting the resistance of force, or load.

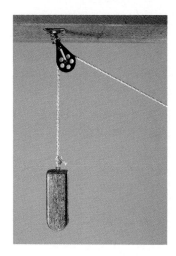
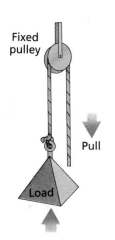
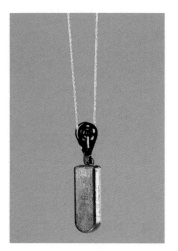
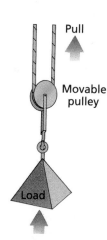

Movable pulleys, however, are attached to the object being moved. Unlike fixed pulleys, they move up and down with the weight being lifted. Movable pulleys are used to multiply force. They can multiply force even more when they are used with fixed pulleys in a *block and tackle,* as shown in Figure 4–13. Block and tackles are sometimes used by automobile mechanics to remove heavy engines from cars. By moving the end of the

Figure 4–14. The fixed pulley (left) is used to change the direction of a force. The movable pulley (right) is used to multiply force.

rope a long distance with a small force, a mechanic is able to lift the motor a small distance with a large force. This shows again the trade-off between force and distance. You can put a small force over a large distance into a machine and get a large force moving a small distance out of a machine.

 ASK YOURSELF

In what industries other than those mentioned might pulleys be useful?

I'm Inclined to Use a Ramp

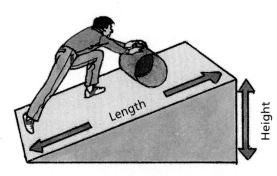

At the house-raising, your neighbors used a ramp to roll the logs to the top of the cabin. A ramp is really an **inclined plane**, a simple machine consisting of a flat, sloping surface. The mechanical advantage of a ramp can be calculated by dividing the ramp's length by the ramp's height. This tells you that a longer, more gently sloped inclined plane has a higher mechanical advantage than a short, steep one. How many other uses can you think of for ramps?

DISCOVER BY *Calculating*

Prop up a book so it forms an inclined plane. Attach a spring scale to a block of wood or other object. Using the spring scale, pull the object up the inclined plane. Record the amount of force needed to pull the object. Compare this force with the amount of force required to pull the object up inclined planes at different angles. How does the effort required to pull the object change with the slope of the inclined plane? ✐

In chopping down trees and processing them into boards and beams, you would use many different kinds of tools, including a wedge. A **wedge** is an inclined plane that is used to push objects apart. The head of an ax is a wedge. Other wedges include the blades of knives, saws, razors, and other cutting tools. Sharpening these tools makes the cutting edge thinner. This increases the mechanical advantage and makes them easier to use. Nails are also wedges and are often used to fasten boards together.

Another kind of fastener is the screw. A **screw** is an inclined plane that is wrapped around a cylinder. Screws are used on many items, such as screw-on jar lids, water faucets, and some kinds of automobile jacks. The mechanical advantage of the screw is much like that of the inclined plane. Looking at the threads on a screw is like looking at many tiny inclined planes. The more gentle the slope of these tiny ramps, the easier it is to turn the screw. That means it has a higher mechanical advantage. Remember though, if it takes less force to turn, it will require that you turn it a greater distance.

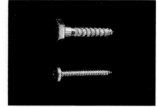

Figure 4–15. A screw is basically an inclined plane wrapped around a cylinder. This illustration shows the mechanical advantage of a screw with gently sloping threads.

gentle slope threads
high MA
small force required
GREAT DISTANCE

steep slope threads
low MA
LARGE FORCE REQUIRED
small distance

DISCOVER BY *Problem Solving* _____

Most of the examples of machines given in this section are machines that enable people to exert more force than their muscles are capable of. Describe one situation in which it is important for humans to exert less force than their muscles normally do. Write your description in your journal. ✎

▶ **ASK YOURSELF**
How are an ax blade, a paring knife, and a screw similar?

Two or More Machines

Most machines are made up of combinations of machines. **Compound machines** are two or more simple machines put together to do work. The ax is a compound machine. You just read that the head of the ax is a wedge. What simple machine is the handle? In the next activity, you'll have a chance to design your own compound machine.

Rube Goldberg, a famous cartoonist, was known for his drawings of imaginary machines like the one shown here. Although humorous, his strange and unusual machines employed many types of simple machines. Study the Rube Goldberg drawing shown here, and list all the simple machines you recognize. Try your hand at designing and drawing your own "Rube Goldberg machine." Which simple machines did you use in your machine? How efficient do you think your machine is?

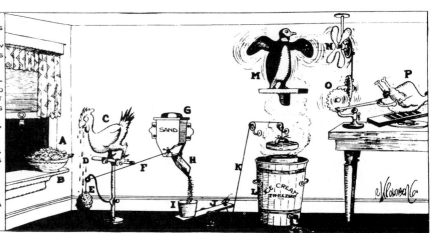

SIMPLE WAY TO CARVE A TURKEY. THIS INVENTION FELL OFF THE PROFESSOR'S HEAD WITH THE REST OF THE DANDRUFF. PUT BOWL OF CHICKEN SALAD (A) ON WINDOW SILL (B) TO COOL. ROOSTER (C) RECOGNIZES HIS WIFE IN SALAD AND IS OVERCOME WITH GRIEF. HIS TEARS (D) SATURATE SPONGE (E), PULLING STRING (F) WHICH RELEASES TRAP DOOR (G) AND ALLOWS SAND TO RUN DOWN TROUGH (H) INTO PAIL (I). WEIGHT RAISES END OF SEE-SAW (J) WHICH MAKES CORD (K) LIFT COVER OF ICE CREAM FREEZER (L). PENGUIN (M) FEELING CHILL, THINKS HE IS AT THE NORTH POLE AND FLAPS WINGS FOR JOY, THEREBY FANNING PROPELLER (N) WHICH REVOLVES AND TURNS COGS (O) WHICH IN TURN CAUSES TURKEY (P) TO SLIDE BACK AND FORTH OVER CABBAGE-CUTTER UNTIL IT IS SLICED TO A FRAZZLE — DON'T GET DISCOURAGED IF THE TURKEY GETS PRETTY WELL MESSED UP. IT'S A CINCH IT WOULD HAVE EVENTUALLY BECOME TURKEY HASH ANYWAY.

Bicycles Built for One or More Bicycles are compound machines consisting of a variety of simple machines. The back wheel of a bicycle is a wheel and axle. Attached to its axle are gears that allow the chain to grip tightly as it turns on one of the various different-sized wheels. The teeth on these gears are actually a series of levers spaced evenly around the outside of each gear. These gears are attached to another gearwheel and axle, which is attached to two pedals. In addition to the gearwheel and axle system, the bicycle also has levers that are used to apply the brakes and to steer the front wheel.

Figure 4–16. The gears on this bicycle are actually small levers attached to a wheel and axle. The brakes consist of two levers that pivot on a single fulcrum.

Winter—At Last   Outside the wind blows, and the piles of snow reach above the windows. Inside your cabin, a fire in the fireplace blazes. Before winter set in, you helped to make the tables and benches you and your family now use. The winter will be long, but there is a lot to do inside the cabin—grinding corn, repairing farm tools, and making candles and clothing. Next spring, you and your neighbors plan to build a schoolhouse. Why not? The resources and machines to do the job are just outside your front door.

 ASK YOURSELF

Identify a compound machine not mentioned in this chapter, and list the simple machines it contains.

SECTION 2   *REVIEW AND APPLICATION*

Reading Critically

1. Categorize the six simple machines as either levers or inclined planes.
2. Is a pulley considered to be a wheel and axle? Why or why not?
3. Explain why an ax is a compound machine.

Thinking Critically

4. Give an example of a device that contains one or more of each of the following simple machines: lever, wheel and axle, and wedge.
5. How do you think a tandem bicycle (a bicycle "built for two") might be similar to and different from a single-passenger bicycle?

INVESTIGATION

Building Simple and Compound Machines

▶ MATERIALS

- bottle caps ● cardboard ● craft sticks ● empty thread spools ● glue
- modeling clay ● paper ● pencils ● rubber bands ● scissors ● shoe
boxes ● stones ● straws ● string ● tape ● any other materials to which
you have access

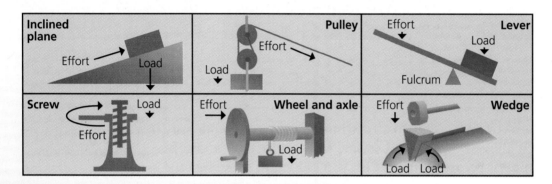

▼ PROCEDURE

1. Make an inclined plane using any of the materials listed above. Compare your inclined plane with those of other groups.
2. Take the inclined plane apart. Now make a pulley, a lever, a screw, and a wheel and axle. Compare your group's simple machines with those of other groups.
3. With the others in your group, decide how to make a compound machine. In your journal, draw a picture of your machine and label the parts. Be creative! Your compound machine can be a machine that already exists or one that you have invented.
4. Build your machine. Then demonstrate to the other groups how your machine makes work easier.

▶ ANALYSES AND CONCLUSIONS

1. How many simple machines are in your compound machine? What are they?
2. Why did you combine the machines in this way?
3. What are the mechanical advantages of your machine?

▶ APPLICATION

Design a compound machine that has all six simple machines in it. Explain what the machine will do and how it will make work easier.

✳ Discover More

If your school has an industrial arts or technology department, ask the teacher to show you some of the large tools, such as a lathe, drill press, and radial arm saw, as well as some of the special-purpose hand tools. Examine each of these compound machines to determine the simple machines that make up their parts.

The Big Idea

Albert Bierstadt, "The Oregon Trail" 1869, 31" × 49", oil on canvas. The Butler Institute of American Art.

Work is the application of force to make something move a certain distance. Power is the amount of work you can do in a certain amount of time. Because there is a limit to the force even a very strong human or several humans can exert, people need machines to help them do their work.

There are six types of simple machines. Simple machines are often combined to make many different kinds of compound machines. All machines, whether simple or compound, have a mechanical advantage—the ratio of force that comes out of a machine compared with the force that is put into the machine. Mechanical advantage is always less than 100 percent because of friction.

For Your Journal

Look back at the ideas you wrote in your journal at the beginning of the chapter. How have your ideas about work and machines changed? Revise your journal entry to show what you have learned. Include information on how machines are useful in your everyday life.

Connecting Ideas

Consider the food grinder and the tricycle shown. Identify these two machines as either simple or compound, and name the simple machine or machines that make each of them work.

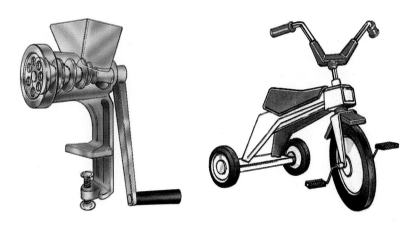

REVIEW

Understanding Vocabulary

1. For each set of terms, define each term and explain how the terms are related.

a) work (101), power (102)

b) simple machine (109), compound machine (115), mechanical advantage (106)

c) lever (110), wheel and axle (111), pulley (112)

d) ramp (114), wedge (114), screw (115)

Understanding Concepts

MULTIPLE CHOICE

2. A screw is an example of what kind of simple machine?

a) wedge

b) inclined plane

c) pulley

d) wheel and axle

3. Work is

a) making an effort.

b) applying force.

c) making something move.

d) applying a force and making something move.

4. When a fixed pulley is used, the force and the load

a) move in opposite directions.

b) move in the same direction.

c) do not move.

d) are always equal.

5. Machines can multiply

a) only force, but not distance.

b) only distance, but not force.

c) neither force nor distance.

d) force or distance.

6. The name of the metric unit that measures 1 joule of work done each second is the

a) watt.   **b)** kilowatt.

c) newton.   **d)** gram.

SHORT ANSWER

7. Briefly explain how levers and inclined planes allow force to be made larger by making the distance through which the force is applied smaller.

8. Explain which part of a pioneer's plow was a wedge.

Interpreting Graphics

9. Look at the art below. What class of lever is the bottle opener? Explain your answer.

10. Look at the art below. What class of lever is the fishing rod? Explain your answer.

Reviewing Themes

11. *Systems and Structures*
Choose a compound machine, such as scissors. List the simple machines that make up the compound machine and explain how the simple machines work together.

12. *Energy*
Explain how energy relates to machine usage.

Thinking Critically

13. In a wheel and axle, the radius of a wheel is like the effort of a straight lever. The radius of the axle is like the resistance arm. What part of the wheel and axle is like the fulcrum of a straight lever? Explain your answer.

14. Think about machinery commonly used today. If the lever were to be eliminated, how many machines would still be in use? Explain your answer.

15. An elevator is a single pulley lifting machine. Explain in your own words how you think this machine works or draw a diagram showing how you think it works.

16. Explain why a merry-go-round like the one shown here is not considered a wheel and axle.

17. In moving the piano, the movers in the top photograph will use more force but will do slightly less work than the movers in the bottom photograph. Explain why.

Discovery Through Reading

Rosta, Paul. "Elevators that Capture Energy." *Popular Science* 237 (December 1990): 98–99. Millions of elevators waste energy. This article discusses an invention that taps and stores energy to run other systems.

ENERGY

*F*ew things are as dramatic or display as many forms of energy as the launch of a space shuttle. The light, heat, the deafening sound of the engines and rockets, and the movement of the huge orbiter are all forms of energy. The amount of energy released during a launch is tremendous.

Launch minus 10 seconds . . . 9 . . . 8 . . . 7 . . . The three launch engines light. The shuttle shakes and strains at the bolts holding it to the launch pad. The computers check the engines. It isn't up to us any more—the computers will decide whether we launch.

3 . . . 2 . . . 1 . . . The rockets light! The shuttle leaps off the launch pad in a cloud of steam and a trail of fire. Inside, the ride is rough and loud. Our heads are rattling around inside our helmets. We can barely hear the voices from Mission Control in our headsets above the thunder of the rockets and engines. For an instant I wonder if everything is working right. But there is no more time to wonder, and no time to be scared.

In only a few seconds we zoom past the clouds. Two minutes later the rockets burn out, and with a brilliant whitish-orange flash, they fall away from the shuttle as it streaks on toward space. Suddenly the ride becomes very, very smooth and quiet. The shuttle is still attached to the big tank, and the launch engines are pushing us out of Earth's atmosphere. The sky is black. All we can see of the trail of fire behind us is a faint, pulsating glow through the top window.

Launch plus six minutes. The force pushing us against the backs of our seats steadily increases. We can barely move because we're being held in place by a force of 3 g's—three times the force of gravity we feel on Earth. At first we don't mind it—we've all felt much more than that when we've done acrobatics in our jet training airplanes. But that lasted only a few seconds, and this seems to go on forever. After a couple of minutes of 3 g's, we're uncomfortable, straining to hold our books on our laps and craning our necks against the force to read the instruments. I find myself wishing we'd hurry up and get into orbit.

Launch plus eight and one-half minutes. The launch engines cut off. Suddenly the force is gone, and we lurch forward in our seats. During the next few minutes the empty fuel tank drops away and falls to Earth, and we are busy getting the shuttle ready to enter orbit. But we're not too busy to notice that our books and pencils are floating in midair. We're in space!

from *To Space & Back*
by Sally Ride with
Susan Okie

For Your Journal

🖍 List as many forms of energy as you can think of.

🖍 What can these forms of energy do?

🖍 How do you use these forms of energy every day?

Forms of Energy

Objectives

Compare potential and kinetic energy.

Calculate the kinetic energy of a moving object.

Relate the amount of kinetic energy and the speed of an object to the force of a collision.

Our universe is rich in different kinds of energy. You know that during a shuttle liftoff, a tremendous amount of energy is released. Sometimes, though, energy can be hard to recognize. Is energy in action when a flag flutters in the breeze? Does energy cause the ice in your soda to melt? Yes. Energy is recognized mostly by what it does.

Energy can be violent, as when thunder claps so loudly the windows shake. But energy can also be gentle, like a dewdrop quietly evaporating on the petal of a rose. You might say energy is an expert at disguise.

Energy Has Many Forms

Suppose you wanted to move your textbook from your backpack to your desk. You couldn't do that without using energy. You would use mechanical energy—supplied by muscle power—to move the book. Energy is found in many forms. Heat is the energy produced by moving atoms and molecules. Light energy travels in waves to reach your eyes. Chemical energy is the energy that binds atoms and molecules together. Your ear receives energy vibrations as sound. Mechanical energy is the energy produced by moving objects. It is this form of energy that you would use to move your book. A Frisbee® flying through the air has mechanical energy. A moving bicycle, flowing water, and a rolling marble also have mechanical energy. What's more, this is only a small sampling of the types of energy that affect your life.

Figure 5–1. The motion of a flying Frisbee® and the leap of the dog that catches it are both forms of mechanical energy.

You know that energy can take many different forms, but what exactly is it? *Energy* is the ability to cause change. Think back to removing the book from your backpack. In that case, the change was in the position of the book and the change was caused by mechanical energy. Let's look at another example.

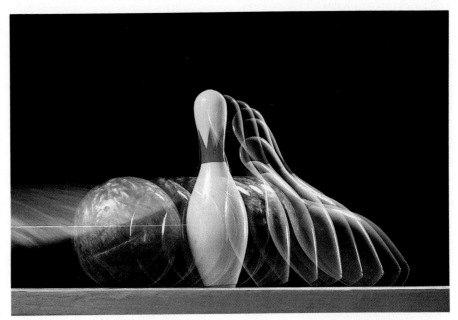

Figure 5–2. The energy of a moving bowling ball causes dramatic changes when it crashes into the pins.

Suppose you get home from school and there's a note asking you to wash the dishes. Sheesh! You are always having to do the dishes. Muttering under your breath, you collect the dirty dishes and stack them up. You run some hot water and add dishwashing detergent. After scrubbing the dishes clean and rinsing them with warm water, you stack them to dry. In this example, energy was used to cause many changes. How many can you name?

Energy is a necessary ingredient for change. The amount of energy used to cause a change is measured in *joules (J)*. You may recall from Chapter 4 that work also is measured in joules. This is because work is a type of mechanical energy.

 ASK YOURSELF

Energy has many forms, and each of the forms causes change. Give an example of your own that shows this is true.

Stored Energy

Think about a rubber band. Just lying around on a desk, a rubber band cannot force anything to move or do any work. However, if you pull back on the rubber band and hold it, it has stored energy. The energy stored in an object due to its position is called **potential energy.** The energy used to stretch the rubber band becomes stored in the rubber band as potential energy.

If you release the stretched rubber band, it snaps back to its original shape. When this happens, the potential energy in the rubber band changes to motion.

The potential energy stored in a stretched rubber band is called *elastic potential energy.* Elastic potential energy can be stored in stretched springs as well as stretched rubber bands. Elastic potential energy can also be stored in compressed springs and bent vaulting poles. In fact, any object that can be forced into a shape that is different from its natural shape can store energy. If the object is able to return to its natural shape, it has elastic potential energy.

DISCOVER BY *Writing*

Imagine that you are the pole shown in the photographs below. Write a poem or a short story in your journal that describes how your energy changes and how you feel as it happens. ✎

Figure 5–3. A pole-vaulter stores energy in the pole as it bends. As the pole straightens, the potential energy is released and the vaulter is launched over the crossbar.

Figure 5–4. As the drop of water falls and as the pitcher moves to throw the ball, potential energy is released.

Another form of potential energy is *gravitational potential energy*. Gravitational potential energy is the energy an object has when it is in an elevated position. A baseball in the air and a drop of water suspended from a faucet both have gravitational potential energy. Anything that can fall or drop has gravitational potential energy.

 ASK YOURSELF

Describe two examples of potential energy, and explain how they are different.

Energy in Motion

Think back to the rubber band example. Remember that when the rubber band is stretched, it has potential energy. When you let go of the rubber band, it snaps back to its original shape and the potential energy changes to motion. What kind of energy is the motion?

The energy an object has due to its mass and its motion is called **kinetic energy.** Anything in motion has kinetic energy. Mathematically, kinetic energy is described as follows:

$$\text{kinetic energy} = \frac{\text{mass} \times \text{speed}^2}{2}$$

Note that this equation uses speed instead of velocity. The amount of kinetic energy in an object is the same no matter what direction it moves. Try the next activity to find out more about kinetic energy.

ACTIVITY

How can the kinetic energy of a rolling cart be measured?

MATERIALS
several books, board, masking tape, meter stick, balance, rolling cart, stopwatch

PROCEDURE

1. Make a ramp with the books and board as shown. Make sure the starting line is far enough from the top so the cart can be placed behind the line.

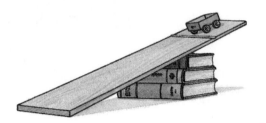

2. Make a table like the one shown.

TABLE 1: KINETIC ENERGY OF A CART					
Length of Ramp	Mass of Cart	Time of Trial			Average
		1	2	3	

3. Measure and record the distance (in meters) between the start and the finish lines. Also measure how much higher the middle of the cart is at the top of the ramp as compared with the cart's middle when it is at the bottom of the ramp.

4. Use the balance to find the mass of the cart in kg. Record the mass.

5. Set the cart so that it is behind the starting line and release it. Time how long it takes for the cart to reach the finish line.

6. Repeat step 5 two more times, and average the results.

7. You already know the kinetic energy of the cart before it begins to roll down the ramp. The cart is not moving, so its speed is zero. Therefore, its kinetic energy is zero. What is the kinetic energy of the cart at the bottom of the ramp? To determine kinetic energy at the bottom, you need to know the cart's mass *(m)*, which you have already measured, and its speed *(v)*.

8. You can find the cart's average speed by dividing the distance traveled down the ramp by the average time taken to travel that distance. You have measured both these quantities. But the cart is accelerating from no speed at the top of the ramp to a speed at the bottom of the ramp that is faster than average. Because it is accelerating smoothly, the speed at the bottom is just twice the average speed.

9. Now that you know the speed at the bottom of the ramp, you can calculate kinetic energy as $1/2\ mv^2$.

APPLICATION

1. Gravitational potential energy is calculated as *mgh,* where *m* is the cart's mass, *g* is the acceleration due to gravity (9.8 m/s/s), and *h* is the height of the ramp. How does the cart's gravitational potential energy at the top of the ramp compare with its kinetic energy at the bottom of the ramp?

2. Suppose you are out riding your bike and you come to the top of a small hill. You coast down and enjoy the ride. Later you come to the top of a large hill. You fly down. Why does your bike go faster on the large hill?

Based on what you learned in the activity, try to answer the following question: How would the crashes be different if one car going 2 km/h and another car going 1 km/h each hit a tree?

Because kinetic energy increases in proportion to the square of the speed, you know that the car going 2 km/h has more kinetic energy and will hit the tree with more force. How much more force will there be at impact? The kinetic energy of the car going 2 km/h is four times as great as the car going 1 km/h: $2^2 = 4$ and $1^2 = 1$. Any time the speed of a vehicle is doubled, the force of impact is four times greater. Now you know why car accidents at higher speeds are much worse than accidents at lower speeds. If everything else is the same, a crash at 60 km/h causes four times the damage that a crash at 30 km/h would cause.

Now you know that the faster an object travels, the more energy it will have as it hits something. Think about what this means if you want to stop *before* you run into something. The amount of force required to make you stop in a certain distance is equal to your kinetic energy divided by the distance. Using your common sense, you can see that the faster you are moving the greater the force necessary to stop you. It all has to do with energy.

Figure 5–5. The damage from this wreck is severe. When two cars hit each other head-on, the damage is the result of the kinetic energy of both vehicles.

ASK YOURSELF

A football player runs twice as fast as an opponent. Is he twice as hard to stop?

SECTION 1 *REVIEW AND APPLICATION*

Reading Critically

1. Use the space shuttle as an example to show the difference between potential energy and kinetic energy.

2. A 50-kg gymnast swings off a bar at 6 m/s. What is the gymnast's kinetic energy as she leaves the bar?

Thinking Critically

3. A friend doesn't believe that energy can change forms. You have a spring and a book on your desk. How can you use these things to prove to your friend that energy can change forms?

4. Baseball players who play third base often field bunted balls with their bare throwing hands. A line drive, which travels about three times as fast as a bunt, might break the player's hand if the ball were caught barehanded. Explain why this is true.

Energy Laws

Illustrate the first and second laws of thermodynamics by tracing energy flow through a series of events.

Apply the first and second laws of thermodynamics to explain energy exchange.

Explain how energy conversions result in energy being wasted as heat.

It's a hot summer day, and you and your friends are outside playing basketball. The group decides to take a break. You run into your house to get something cold to drink. But before you reach for something to drink, you stand in front of the open refrigerator door and enjoy the cool air. If you stood there long enough with the door open, the entire room would become cooler, right? Wrong! You may not know it, but right before your eyes, the laws of energy are at work. Understanding these laws makes most technology possible—even refrigerators. Read on to find out more.

Energy Is Conserved

Figure 5–6. When these children jump on their pogo sticks, they are converting potential energy to kinetic energy and back again.

A pogo stick is basically a pole on which foot-rests are supported by a strong spring. When you jump on a pogo stick, the spring compresses and changes kinetic energy into potential energy. Then the spring begins to expand and converts potential energy back into kinetic energy. The amount of kinetic energy produced by the spring, plus a small amount of kinetic energy changed into heat energy, is equal to the amount of potential energy stored by the spring. This is an example of the law of conservation of energy.

The law of conservation of energy states that the total amount of energy in the universe always stays the same. Another way to say this is that energy cannot be created or destroyed. It can be changed from one form to another, but the total amount of energy never changes. The law of conservation of energy is also called the *first law of thermodynamics.*

Thermodynamics in Action   The operation of the pogo stick shows how elastic potential energy is converted into kinetic energy and vice versa. However, the pogo stick also illustrates how gravitational potential energy is converted into kinetic energy. As the spring converts its potential energy into kinetic energy, the person and the pogo stick are moved up and into the air. As the person rises, the kinetic energy is being converted back into potential energy. This time the potential energy is in the form of gravitational potential energy. When the person reaches maximum height, all of the kinetic energy has been converted into gravitational potential energy. As the pogo stick moves downward, this potential energy is converted back into kinetic energy.

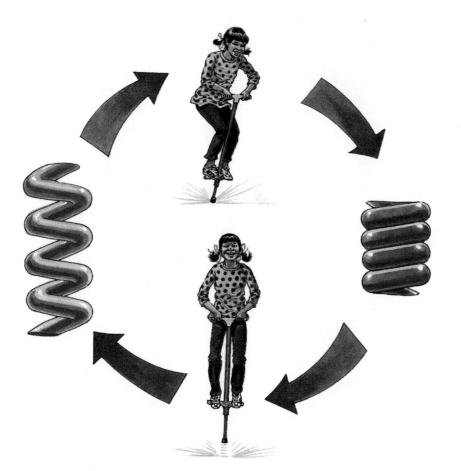

Figure 5–7. Energy is changed from one form to another in cycles.

Remember that the total amount of energy always stays the same, but some energy is given off as heat. In the pogo stick example, the person on the pogo stick gives off some heat, and the spring, as it is compressed and released, gives off some heat.

Figure 5–8. Even though you feel cool air when you open a refrigerator, the kitchen does not become cooler. The heat removed from inside the refrigerator is released into the kitchen by the refrigerator's working parts.

Cooler or Hotter?

You are probably still wondering about the open refrigerator and why it can't cool the entire room. The refrigerator works by pumping heat from the inside to the outside of the refrigerator. When you open the door, cool air passes out of the refrigerator and is replaced by warm air from the room. Any heat that goes into the refrigerator is also pumped out of the refrigerator.

You do feel cooler standing in front of the open refrigerator, but the room will not get cooler. In fact, the room will get slightly hotter. This is because some of the refrigerator's parts, such as the motor and compressor, change some energy into heat during the energy exchange. This energy is wasted, but it is not lost. This is an application of the law of conservation of energy.

ASK YOURSELF

Why is it important for scientists and engineers to understand the law of conservation of energy?

Less and Less

The track meet is about to start. Seven women are poised at the starting blocks. The starting gun fires! The women sprint away from the blocks and race 100 m, running as fast as they possibly can. At the end of the race, the winner bends over and puts her hands on her knees to catch her breath. Her face is glistening with sweat.

Figure 5–9. Runners use energy to race. This use of energy results in waste heat.

How is this track meet related to the conservation of energy? Any time energy changes from one form to another—such as when potential energy from food changes into motion—some of the energy changes into heat. The runners' bodies heat up, and they sweat to cool down. This is the basis of the *second law of thermodynamics*. Whenever there is an energy conversion, heat is produced that cannot be used to do useful work. The unuseable heat is called *waste heat*. Why do you think it is called "waste"?

Applied to machines, the second law of thermodynamics says that no machine is 100 percent efficient. The amount of work a machine does is always less than the amount of energy put into it. Some of the input energy always becomes waste heat, even if all friction is eliminated. Even the most carefully designed system cannot convert all input energy into work. What would happen if you could invent a machine that was 100 percent efficient?

You can observe waste heat when you use almost any machine. You can even observe it using a bicycle pump. See for yourself how this happens.

Figure 5–10 Not all of this person's work is being used to pump up the bicycle tire. Some is converted into waste heat.

DISCOVER BY Doing

Use a bicycle pump to pump air into a very deflated bicycle tire. Before you start pumping, feel the bicycle tire and the outside of the pump. Make a note of the temperature of each one. After you finish pumping air into the tire, feel the tire and the pump. What happened? Explain why. ✎

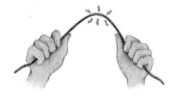

You can also observe waste heat by bending wire from a clothes hanger back and forth. As you bend the wire, it becomes warm. In fact, if you bend it enough to break it, it becomes quite hot. The wire is warmed by waste heat. **CAUTION: If you try this, be careful with the broken ends of the hanger.**

Figure 5–11. Bending a coathanger back and forth produces waste heat.

▼ ASK YOURSELF

How do people give off waste heat?

Perpetual Motion

Many people throughout history have tried to make perpetual motion machines. A perpetual motion machine is one that makes at least as much energy as it uses. In other words, a perpetual motion machine, once started, would continue operating forever with no loss of motion, no waste heat, no friction, and no additional energy input. No one has ever succeeded in making one of these machines because no one has yet found a way to get around the second law of thermodynamics. Even complex natural systems—such as the solar system—that appear to run without energy input have been found to be running down very slowly.

Figure 5–12. For centuries, inventors have attempted to design perpetual motion machines. Here are two examples. One from ancient Greece and one from a modern museum exhibit.

 **ASK YOURSELF**

Why can't there be a perpetual motion machine?

Energy and Life

Even living systems obey the second law of thermodynamics. Plants, animals, and all other creatures require a source of energy, such as food. Animals get food from eating plants or eating other animals. Plants make their own food using sunlight, water, and carbon dioxide. Regardless of the type of living system and how it gets its energy, each produces heat and motion as a result of that energy.

Think back to the 100-m race at the track meet. The women get the energy to run the race by eating nutritious foods. When the race is over, they replace their energy by eating again.

Look at the picture of the desert rabbit. Where does it get its energy? Where does the cactus get its energy?

The second law of thermodynamics is fairly simple mathematically, but it is difficult to explain how it works—especially when explaining living systems. The explanation here uses only the simplest examples. Like all the other laws of nature, the second law of thermodynamics describes *how* things happen; it does not explain *why* things happen.

Figure 5–13. Animals convert the nutrients from plants or other foods into heat and motion.

 ASK YOURSELF

What would happen to the rabbit if it couldn't find a supply of energy?

SECTION 2 *REVIEW AND APPLICATION*

Reading Critically

1. Relate the action of a bouncing ball to the laws of thermodynamics. Assuming it is not caught, what finally causes the ball to stop bouncing?

2. Give two examples that illustrate the second law of thermodynamics.

Thinking Critically

3. The term *thermodynamics* is formed from the Greek words *therme* meaning "heat" and *dynamis* meaning "power, strength, or movement." How does the meaning of *thermodynamics* apply to energy conservation? How does it apply to energy that is wasted?

4. Trace the forms of energy and their changes in a simple action, such as pedaling a bicycle from one place to another. Be sure to include the input energy. Show where energy is conserved, and describe how some of the energy changes to waste heat at each change in the form of energy.

SKILL *Communicating Using a Line Graph*

▶ MATERIALS
- thin books (10 to 20) ● stiff cardboard, 15 cm × 40 cm ● ruler or meter stick
- cart ● tape ● graph paper

▼ PROCEDURE

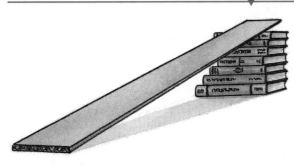

1. Set up a ramp as shown. You will do five trials to gather the data for your graph. Using different numbers of books, make the ramp a different height each time. Record the height of the ramp in your data table. Let the cart roll down the ramp, and then measure and record how far it travels beyond the bottom of the ramp.

2. Use the graph shown as a guide to make the graph of your data.

3. Mark points for the distances you measured. Then make a smooth line, crossing as many points as you can.

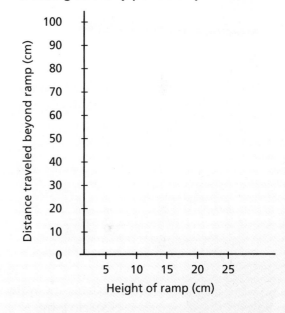

▶ APPLICATION
Look at your graph. What pattern do you observe? How does increasing the height of the ramp affect the distance the cart rolls? Use your graph to predict how far the cart might roll if you started it at a height of 30 cm.

✳ *Using What You Have Learned*
Are some patterns easier to understand when the data is shown on a line graph? Why?

Heat

Distinguish between heat and temperature.

Describe the properties of thermal expansion.

Infer how heat is transferred when given an example.

When you think of the sun, you probably think of its blazing heat. Considering that the sun is 150 million kilometers away, how does this heat reach the earth?

Heat Is Energy

It is a cold winter day. The snow lies deep on the ground, and school has been cancelled. Children are outside swooshing down hills on their sleds. Inside, it is warm and cozy. Why? It's warm because the house is heated by a furnace that pumps warm air into all the rooms. This heat is a form of energy.

Heat is used and is produced when many kinds of work are done. When you shovel the snow from the sidewalk, you give off heat. You probably know that heat and work are related, but what is the connection? Try the next activity, and see for yourself.

ᴅɪꜱᴄᴏᴠᴇʀ ʙʏ *Doing*

Fill a bowl with water, and allow the water to come to room temperature. Carefully measure and record the temperature of the water. Then, using an electric mixer set on high speed, beat the water for three minutes. Stop the mixer and recheck the water temperature. Has the temperature changed? Explain why. ✐

Many scientists have had questions about the relationship between work and heat. In the late 1700s, an American scientist named Benjamin Thompson, better known as Count Rumford, noticed that the barrels of cannons became hot as the holes were drilled in them. At that time, people thought heat was a substance called *caloric* that flowed from object to object. However, Thompson noticed that in drilling a cannon, there was no source of heat. He concluded that the work done in making the holes somehow changed into heat. Therefore, there had to be a connection between heat and work.

Several years later, English physicist James P. Joule concluded that heat and work were different forms of the same

Figure 5–14. When a substance is heated, its molecules move faster and farther apart; it becomes less dense. A hot-air balloon rises because the hot air inside is less dense than the air outside.

thing. Because of his work, the unit of energy in SI is named after him.

Now you know that heat is a form of energy, but what causes it? To help you understand what causes heat, think about this. As an object becomes hotter, its atoms and molecules move faster. As an object becomes colder, its atoms and molecules move slower. So you can say that heat is due to the motion of molecules or atoms. Specifically, the **heat** of an object is the total kinetic energy of the random motion of its atoms and molecules.

Research about heat and work didn't stop with James Joule. Today, scientists are still trying to understand more about heat and how heat can be used to do work. One of these scientists is Annie Easley, who works for NASA. Easley's work has centered on identifying how energy changes from one form to another. She applies what she finds out to improve existing technology. Some of her discoveries have been used to improve the operation of electric-powered vehicles used by NASA.

Benjamin Thompson (Count Rumford)

James Joule

 ASK YOURSELF

Why are joules used to measure both heat and work?

Temperature and Heat

If someone asked you the difference between temperature and heat, what would you say? Maybe the following statements will help you answer the question.

A large pot of boiling water has more heat than a small pot of boiling water, even though their temperatures are the same. A small pot of boiling water has the same temperature as a large pot of boiling water, even though the amount of water is different.

Annie Easley

Figure 5–15. Temperature is average kinetic energy, and heat is total kinetic energy.

Still stumped? The difference between temperature and heat has to do with the movement of atoms and molecules—the kinetic energy—of substances. For example, when a pot of water is boiling, the average kinetic energy of the atoms and molecules is the same regardless of the size of the pot. However, a large pot of boiling water has many more water molecules than a small pot. Therefore, the large pot contains more heat energy. Likewise, you have to add more energy to the molecules and atoms in the large pot to give them the same average kinetic energy as the molecules and atoms in the small pot. That is why it takes a large pot of water longer than a small pot of water to begin boiling, even when the burner is set on high under both. There are more molecules to heat in the large pot.

How Hot Is It? Many people confuse the meanings of temperature and heat, but now you know they're not the same. Temperature is *not* a measure of how much heat a substance contains. **Temperature** is the measure of the average kinetic energy of the moving atoms and molecules of a substance. Compare this definition to the definition for heat—the *total* kinetic energy of the moving atoms and molecules of a substance. To put it more simply, think of temperature as measuring how fast the average atom or molecule in a pot vibrates. Heat measures both how fast an atom or molecule vibrates and also how many molecules and atoms are vibrating.

Figure 5–16. The sun has a very high temperature, but the area around it doesn't have much heat because there are very few molecules in space.

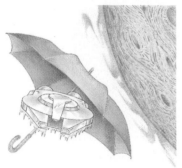

It's Not the Heat . . . When you think of heat and temperature as the total and the average kinetic energies of a substance, you can understand many seemingly strange things. For example, suppose you were in a spacecraft and you traveled in a path very close to the sun. You would certainly get very hot, right? Not necessarily! If you could shade your spacecraft from the light, you could actually freeze. This is true because, although the temperature near the sun is quite high, there is very little heat. The average kinetic energy of the atoms and molecules near the sun is very large; thus, the temperature is high. However, there are very few atoms and molecules in space. For this reason, the amount of heat near the sun is low.

How does the number of molecules in a substance affect temperature?

Thermal Expansion

Have you ever noticed that beverage manufacturers do not fill glass bottles to the very top? Why do you think this is? Here's a hint: As a substance becomes hotter, its atoms and molecules move faster and the substance expands.

Now that you've had a chance to think about the question, you have probably figured out the answer. If a bottle were filled to the top with a beverage and then warmed, the liquid would expand and shatter the glass.

You have just discovered one of the physical properties related to heat: thermal expansion. *Thermal expansion* is an increase in the size, or volume, of a substance due to an increase in the motion of its molecules and atoms.

When a substance is heated, the kinetic energy of the atoms increases. This increase in kinetic energy causes the atoms to move farther apart. The farther apart the atoms move, the greater the thermal expansion. Exactly how far apart the atoms move determines the thermal expansion of the particular substance. Thermal expansion is a characteristic property of substances.

Figure 5–17. As this metal strip inside the thermostat warms or cools, it changes shape. This change turns the heater circuit on and off.

Turn Up the Heat Understanding the thermal expansion of substances can be very useful in technology. For example, a thermostat that controls the temperature in a house relies on thermal expansion to turn the heat on and off automatically. The thermostat contains a metal coil made of two different metals fastened together. The metals expand at different rates. When the temperature increases, the outside metal expands more than the inside metal. This action causes the coil to curl up on itself and trip a switch, which turns on the heat.

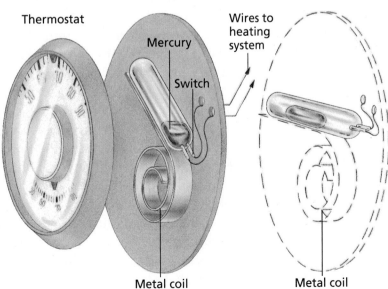

Thermostat

Mercury

Switch

Wires to heating system

Metal coil

Metal coil

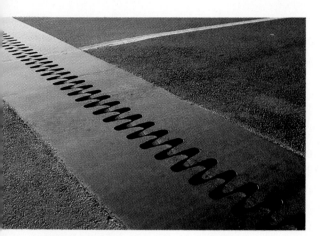

Bridging the Gap   Knowledge of thermal expansion is also helpful to engineers. When building a bridge, engineers know that the metal and concrete in the bridge are going to expand when the weather is warm and contract when the weather is cold. To prevent the bridge from buckling and collapsing, expansion joints are inserted at regular intervals to allow the bridge to expand and contract without damage. An expansion joint is shown here.

Figure 5–18. Expansion joints such as this one allow structures to expand and contract.

ASK YOURSELF

Why is understanding the thermal expansion of substances useful in technology?

Movement of Heat

You have just made a cup of hot chocolate. You stir it and leave the spoon in the cup. In a short while, the spoon gets warm. This happens because the hot chocolate transfers some of its heat to the spoon. This is just one example of how heat moves. Heat energy moves in three different ways: conduction, convection, and radiation.

Conduction   If you saw a quarter lying on a hot stove, would you use your hand to pick it up? Of course not! But *why* wouldn't you pick it up? The answer has to do with conduction.

Conduction is the transfer of heat energy from one substance to another by direct contact. Conduction occurs when atoms or molecules bump into each other. When the fast-moving molecules in a hot substance hit the slow-moving molecules in a colder substance, the slow-moving molecules speed up and the fast-moving molecules slow down. Note the heat transfer always goes from the hotter substance to the cooler substance. That's because heat energy is transferred from the faster molecules to the slower molecules. As more and more of the slow-moving molecules go faster, the substance heats up.

Because conduction is due to colliding molecules, it can occur only between objects that are touching. Think back to the spoon in the hot chocolate. Heat is transferred by conduction from the hot liquid to the spoon. If you were to hold the cup of hot chocolate in your hands, your hands would warm up too. This is because heat is conducted from the hot chocolate to the cup, and then from the cup to your hands.

You can feel the heat from the hot chocolate because the cup allows the conduction of heat. Some materials and substances allow heat to travel more easily through them than others. Materials and substances that readily allow the transfer of heat are called **conductors**. Many metals, such as copper and aluminum, conduct heat well. If the hot chocolate were in a metal cup, the cup would probably be too hot to hold. To find out more about conduction, try the following activity.

DISCOVER BY *Doing*

CAUTION: Put on safety goggles, a laboratory apron, and a glove before trying this activity. Get a burner, a metal rod, a candle, and some matches. Light the candle and put four spots of wax, equally spaced, on the metal rod. Hold the rod in your gloved hand, and place the end of the rod into the flame. Record how long it takes for each spot of wax to melt. Why does the wax melt in this pattern? How does this example demonstrate conduction? ✐

Some materials limit the amount of heat that passes through them. These materials and substances are called **insulators.** Air is an example of a very good insulator. If the hot chocolate were in a cup made of plastic foam, which has many small air pockets, the cup would transfer less heat and only be warm to the touch.

Look at the picture of the two students in the lab. Both are heating rods in a flame. The girl's rod is glass; the boy's is aluminum. The boy is wearing a glove. Why?

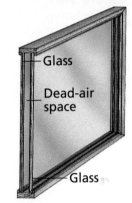

Figure 5–19. This house has special windows that reduce heat loss in the winter. The air space between the panes of glass acts as an insulator.

Figure 5–20. Study this photograph closely. Are these students using proper safety precautions? Explain your answer.

143

Figure 5–21. Convection is responsible for moving heat from the aquarium heater to the rest of the water.

Convection   Look at the picture of the aquarium. It contains tropical fish that must be kept in water that stays at a constant temperature. To maintain this temperature, a small heater is installed in one corner. Because of convection, this small heater can keep the entire aquarium at a constant temperature. **Convection** is the transfer of heat in liquids and gases as groups of molecules move in currents.

The water surrounding the aquarium heater is heated by conduction. However, as the water gets hot, its molecules move faster and take up more space. This action causes the warm water around the heater to become less dense, because the molecules are farther apart. As a result, the water is pushed upward by the cooler, more dense water in the aquarium. Warmer water, like warmer air, rises. As the cooler water takes its place near the heater, it is also heated. Meanwhile, the warm water that was pushed away from the heater begins to lose kinetic energy to the surrounding cooler water. As this cycle continues, a *convection current* is set up due to the changing densities of the water.

Under certain conditions, you can see convection currents. During cold weather, for example, the air over a hot spot, such as a heated building, will rise and seem to shimmer. This shimmer is caused by the different ways light travels through hot air and cold air. Even during the summer, convection currents can be seen as a shimmer above hot roads. Sometimes this shimmer appears from a distance to be water on the road—a *mirage*.

Figure 5–22. The shimmer above this road is caused by convection currents. Winds and storms are caused by convection currents in the earth's atomosphere.

A type of convection current is also responsible for the weather. Air that is over warm surfaces near the ground is warmer and less dense than air above. The warm air then rises and the cold air sinks. This forces the bottom layer of the atmosphere, the one nearest the earth, to turn over. This overturning is a type of convection, which creates the weather systems of the world. In a similar way, convection currents also cause some ocean currents and the movement of water in ponds and lakes. This helps to mix the nutrients in the water and, therefore, support more life.

Radiation It is a warm summer day. You are outside, and the sun feels hot on your skin. How does the heat from the sun travel 150 million kilometers through space to Earth? It travels by radiation. **Radiation** is the transfer of energy by electromagnetic waves. The movement of heat by radiation does not require matter. Both conduction and convection depend on molecules to carry heat.

The word *radiation* can also be used to refer to the waves and particles given off by radioactive materials such as uranium and plutonium. Therefore, you must pay careful attention to how this word is used whenever you see it.

There are many examples of heat transfer by radiation. For example, if you warm your hands by a fire in a fireplace, the heat travels to your hands by radiation. For another example of radiation, consider again the cup of hot chocolate. If you place your hand about 1 cm to the side of the cup, you can still feel the warmth from the hot chocolate.

ASK YOURSELF

When you walk barefoot on the street in the summer, the pavement can be so hot that it burns your feet. How is heat transferred from the sun to your feet in this example?

SECTION 3 *REVIEW AND APPLICATION*

Reading Critically

1. Describe how heat is transferred by convection.
2. Why is thermal expansion considered to be a characteristic property of substances?

Thinking Critically

3. Someone asks you how warm you are, and you tell them your temperature. Explain why just telling them your temperature does not give enough information to answer the question.
4. When entering a burning building, firefighters stay low to the floor. Explain why this is so.

INVESTIGATION

Predicting Temperature Change

▶ MATERIALS
- plastic-foam cups (3) ● thermometer ● graduated cylinders, 50 mL (2)
- cold water ● hot water ● stirring rod

▼ PROCEDURE

1. Copy the table shown.

TABLE 1: TEMPERATURE CHANGES					
Cold Water		**Hot Water**		**Mixture**	
Amount	Temperature	Amount	Temperature	Amount Predicted	Temperature Actual
20 mL		20 mL			
20 mL		40 mL			
40 mL		20 mL			

2. Label the cups "A," "B," and "C."

3. Place 20 mL of cold water in cup A and 20 mL of hot water in cup B.

4. Measure and record the temperature of the water in each cup.

5. Predict what the temperature will be if you mix the water in cups A and B. Record your prediction.

6. Pour the water from each cup into cup C, and stir with the stirring rod.

7. Measure and record the temperature of the mixture.

8. Repeat steps 3–7, but change the amounts of water used to 20 mL of cold water and 40 mL of hot water.

9. Repeat steps 3–7, but change the amounts of water used to 40 mL of cold water and 20 mL of hot water.

▶ ANALYSES AND CONCLUSIONS

1. How did your predictions compare with the actual temperatures? Explain any differences.

2. How did the temperature of the mixture compare with the starting temperatures when equal volumes of hot and cold water were mixed?

3. How did the temperature of the mixture compare with the starting temperatures when unequal volumes of hot and cold water were mixed?

4. How can you predict the final temperature of a mixture of hot and cold water?

▶ APPLICATION

Suppose you want to warm up a child's wading pool, which holds 200 L of water. Now it has 120 L of water at 20° C. You want the pool to be 30° C. How much hot water, at what temperature, would you have to add? Would you have to take out some of the cold water before you added the hot water?

Discover More

Repeat this investigation using a liquid other than water. Does the temperature of the liquid change in the same way as water did? Explain your results.

*H*IGHLIGHTS

The Big Idea

Energy is the ability to change something or to do work. Energy is involved anytime the position, appearance, or makeup of something changes. All changes involve energy.

Energy is found in many forms and can change from one form to another. Kinetic energy can change to potential energy, chemical energy to heat energy, electrical energy to light energy, and so on. Energy can also be stored and transferred. Energy is transferred when objects move.

Every time energy is transferred, some of the energy becomes waste heat. Heat energy can be used to do work. Heat can be transferred by conduction, convection, or radiation.

For Your Journal

Look back at the ideas you wrote in your journal at the beginning of the chapter. How have your ideas changed? Revise your journal entry to show what you have learned. Be sure to include information on how energy is used and how it changes form.

Connecting Ideas

This concept map shows how the big ideas of this chapter are related. Copy the map into your journal, and fill in the blanks to complete it.

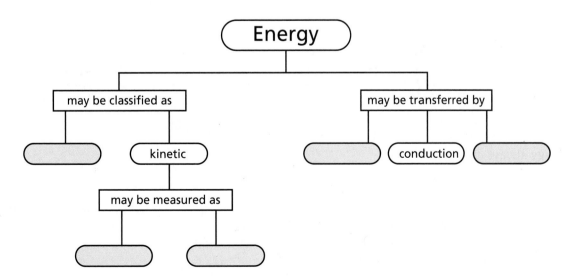

REVIEW

Understanding Vocabulary

1. For each set of terms, explain the similarities and differences in their meanings.
   a) potential energy (125), kinetic energy (127)
   b) heat (139), temperature (140)
   c) conductor (143), insulator (143)
   d) conduction (142), convection (144), radiation (145)

Understanding Concepts

MULTIPLE CHOICE

2. Which of the following actions does *not* describe potential energy being changed into kinetic energy?
   a) releasing a stretched rubber band
   b) a child sliding down a slide
   c) a spring being compressed
   d) an apple falling from a tree

3. A washer is a round, flat piece of metal with a round hole in it. Suppose you heat a washer. The hole in the washer will
   a) get larger.
   b) get smaller.
   c) stay the same size.
   d) There is no way to know.

4. Starting at rest, a skateboarder rolls downhill. At the bottom, the speed is 10 m/s. If the rider's mass is 30 kg, what is the kinetic energy of the rider?
   a) 300 J   b) 1000 J
   c) 1500 J   d) 3000 J

5. Every time energy changes from one form to another, some of the energy always changes into
   a) kinetic energy.
   b) potential energy.
   c) heat energy.
   d) mechanical energy.

6. Joules could be used to measure
   a) the mass of a car.
   b) the energy produced by a car engine.
   c) the power produced by a car engine.
   d) the speed of a car.

SHORT ANSWER

7. Use the two laws of thermodynamics to briefly explain why it is impossible to cool a room by leaving a refrigerator door open.

8. Explain the energy conversions that apply to a child on a swing. Why does the child need energy input from time to time?

Interpreting Graphics

9. Look at the figure below. Which car has the greatest potential energy? Explain why.

10. Look at the figure below. Which car has the greatest potential energy? Explain why.

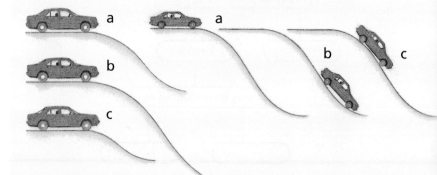

Reviewing Themes

11. *Energy*
During the transfer of energy from one form to another, or from one place to another, waste heat is given off. Explain this occurrence using the laws of thermodynamics.

12. *Systems and Structures*
Using an example other than a pogo stick, explain how objects and people interact with energy to form a repeating cycle.

Thinking Critically

13. The food web is a key idea in biology. Producers (plants) in the food web are eaten by herbivores, which are in turn eaten by carnivores. All three—plants, herbivores, and carnivores—are recycled by decomposers into nutrients for plants. Briefly explain the food web idea in terms of energy conversion. Do not forget to include an energy source(s) for the plants.

14. A greenhouse is a structure made mostly of glass, which allows light in but limits the flow of air in and out. Why does the air in a greenhouse get hot during the daytime?

15. The picture below shows several ways in which heat can be transferred. Identify each transfer, and explain how it occurs.

16. A trick many campers use for rapidly baking potatoes in a bed of coals involves sticking a large nail through the potato, wrapping the potato in foil, and making sure it is surrounded by hot coals. Explain why this method would cook a potato more rapidly than just wrapping it in foil and putting it in an oven at the same temperature as the coals.

17. The popping of corn is a very interesting phenomenon. Popcorn kernels "pop" when they are heated because each kernel contains a small amount of water within its hard shell. Explain in scientific terms why popcorn pops.

Discovery Through Reading

"Can a machine run forever?" *Current Science* 76(August 1991):10. This article tells about a modern attempt at making a perpetual motion machine.

ENERGY SOURCES

Coal was the energy source that fueled the engines and factories of the industrial revolution. However, like other types of energy, coal had its costs. Coal miners did back-breaking work, in constant fear of explosions and cave-ins. Until the 1930s, children did the work, too.

The job of a breaker-boy, or slate-picker, required little skill. The boys sat on narrow seats over chutes, into which coal was dumped and carried to the washers. Their job was simply to pick out the pieces of slate, rock, or other debris, leaving nothing in the chutes but pure coal. The work was not particularly difficult but it was exceedingly tiresome. The boys became saturated with coal dust; their hands were scratched and bruised; and their fingernails were worn off by the constant rubbing of the discarded slate. Cut and crushed fingers were not uncommon. Although the legal age for employment generally was twelve years, many of the boys, some as young as eight, exaggerated their age

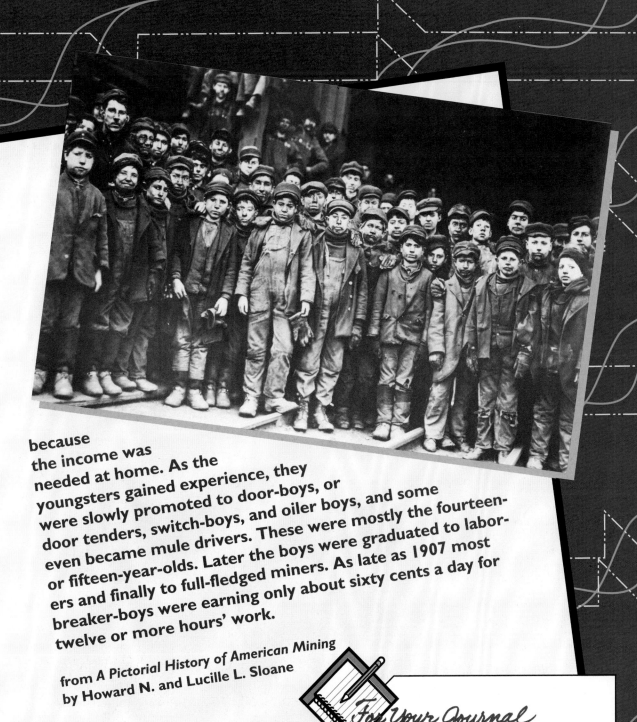

because
the income was
needed at home. As the
youngsters gained experience, they
were slowly promoted to door-boys, or
door tenders, switch-boys, and oiler boys, and some
even became mule drivers. These were mostly the fourteen-
or fifteen-year-olds. Later the boys were graduated to labor-
ers and finally to full-fledged miners. As late as 1907 most
breaker-boys were earning only about sixty cents a day for
twelve or more hours' work.

from *A Pictorial History of American Mining*
by Howard N. and Lucille L. Sloane

For Your Journal

- List the energy sources you use every day.
- How can energy be obtained from the sun, wind, and oceans?
- What tradeoffs are involved in using different sources of energy?

Energy from Fossil Fuels

Identify *the three major types of fossil fuels.*

Compare *and* **contrast** *the physical characteristics, uses, and locations of coal, petroleum, and natural gas.*

Define *and* **identify** *nonrenewable resources.*

There you are, standing outside the public library, ready to begin your research paper on energy use. You push on the doors, but they are locked. The library not open on a Saturday? Through the crack in the doors, you smell a musty, closed-up odor. How long has the library been closed? You turn your head and notice the rusted cars that appear anchored to their parking spots. There are no traffic sounds, no blaring radios, and no human voices.

The footsteps of an approaching stranger startle you. "Conduct your research wisely," he cautions. "The energy choices and decisions you make now may affect future generations. Without energy, the future will be as bleak and silent as this landscape." The stranger shuffles away. Suddenly, you throw off the covers and sit up in bed. What a nightmare!

The Energy Research Begins

You've put off starting your research paper for as long as possible. However, the nightmare has given you the motivation to begin work on it. Your topic is to identify current energy sources and to describe them. You are also supposed to make recommendations about energy choices for the future.

A Fossil Is a Fuel? You discover that the most commonly used energy source today is supplied by burning three major fossil fuels—coal, petroleum, and natural gas. They are called **fossil fuels** because they are formed from plant and animal material that was buried in sediment millions of years ago. As time passed, bacterial action, heat, and pressure converted these fossils into hydrocarbons—molecules consisting of carbon and hydrogen. When hydrocarbons are burned, they release heat, water, carbon dioxide, and some pollutants.

Not Everything Old Becomes New Again

Fossil fuels take millions of years to form. Because they take so long to form, fossil fuels are considered to be nonrenewable resources. **Nonrenewable resources** are those that cannot be replaced after they are used. You go to the library, check out several books about fossil fuels, and head for home.

Figure 6–1. This piece of coal contains the fossil of a fern that lived millions of years ago.

 ASK YOURSELF

Why are fossil fuels considered nonrenewable resources?

Coal

The first fossil fuel you read about is coal. Coal is a dark brown or black rock that can be burned to release energy. It is used most commonly as a fuel for electric-energy plants and for industrial processes, such as iron smelting, that require high temperatures. Coal is found in layers of sedimentary rock thought to be the sites of old peat bogs that were covered with earth and compressed for millions of years. When the layers in coal are split, fossils of plants and animals are often found.

Coal is found in many areas of the United States in layered deposits called *seams*. Coal is removed by two principal methods, depending on the location of the seam in the earth. The first method, strip mining, is used if the seam is near the surface. *Strip mining* involves stripping away the top layer of soil and then removing the layers of coal underneath. This leaves huge scars on the earth's surface and should be—but most often isn't—followed by reclamation. *Reclamation* is the process of restoring the land to its original fertile condition.

Figure 6–2. Peat that has been cut and stacked. Peat can be burned for fuel.

Researching _____

Find information on a strip mine that has been reclaimed. How was the land restored to its original condition? What can the land now be used for? How much did the reclamation cost? You may wish to prepare a poster that shows the results of your research. ✎

Coal located deeper in the earth is removed by *deep mining,* the second method of removing coal. In this process, deep shafts are dug into the earth to reach the coal. The specialized machinery shown in this picture is used to dig tunnels directly into the seam. As the coal is dug away, it is taken to the surface of the mine.

Figure 6–3. In strip mining (left), the top layer of soil is removed to expose the coal. The coal is dug and then carried away in large trucks. In deep mining (right), specialized equipment is used to remove seams of coal from deep beneath the earth's surface.

DISCOVER BY *Researching* _____

Find information on deep mining. What are the dangers involved in this type of mining? What are the advantages and disadvantages of deep mining? Share your findings with your classmates. ✎

▶ **ASK YOURSELF**

What are the different methods used to obtain coal from the earth?

Petroleum

The second fossil fuel you research is petroleum. Black gold is one name for petroleum. It is also called crude oil, or simply oil. Petroleum is a dark brown, thick liquid found in underground pools. These pools are located within layers of porous sedimentary rock, such as sandstone. Oil is collected by drilling wells down to the deposits and pumping the oil to the surface. Some oil wells are drilled as deep as seven kilometers into the earth.

Unlike coal, oil cannot be easily used in the same form in which it is taken from the earth. It must first be taken to a **refinery**—a large industrial plant that separates crude oil into products such as gasoline, diesel fuel, heating fuel, petroleum jelly, and asphalt. Oil is separated into various products in a *fractionation tower,* as shown in this drawing.

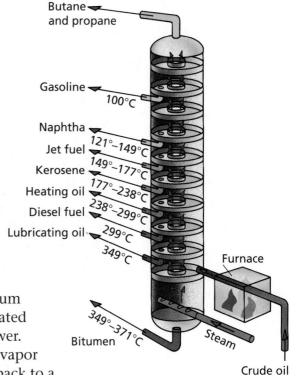

Figure 6–4. Refineries (left) consist of many fractionation towers. The towers distill crude oil into many useful products, as shown on the right.

Within a fractionation tower, the petroleum undergoes fractional distillation. The oil is heated into a gas, which rises in the fractionation tower. Since the products contained in the crude oil vapor have different boiling points, they condense back to a liquid state at different temperatures and levels in the tower. Each product is then pumped from its condensation level in the tower to a separate holding tank.

During the distillation process, many petroleum products undergo still another process called *cracking.* In this process, heat and chemicals are used to break apart the most dense molecules. This process is used primarily to produce more gasoline than is possible through fractional distillation alone.

ASK YOURSELF

What are the two processes used in refining crude oil?

Natural Gas

The third fossil fuel is natural gas. The flame on a gas stove or Bunsen burner is created by burning natural gas. Natural gas is a colorless gas consisting mostly of methane (CH_4). Methane is commonly found with petroleum deposits, but it is also found by itself. Like deposits of petroleum, natural gas deposits are reached by drilling wells. Natural gas is an important fuel because it is relatively easy to collect from the earth, and, compared with some other fossil fuels, it produces little pollution when burned. It is collected and transported by pipelines to processing plants and then distributed to consumers. Natural gas is most often used for heating and generating electricity.

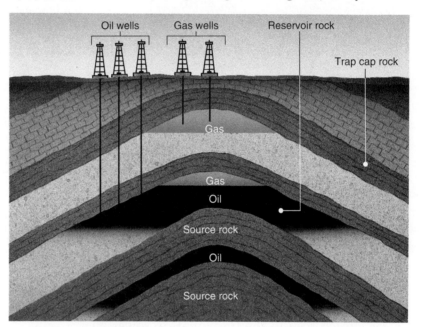

Figure 6–5. Both petroleum and natural gas deposits are reached by drilling wells.

Although natural gas has traditionally been considered a fossil fuel, some scientists have recently challenged this classification. These scientists now hypothesize that natural gas comes from the rocks and dust from which the earth was *originally* made. The facts that some meteorites contain hydrocarbons and that simple hydrocarbons like methane have been discovered in interstellar space are used as evidence to support this hypothesis. One prediction from the hypothesis is that there are huge quantities of natural gas deep in the earth, much deeper than levels where fossils are found.

 ASK YOURSELF

Why do some scientists consider natural gas a fossil fuel, while others do not?

No Easy Answers

Whew! Doing research is tiring work. You consult your outline and begin to add more information to it. Now that you've finished reading about the fossil fuels used today, would you recommend that fossil fuels be used as the primary source for the earth's future energy needs? Why or why not? What are your other choices? Complete the activity to get more information.

 ᴅᴇʀ ʙʏ Researching ─────────────────

Plant and animal materials that were recently living are referred to as *biomass.* For example, firewood is a form of biomass. Research what other types of biomass people have used to supply their energy needs. What are the pros and cons of using forms of biomass? Record your research in your journal. Take turns reading your research findings aloud. ✎

Figure 6–6. The burning firewood is an example of biomass.

After you look over your research notes, you conclude the bottom line is that the supply of fossil fuels is nonrenewable. In addition, the burning of fossil fuels releases pollutants. Future energy choices must include clean, renewable resources. You continue your research—perhaps solar energy and wind energy would be a good place to start.

◣ **ASK YOURSELF**

What are the drawbacks involved in using fossil fuels?

SECTION 1 *REVIEW AND APPLICATION*

Reading Critically

1. What are the similarities and differences among the three fossil fuels?

2. Identify the nonrenewable resources discussed in this section.

Thinking Critically

3. Some people think the United States should use its own fossil fuels rather than purchase such fuels from other countries. Others think we should continue to purchase fossil fuels from other countries. Which do you favor, and why?

4. Many decisions related to energy production and consumption determine how tax money is spent. Briefly discuss both the pros and the cons of the following energy-related tax issue: Interstate highways were strongly subsidized with taxes and railroads were not.

INVESTIGATION

Determining Heat Produced by Fossil Fuels

▶ MATERIALS

- small can
- thermometer ● ring stand ● clamp ● candle
- balance ● calculator ● lighter

CAUTION: Be careful when you light the candle.

▼ PROCEDURE

1. Make a data table.
2. Set up the equipment as shown. If your candle is not already stuck to the can lid, light the candle and attach it to the lid with a few drops of melted wax.
3. Measure 100 mL (100 g) of water into the can. Put the thermometer in the water, and record the water's temperature.
4. Determine the mass of the candle and can lid.
5. Place the candle under the can of water so that the top of the wick is about 1 cm from the bottom of the can, as shown in the illustration.
6. Light the candle. After 15 minutes, record the water temperature and blow out the candle. How much did the temperature rise?

7. Again determine the mass of the candle. How much wax was burned?
8. Calculate the total number of joules of heat energy

produced (100 g × temperature rise (°C) × 4.2 J/g · °C).
9. Calculate the number of joules produced by 1 g of wax.

▶ ANALYSES AND CONCLUSIONS

If all the heat produced by burning the candle was not used to heat the water, where did some of the heat go? Explain your answer.

▶ APPLICATION

A small car gets about 10.5 kilometers per liter of fuel when traveling at 90 km/h on a straight, level highway. If each liter of fuel is 800 g and has the same heat of combustion as candle wax, how many joules of energy are expended each hour while traveling at that speed?

✳ Discover More

Try the experiment again using different fuels, such as a peanut or potato chip. Be sure to use the proper safety precautions.

Energy from the Sun and the Wind

The media specialist hands you a package of information you requested. A photograph at the beginning of an article shows a sleek racing car. You wanted an article about solar energy, not automobiles! Shrugging your shoulders, you begin to read:

Spectators watched as the car, making only a high-pitched hum, glided down the road. Long and slender, the car looked like a cannelloni on wheels! Obviously, this was no ordinary car. It was the Sunvox I solar-powered race car—winner of the cross-country portion of the Second Annual Sun Day Challenge. The car's driver, Tamara Polewick, a 19-year-old engineering student from Dartmouth College, may have a "sunny" future ahead of her—either as a race car designer or driver.

Figure 6–7. A car of the future

Solar Energy

Putting aside the article, you look at the other materials and continue your research on solar energy. Each day, you learn, the sun bathes the earth with as much energy as would be contained in a trainload of coal more than 2 million kilometers long. Energy from the sun is called **solar energy.** Solar energy is a renewable resource. It travels through space as sunlight and provides the earth with both light and heat. Solar energy drives the winds. Solar energy is responsible for the water cycle (the evaporation and precipitation of water that results in rain and snow) that eventually drives hydroelectric generators.

The sun provides the energy necessary for plant growth. Without plants, which support the food chain, there would be no life on Earth and fossil fuels would never have formed. Even coal, petroleum, and natural gas can be thought of as forms of concentrated, stored solar energy.

Life on Earth is dependent on solar energy for its very existence. More sunlight strikes Earth than is needed or is used. This excess is an important energy source and is currently used to meet some energy needs.

 **ASK YOURSELF**

What is solar energy?

Solar Heating

Humans can harness solar energy for many uses. One major use of solar energy is for heating homes and buildings. Many different systems for converting sunlight to heat are now in operation. Most of the systems use either passive solar heating or active solar heating. These two systems differ in the ways that they collect and transfer heat.

Soaking Up Some Rays   Homes that are heated as a result of the simple absorption of solar energy use **passive solar heating.** In this method of heating, no mechanical devices are used to transfer heat from one place to another. The drawing shows an example of a home with a passive solar heating system. In this system, large glass panels trap heat in the home. Natural convection currents carry the heat throughout the rooms. Excess heat is absorbed by a thick concrete floor and wall and is released into the home at night or during cloudy periods.

Figure 6–8. The sun is the earth's ultimate source of energy.

Figure 6–9. In passive solar heating, the heat moves naturally through the home. In the diagram shown here, excess heat from the sun is stored in the masonry wall.

Concrete wall

Collecting the Rays   More complex solar-heating systems use **active solar heating.** This method of heating uses mechanical devices such as fans to move heat from one place to another within a building. Most active solar-heating systems include *solar collectors,* which are devices that gather sunlight and convert it into heat. The photograph shows a home that has many solar collectors installed on its roof.

Figure 6–10. The solar collectors on this roof collect energy from the sun.

Most active solar-heating systems use either air or water as the working fluid to transfer heat. In most cases, pumps or fans move the working fluid through the solar collectors. If the working fluid is air, it can be routed directly from the collectors into the home, where it heats the rooms. When water is used, a *heat exchanger* is necessary to transfer the heat from the water to the air in the rooms. A radiator is an example of a heat exchanger. When hot water flows through the radiator, the fins on the radiator become hot and heat the surrounding air.

In your notebook, you jot down your ideas about solar energy. Perhaps you should recommend that solar energy be the primary energy resource for the future. After all, the sun is expected to shine for billions of years, so it will certainly provide enough nonpolluting energy. But how can you estimate the amount of power the sun produces? You decide to conduct an experiment.

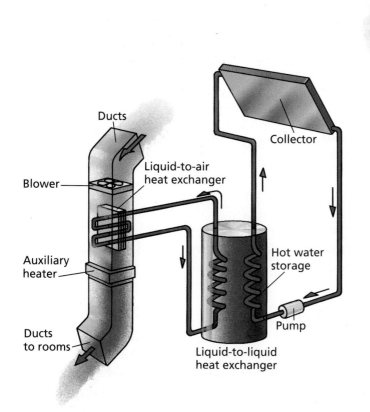

Ducts

Collector

Liquid-to-air heat exchanger

Blower

Hot water storage

Auxiliary heater

Ducts to rooms

Pump

Liquid-to-liquid heat exchanger

Figure 6–11. This diagram shows the operation of one type of active solar-heating system. This particular system uses water as its working fluid.

ACTIVITY

How can you estimate the sun's power?

MATERIALS

piece of thin metal, 2 × 6 cm; graphite suspended in alcohol; meter stick; thermometer in 1-hole stopper; jar and lid with hole; 100-watt lamp

PROCEDURE

1. Carefully prepare your collector as shown in the diagram.

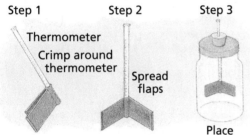

Step 1 — Thermometer / Crimp around thermometer / Metal strip

Step 2 — Spread flaps / Paint these sides with graphite

Step 3 — Place thermometer in jar

2. Place the jar collector outside in the sun. The graphited surfaces of the metal should face the sun. Tilt the jar until the shadow cast by the metal flaps is as large as possible.

3. As soon as it stops rising, record the collector's maximum temperature and return to the classroom.

4. Allow the collector to cool, and then place it 25 cm from a 100-watt bulb. The graphite side of the metal should be facing the bulb.

5. Allow the collector to reach the same maximum temperature it reached while in the sun. If necessary, move the collector closer to or farther from the bulb to achieve the same temperature. If you moved the collector, measure the distance from the bulb to the jar.

6. Using the formula below, calculate the sun's power versus the power of a 100-watt light bulb. The distance from the sun is 1.50×10^{11} m. If you had to move the jar in step 5, replace 0.25 m with your measurement.

$$\frac{\text{power of the sun}}{(1.50 \times 10^{11} \text{ m})^2} = \frac{100 \text{ watts}}{(0.25 \text{ m})^2}$$

APPLICATION

1. Based on your calculations, what is the power of the sun?

2. The value for the sun's power, based on careful measurements, is 3.40×10^{25} watts. How close did your calculation come to this value? What could explain any difference?

3. If we could tap it, do you suppose this is enough solar power for the earth's needs in the foreseeable future? Explain.

ASK YOURSELF

Compare the two basic solar-heating systems that can be used to heat homes and buildings.

Wind Energy

You'd never thought of wind energy as something that comes from the sun, but the facts are there in front of you. The book you're reading explains that winds are convection currents that are formed when the sun heats some parts of the atmosphere more than others. Wind can transfer some of its kinetic energy to other objects. Sailboats, for example, use kinetic energy from the wind to propel them forward.

Figure 6–12. This windmill in the Netherlands uses the energy of the wind to pump sea water away from the land.

Wind Turbines   The energy of wind can be converted into electricity with **wind turbines.** A typical wind turbine consists of a fanlike turbine attached to an electric generator. When wind strikes the turbine, the wind makes the turbine spin. This spins the generator to produce electricity. In some regions, many wind turbines are grouped together in installations called *wind farms.* A single wind farm may consist of more than 30 wind turbines. Wind farms in the United States are located in California, Vermont, Oregon, Montana, and Hawaii.

Energy for Sale   Wind turbines can also be used to generate electricity for individual homes. Many homeowners have installed wind turbines to produce their own electric energy. They can use the electricity from their wind turbines directly, or they can sell it to the electric company serving the area. When energy is sold to the electrical company, it simply flows into the company's electric grid system. Try the next activity to find out more about wind turbines.

Figure 6–13. Many wind turbines grouped together are called wind farms. This wind farm is in California.

Obtain two electric fans and an ammeter. Place the fans on a table so they face each other and are only a few centimeters apart. Plug in one of the fans, and allow it to blow on the other fan.

The wind that the first fan creates should cause the other fan to start spinning, even though it is not plugged in. Connect the ammeter across the prongs of the disconnected plug to measure the current the "wind turbine" produces. ✐

Like all other sources of energy, wind turbines have advantages and disadvantages. Wind energy, like solar energy, is renewable and nonpolluting. But wind turbines work only when the wind is blowing and can be noisy during operation.

Tough Questions, Tough Decisions   Energy supply presents challenging questions. In a rough draft of your paper, you've already decided people should not depend completely on fossil fuels. Should you recommend solar energy or wind energy as the primary energy supply?

The advantages and disadvantages of each are many. Weather and climate systems must be very carefully considered. If there are long cloudy periods, solar methods might not capture enough energy to support society's needs. Perhaps the solution is to recommend the use of more than one type of energy source. After all, the healthiest ecosystems are those with the largest number of different species in them. Might that principle apply to human energy systems as well?

 ASK YOURSELF

How can wind be converted into electricity?

SECTION 2   *REVIEW AND APPLICATION*

Reading Critically

1. Which type of solar-heating system uses solar collectors?

2. What device converts the kinetic energy of the wind into electricity?

Thinking Critically

3. Compare passive and active solar heating. Describe a situation in which each type of heating system would be more practical.

4. Where do you think solar energy would be most useful? least useful?

SKILL Communicating Using a Diagram

▶ **MATERIALS**
- paper
- pencil

Observe the diagram. It shows how a solar hot-water system works. Notice that all the important parts of the diagram are labeled. Arrows are used to show how the water and heat travel through the system. The diagram lets you see at a glance how the system works.

▼ **PROCEDURE**

1. Make a diagram to explain one of these processes: how your home is heated, how a windmill is used to generate electricity, or how a water wheel can be used to generate electricity.
2. To make a good diagram, you need to understand what you want to show. You may have to do some research before you begin. If you do, make some sketches as you read about your topic. Try out different ways of making your diagram.
3. When you finish your diagram, show it to another person. Ask him or her to explain what the diagram shows.

▶ **APPLICATION**
In what other parts of your life might diagramming skills be useful?

✳ **Using What You Have Learned**
Diagrams are important for sharing ideas. Get together with a group, and make a list of tips for making good diagrams. Share your list of tips with other groups.

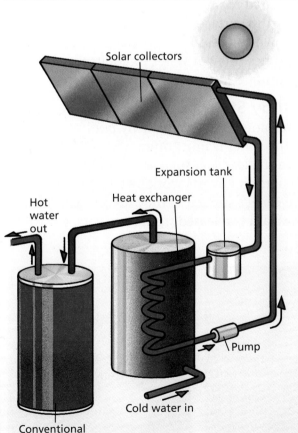

Solar collectors

Hot water out

Heat exchanger

Expansion tank

Pump

Cold water in

Conventional hot water tank

Other Energy Resources

Objectives

Tell *how geothermal energy is converted into usable energy.*

Compare *and* **contrast** *methods of extracting energy from water.*

Explain *how energy is converted from one form to another in alternative methods of energy production.*

At last, you believe, you are on the right track. In your research paper, you intend to suggest that energy choices for the future should include a combination of fossil fuels, solar energy, wind energy, and other energy resources. What other energy resources are there? You check out a video from the library that may have some useful information.

Geothermal Energy

The first image you see on the video is a geyser. The narrator explains that deep in the earth's core, radioactive nuclei produce heat as they decay. Because the crust prevents most of the heat from escaping, it builds up. This heat is a source of energy, called **geothermal energy.** In some places, geological forces push large masses of melted rock, or *magma*, to within four to six kilometers of the earth's surface, forming geothermal *hot spots*. As water seeps down to these hot spots, the water becomes very hot. Sometimes this hot water forces its way back to the earth's surface, forming geysers or hot springs.

Turbines Powered by Steam In most cases, the fluid that turns generator turbines is steam made by boiling water. The fuel used to boil the water is most often a fossil fuel. In some cases the fuel is a nuclear fuel such as uranium. The steam turbines in electric generating plants work just as well on steam from under the earth's surface as they do on steam from a boiler. Therefore, if steam is released from hot spots, electricity can be generated.

Figure 6–14. Geysers are produced by geothermal energy.

Figure 6–15. Turbine generators (left) are used to produce electricity. The turbines (right) convert steam into rotary motion to turn the generator.

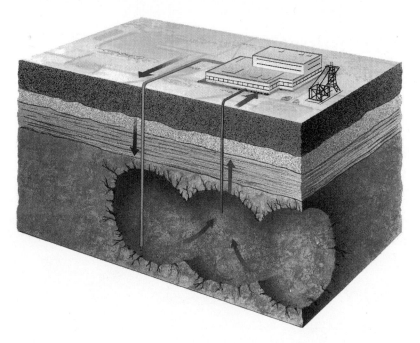

Figure 6–16. A geothermal electric plant receives high-pressure steam from a geothermal hot spot located beneath the earth's surface.

In some cases, hot spots do not already contain water and steam. In these situations, water is pumped down to the hot spots to produce the necessary steam to operate the turbines.

Geothermal Electric Plants   Many countries, including the United States, the Commonwealth of Independent States, Italy, Australia, Japan, and Iceland, already have electric plants that operate on geothermal energy. Some countries, such as Iceland, use geothermal energy directly rather than converting it to electricity. Such direct uses include heating homes, buildings, and household water. The hot water from the earth is even used to heat swimming pools.

Like solar or wind energy, geothermal energy is free, if you don't count the cost of trapping and processing it. Unfortunately, it is only available in a limited number of locations. Many of these are not near places where many people wish to live. Also, the water from some hot spots contains high concentrations of dissolved minerals that are slightly radioactive. This radioactive water is less dangerous than spent fuel from a nuclear reactor, but it must be disposed of carefully.

Mining the Earth's Heat   If geothermal energy is to be used as a future energy resource, hot spots must be close enough to the surface to allow inexpensive "mining" of the earth's internal heat. This occurs most commonly where tectonic plates push into each other or pull apart.

ASK YOURSELF
What produces geothermal energy?

Hydroelectric Energy

The video continues with footage of a gigantic concrete structure. The narrator identifies the structure as a hydroelectric dam. In areas where deep bodies of water are collected behind dams, the fluid turning the turbine is water falling through a dam. Electricity produced in this fashion is called *hydroelectric energy*. The drawing on this page shows the basic arrangement of a hydroelectric dam.

Figure 6–17. The Grand Coulee Dam in the state of Washington is 165 m tall and 1272 m wide. It is so large it can easily be seen from orbit with a simple telescope! In a hydroelectric dam, water flows from a reservoir through turbines. Each turbine, in turn, operates a generator to produce electricity.

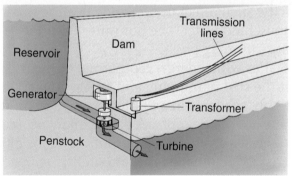

 ASK YOURSELF

How is electricity produced by a hydroelectric dam?

Ocean Tide Energy

Water flowing down a river can be used to produce hydroelectric energy if that river can be trapped behind a dam. However, this is not the only way to use the energy of moving water. Another method involves the ocean tides, which also represent large masses of moving water. In a few locations on Earth, dams can be used to trap water within a bay area during high tide. Then at low tide, the water is allowed to drain through turbines located in the dams to produce hydroelectric energy.

Water Temperature Energy? The oceans also represent a source of energy that is not based on water movement. Instead, this energy source is based on water temperature. The water at the surface of the oceans absorbs large amounts of solar energy. As a result, surface water is much warmer than deeper water. You have probably experienced this difference while swimming. You can even feel it in a swimming pool as you dive to the bottom. A process that uses the differences in ocean water temperature to produce electricity is called *ocean thermal energy conversion,* or *OTEC*.

Figure 6–18. On the Rance River in France, ocean water is trapped behind a dam at high tide. At low tide, the water is used to operate turbine generators.

The OTEC System The OTEC system works much like a refrigerator. It uses warm surface water to heat a special fluid, such as ammonia or Freon, that has a very low boiling point. When surface water heats the fluid, the fluid begins to boil. This boiling action creates pressure that turns turbine generators, producing electricity. Once the fluid is used in this way, it is routed down to a cold layer of the ocean. Here the fluid is cooled and converted back into a liquid, and the cycle begins again.

One of the advantages of the OTEC system is that it is basically pollution free. It uses solar energy to heat the surface of the ocean and can operate 24 hours a day. Unfortunately, OTEC plants are very expensive to build. Also, since they are built offshore, the energy they produce is hard to transport to cities where it will be used. It is possible to convert the energy into microwave radiation, beam it ashore, and then convert it to electricity. But that would add a lot to the cost. Moreover, scientists are not sure how OTEC energy factories will affect the marine life of the area.

Figure 6–19. This illustration shows what an OTEC plant might look like.

ASK YOURSELF

How can tides be used to produce energy?

Nuclear Energy

The final portion of the video includes a discussion of nuclear energy—energy stored within the nuclei of atoms. Nuclear energy is very concentrated, but it is also very hard to release in a controlled fashion. Complex devices called **nuclear reactors** are used to convert nuclear energy into heat. The heat from nuclear energy is used to boil water into steam. The steam is then used to spin a generator turbine that produces electricity.

Splitting Up   All operating nuclear energy plants use nuclear fission. *Fission* means "splitting." In nuclear fission, the nucleus of a uranium or plutonium atom splits when a free neutron crashes into it. When each nucleus is split, it releases heat energy, smaller atoms, and additional neutrons. The additional neutrons go on to split other nuclei to cause a *chain reaction*. This chain reaction must be carefully regulated to keep it under control.

Nuclear reactors are designed in many different ways. A simple diagram of one kind of nuclear reactor is shown on this page. Most reactors, regardless of their design, contain the same basic parts.

The nuclear chain reaction takes place and is controlled within the *reactor core*. The reactor core contains uranium (or plutonium) fuel pellets that have been sealed inside long rods called *fuel rods*. The uranium inside these rods is the fuel for the entire nuclear reactor.

Figure 6–20. Nuclear reactors, such as this one, use fission reactions to produce electricity.

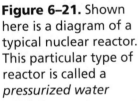

Figure 6–21. Shown here is a diagram of a typical nuclear reactor. This particular type of reactor is called a *pressurized water reactor.*

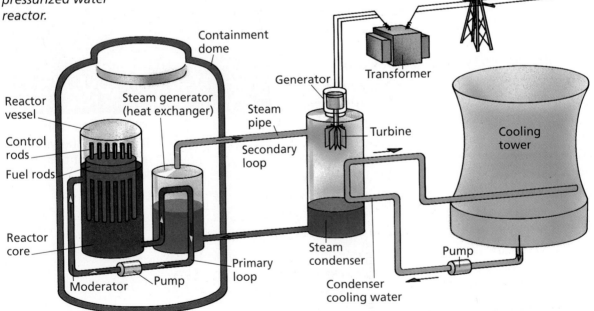

Another set of rods is also located inside the reactor core. These rods are called *control rods* because they are designed to control the nuclear chain reaction. The control rods are filled with materials such as boron and cadmium that can absorb free neutrons. When the control rods are lowered into the reactor core, more neutrons are absorbed, so a smaller number of nuclei are split. This, in turn, slows the chain reaction, and less heat is produced. When reactor technicians "power up" or "power down" a nuclear reactor, they are raising or lowering the control rods.

Is There a Moderator in the House?

In addition to fuel rods and control rods, the reactor core also contains a moderator. The moderator is a substance such as water or graphite that can slow down the speed of free neutrons. If neutrons are moving too fast, they will not cause the uranium or plutonium nuclei to split. In order to have many of the neutrons travel slowly enough to cause fission, they must bounce off something that can absorb some of their kinetic energy. Neutrons transfer some of their kinetic energy to the atoms of a moderator in collisions. Most commercial nuclear energy reactors in the United States use water as a moderator.

The water in nuclear reactors also acts as a coolant. It flows around the fuel rods, where it absorbs heat. If left uncooled, the reactor core could get so hot that it would actually melt, causing radiation to escape. This situation is called a *meltdown.* Partial meltdowns have occurred at Chernobyl in the Ukraine and at Three Mile Island in Pennsylvania.

Reactors also contain *shielding* to prevent radiation from escaping the core. Some shielding materials reflect stray neutrons back into the core. Others absorb radiation to protect the structure of the reactor and to keep radiation from escaping. Radiation can cause biological damage to the people who work at the reactor.

The *containment dome,* which covers the entire reactor, is designed to prevent radiation from escaping into the environment even if an accident occurs in the reactor. The Three Mile Island accident caused little damage to the environment because the radiation was held in the reactor by the containment dome. By contrast,

Figure 6–22. New fuel rods are shown here being installed into a reactor core. Once the rods are installed, they will not need to be replaced again for approximately one year.

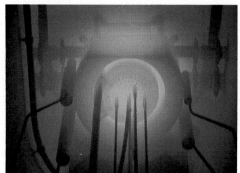

Figure 6–23. The liquid around the fuel and control rods in a reactor core is the moderator.

the Chernobyl reactor did not have a containment dome, and the Chernobyl accident released large quantities of radiation into the environment.

The Solution? Should you recommend full implementation of nuclear fission as an energy alternative? Although nuclear fission can produce tremendous amounts of energy, fissionable materials are expensive and not readily available. In addition, there is currently no accepted method for storing or disposing of radioactive waste from a nuclear reactor.

Through all of your research, you've noticed a pattern. There are advantages and disadvantages to each energy resource. You have read over and over again that the success or failure of humanity depends on its energy supply. But the energy supply must fit both the needs of the people and the resources of the planet. As you've discovered, there are many different energy resources available, but the decision to use each one must be made with care.

DISCOVER BY Researching

Some alternative energy sources that have not been discussed include (a) hydrogen fuel from ocean water, (b) solar energy used to operate steam engines to produce electricity, and (c) solar energy collected in space and transmitted to Earth with microwaves. Conduct research on one of these or another alternative energy source, and prepare a poster. In your poster, include the tradeoffs associated with each energy source. ✎

 ASK YOURSELF

How do nuclear reactors convert nuclear energy into heat?

SECTION 3   *REVIEW AND APPLICATION*

Reading Critically

1. Choose one alternative energy source, and explain the scientific process by which it produces useful energy.
2. What are the advantages and disadvantages of geothermal energy?

Thinking Critically

3. Of the alternative energy sources described in this section, which one do you think is the best choice for widespread use? Explain your answer.
4. In what ways are the methods for producing electricity from conventional, nuclear, and alternative energy sources alike? In what ways are they different?

HIGHLIGHTS

The Big Idea

Heat, light, sound, and electricity—these are all forms of energy. Without energy, everything on Earth would stand still. Nothing moves without it.

Most of the energy people use comes from nonrenewable sources such as fossil fuels. Because fossil fuels are nonrenewable resources, energy alternatives must be developed. Everyone is concerned about energy sources for the future. Scientists are exploring new sources—renewable ones that won't run out. They are also working on new technologies to make energy from these sources continuously available.

For Your Journal

Think about what you have learned in this chapter about types of energy sources. Look back at what you have written in your journal, and revise or add to the entries to reflect your new understanding about energy sources.

Connecting Ideas

Copy this concept map into your journal. Complete the map by filling in details where needed. Remember that a concept map tells you not only which concepts are hooked together, but also what their relationships are.

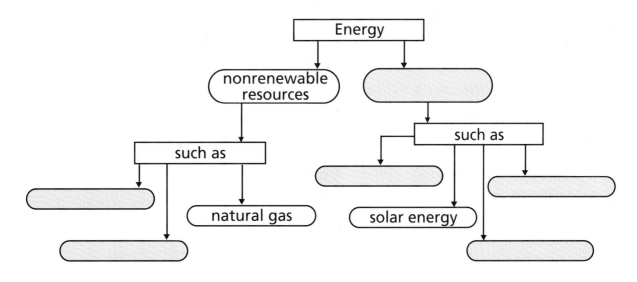

REVIEW

Understanding Vocabulary

1. Classify each term into one of two categories: types of energy or methods of converting energy.
   a) solar energy (159)
   b) wind turbines (163)
   c) passive solar heating (160)
   d) active solar heating (161)
   e) geothermal energy (166)
   f) nuclear reactors (170)

2. Compare and contrast these types of industrial plants: nuclear reactor (170), refinery (155), hydroelectric dam (168), OTEC system (169).

Understanding Concepts

MULTIPLE CHOICE

3. Which term does not relate to coal?
   a) reclamation   **b)** seam
   c) refinery   **d)** strip mine

4. A type of energy production not dependent on location is
   a) tidal.   **b)** solar.
   c) wind.   **d)** nuclear.

5. Which of the following is not a fossil fuel?
   a) solar energy   **b)** coal
   c) natural gas   **d)** petroleum

6. Which part is not found in a nuclear reactor?
   a) fractionation tower
   b) containment dome
   c) fuel rods
   d) control rods

7. Which energy source is nonrenewable?
   a) geothermal
   b) oil shale
   c) solar
   d) wind

SHORT ANSWER

8. Which types of energy collection alter the surface of Earth, and in which ways?

9. Why would solar energy be less efficient during the winter than during the summer?

Interpreting Graphics

10. Follow the path of the steam in this turbine engine. Explain how the blades are turned by the steam.

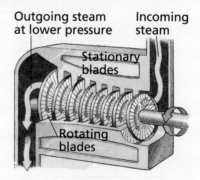

Outgoing steam at lower pressure   Incoming steam

Stationary blades

Rotating blades

Reviewing Themes

11. *Energy*

Starting with the sun shining on the ocean, trace the number and types of energy changes that occur during the OTEC system of energy production. Follow the process all the way to a light bulb in your house.

12. *Technology*

What are the advantages and disadvantages of the OTEC system?

Thinking Critically

13. Most of the solar energy hitting Earth becomes thermal energy. However, the average temperature of Earth does not increase with time. What are some possible reasons for this?

14. In 1775 French explorers in the Ohio Valley saw what they called "pillars of fire." Explain what they saw.

15. How could acid rain damage forests and lakes that are hundreds of kilometers from the nearest coal-burning power plant?

16. Suppose it costs $1,500 to install solar collectors on the roof a house. The homeowner's $100 monthly heating bill goes down 50 percent after installing the solar collectors. How long does it take for the savings to equal the cost of installation? Explain your answer.

17. One proposal for using solar energy is to build massive collectors in space. The large solar collectors would then receive energy from the sun as they orbit the earth. The energy would then be transmitted to the earth in the form of waves. What do you think would be the advantages and disadvantages of this proposed system? Do you think this system will ever become a reality? Explain.

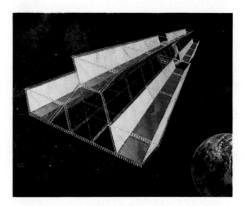

Discovery Through Reading

Haybron, Ron. "Sun Lover." *Discover* 11 (July 1990): 30. Read about a solar dynamic generator planned for the *Freedom* space station.

ENERGY AND THE ENVIRONMENT

John Muir Papers, Holt-Atherton Special Collections, University of the Pacific Libraries, © 1984, Muir-Hanna Trust

The naturalist John Muir spent his life exploring and describing the natural beauty of the American West. He was troubled that people saw in nature only "resources" to be exploited. In the last ten years of his life, Muir battled plans to dam the Hetch Hetchy valley, then a part of Yosemite National Park in California. Muir wrote of the valley's awesome beauty.

The artist . . . wandered day after day along the river and through the groves and gardens, studying the wonderful scenery; and, after making about forty sketches, declared with enthusiasm that although its walls were less sublime in height, in picturesque beauty and charm Hetch Hetchy surpassed even Yosemite.

That anyone would try to destroy such a place seems incredible; but sad experience shows that there are people good enough and bad enough for anything. The proponents of the dam scheme bring forward a lot of bad arguments to prove that the only righteous thing to do with the people's parks is to destroy them bit by bit as they are able . . . Landscape gardens, places of recreation and worship, are never made beautiful by destroying and burying them.

from *The American Wilderness*
by John Muir

For Your Journal

- What are our natural resources?
- How does the burning of fossil fuels affect the environment?
- What can you do to reduce the amount of energy you use?

Environment at Risk

List and **discuss** the sources of air pollution, acid rain, and nuclear waste.

Compare and **contrast** the arguments for and against storing nuclear waste underground.

Evaluate the pros and cons of various pollution-control methods.

On everything from billboards to T-shirts you see the slogans: Save the Planet, Save Our Earth, Preserve the Planet, Keep Earth Green, and on and on. What's happening? We have come face to face with our environment—an environment at risk. The issues related to the environment are enormous—air pollution, acid rain, nuclear waste, energy conservation, energy economics, and energy politics. You are probably willing and eager to "Save the Planet," but you need information about the issues. This information will be your survival tool—a tool with which you *can* help the environment.

Air Pollution

Nearly 90 percent of the United States' energy requirements are supplied by fossil fuels. Unfortunately, the use of fossil fuels creates air pollution. The burning of fossil fuels produces nitrogen oxides, unburned and partially burned hydrocarbons, sulfur oxides, carbon oxides, and smoke particles. These pollutants affect the atmosphere in several ways.

Nitrogen oxides and partially burned hydrocarbons react with sunlight to produce *smog*. Smog looks like dirty fog and is a health hazard. The word itself comes from a combination of the words *smoke* and *fog*. In cities such as Los Angeles, high concentrations of smog can even lead to the cancellation of outdoor athletic events and sometimes city residents are advised to stay indoors.

Figure 7–1. Smog, visible in the photograph on the left, is a serious health hazard. The photograph on the right shows the same city on a low-smog day.

Incomplete combustion of fuels also produces carbon monoxide, a colorless and odorless pollutant that is poisonous to humans and other animals. Carbon monoxide reduces the ability of red blood cells to carry oxygen in the body. It can stick to hemoglobin in the same way oxygen does, blocking the oxygen. Carbon monoxide is not usually a problem because it is produced in low concentrations and is changed in the atmosphere to carbon dioxide. However, when air circulation is poor, such as in closed garages, in long tunnels, or during unusual weather conditions, the effects of carbon monoxide can be deadly.

Reducing air pollution is difficult. For example, devices used to clean polluted air from factory smokestacks often do so by trapping the pollutants in water. However, this causes water pollution. When pollutants are removed from the water, they often become a solid-waste problem. There are no easy solutions. Often, though, one option is better than another. The solid waste from the water and air can be stored in one place and be kept track of. When it is still in the air or water, it can make its way anywhere on the planet and is uncontrolled.

Figure 7–2. To avoid carbon monoxide poisoning, this mechanic attaches an exhaust hose to the tailpipe of a car. The hose carries exhaust fumes out of the garage.

A Hot Time on the Old Planet

In addition to causing air pollution, the burning of fossil fuels causes other problems. Have you noticed how hot it gets inside a parked car on a sunny day with the windows rolled up? Something similar happens on Earth. Carbon dioxide in the earth's atmosphere acts like a glass globe surrounding the planet. The carbon dioxide allows sunlight to pass inward to the earth but reduces the flow of heat energy outward into space. More carbon dioxide in the air traps more heat energy. This phenomenon is called the **greenhouse effect** because the carbon dioxide acts much like the glass of a greenhouse. To find out more about the greenhouse effect, try the next activity.

DISCOVER BY *Doing*

Place a dark cloth on the bottom of a small, empty aquarium. Now lay a thermometer on the dark cloth. Put the aquarium in direct sunlight, and cover it completely with plastic wrap or a glass lid. How does the temperature in this "greenhouse" compare with the temperature outside? In your journal, write an explanation for this, including a diagram.

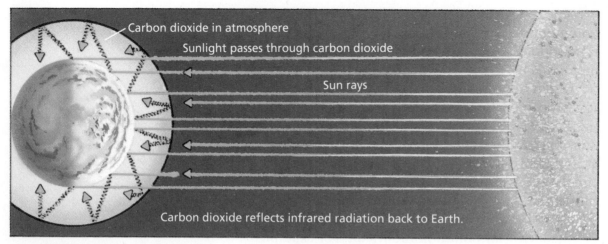

Carbon dioxide in atmosphere

Sunlight passes through carbon dioxide

Sun rays

Carbon dioxide reflects infrared radiation back to Earth.

Figure 7–3. Excess carbon dioxide in the atmosphere might cause global warming due to the greenhouse effect.

Scientists hypothesize that the greenhouse effect may result in an increase in the average temperature of Earth's atmosphere. This heating is called *global warming*. One predicted effect of global warming is a large change in wind and rainfall patterns, which could lead to crop failures. What other changes might occur if the planet warmed too much?

One way to reduce the greenhouse effect and global warming involves reducing fossil-fuel consumption or using alternative fuels that produce fewer pollutants. The major alternative fuels that are now under consideration as the "fuels for the future" include natural gas, propane, ethanol, and methanol. Each fuel has its advantages and disadvantages. Over the next decade, federal Clean Air Act rules will force more vehicles—from school buses to garbage trucks—to burn alternative fuels.

Holes in the Shield   You know that the earth is surrounded by atmosphere, which is divided into distinct layers, as shown in Figure 7–4. Near the top of the stratosphere is a layer called the *ozone layer*. Ozone is a form of oxygen. The ozone layer protects Earth from the sun's ultraviolet radiation by absorbing it. This protection is important because too much ultraviolet radiation can cause problems like skin cancer, impairment of vision, and even death.

There are several ways the ozone is being depleted. For example, nitrogen oxides, produced by high-flying airplanes, can damage the ozone layer by reacting with the ozone. The ozone layer is also damaged by compounds called *chlorofluorocarbons* or *CFCs*. Freon, an example of CFCs, is used in refrigeration and air-conditioning units and in some manufacturing processes. When Freon is released into the air, it works its way up into the atmosphere and destroys the ozone layer. Nitrogen oxides and

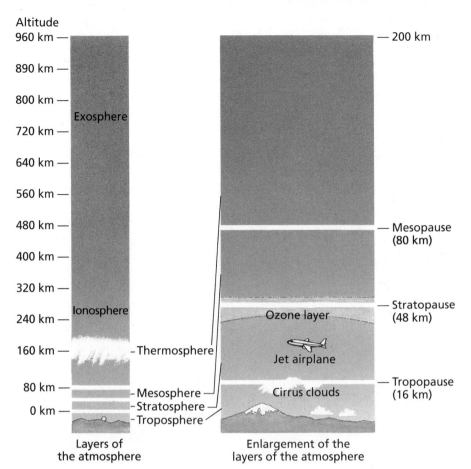

Figure 7–4. Layers of the earth's atmosphere

Altitude

960 km — Exosphere
890 km —
800 km —
720 km —
640 km —
560 km —
480 km —
400 km — Ionosphere
320 km —
240 km —
160 km — — Thermosphere
80 km — — Mesosphere
0 km — — Stratosphere
— Troposphere

Layers of the atmosphere

— 200 km
— Mesopause (80 km)
— Stratopause (48 km)
Ozone layer
Jet airplane
— Tropopause (16 km)
Cirrus clouds

Enlargement of the layers of the atmosphere

Freon destroy the ozone layer faster than it can form through natural processes.

How can the production of CFCs be lessened? American manufacturers have been ordered to stop production of all ozone-depleting chemicals by the end of 1995. Scientists are searching for replacements that do not harm the environment. In the meantime, automobile air-conditioner repair shops have begun using a machine that can recycle the refrigerant, so none is vented to the atmosphere. So far, more than 200 000 recycling machines have been sold.

Figure 7–5. As you can see from these photographs, the ozone hole over the Antarctic has grown larger.

23 Sep., 1979   23 Sep., 1980   23 Sep., 1981   23 Sep., 1982
23 Sep., 1989   23 Sep., 1990   23 Sep., 1991   23 Sep., 1992

100   Total DU   500

ASK YOURSELF

What is air pollution, and how can it be stopped?

Raindrops Falling on My Head

Rain that contains acid is called **acid rain.** Acid rain is formed when air pollutants (sulfur oxides and nitrogen oxides) combine with oxygen and water vapor in the atmosphere to form weak sulfuric and nitric acids. This acidic water can travel as clouds for hundreds and even thousands of kilometers before falling to the earth as acid rain. You can find out whether acid rain is a serious problem in your neighborhood by doing the next activity.

Figure 7–6. Acid rain can destroy forests (left) and kill fish and other organisms living in lakes (right).

DISCOVER BY Doing

Test the acid content of water. Take a water sample from the tap, and dip one strip of blue litmus paper and one strip of red litmus paper into the sample. Then add a couple of drops of vinegar to the water sample, and repeat the procedure. How would the results compare to those of normal rainwater and acid rain? Collect rain samples and test them. How do they compare? ✐

It's Raining Acid   Acid rain affects our environment in many ways. As runoff, it might make lakes and rivers acidic, killing fish and other aquatic organisms. Some lakes become so acidic that not even bacteria can live very well in them. The water in these lakes is very clear because none of the tiny organisms that cause water to look cloudy can survive. Acid rain also reacts with soil and rocks to dissolve ions of aluminum and other metals, which can be harmful to living organisms. When washed into rivers and lakes, these dissolved metals pollute the water.

Figure 7–7. Damage to the Statue of Liberty was caused largely by the effects of acid rain.

Acid rain is also harmful to land organisms. Some food crops have been harmed by the effects of acid rain. Acid rain is killing huge forests in Canada, Europe, and the northeastern United States. Much of the acid rain that is destroying Canadian forests comes from polluted air in the United States. Much of the acid rain that is killing Swedish and Norwegian forests comes from Germany and Poland. As you can see, acid rain is a global problem that can be solved only with international cooperation.

Acid Rain Damages Structures   In addition to killing living organisms, acid rain damages buildings, bridges, statues, and other structures. Much of the damage done to the Statue of Liberty was caused by acid rain. The total repair cost for the Statue of Liberty was $30 million. Structures made of limestone or marble react with acid rain in a manner similar to the reaction that occurs between sodium bicarbonate and the acid in vinegar.

The only way to control acid rain is to reduce the emission of sulfur oxides and nitrogen oxides into the air. The main source of sulfur oxides is the burning of coal and other fossil fuels in electric-energy plants. As much as 50 percent of the

smog and 90 percent of the carbon monoxide in the air pours out of automotive tailpipes. Nitrogen oxides are produced by internal combustion engines. When fuel is burned at high temperatures, oxygen reacts with nitrogen in the air as well as with the fuel, producing nitrogen oxides. Making engines smaller and less powerful is one way to reduce nitrogen oxides. But there is something else that is in place now that works, the pollution-control systems on most automobiles.

In addition, in November 1992, new regulations required specific urban areas to implement tough emission testing for cars and light trucks. The simple tailpipe gauge that measures exhaust while engines idle has been replaced with a new high-tech treadmill device. It collects exhaust while the car idles, accelerates, and brakes. The information is fed into computerized equipment that is so sensitive that millions of cars that formerly passed emissions tests will now probably fail. The vehicles that fail must have repairs made and attempt the test again. The new regulations may help reduce pollution from vehicles by 30 percent in many cities.

Figure 7–8. Most passenger vehicles sold in the United States have a catalytic converter—a device used to remove pollutants from automotive exhaust. Catalytic converters, such as this one, help reduce carbon monoxide, nitrogen oxides, and hydrocarbons.

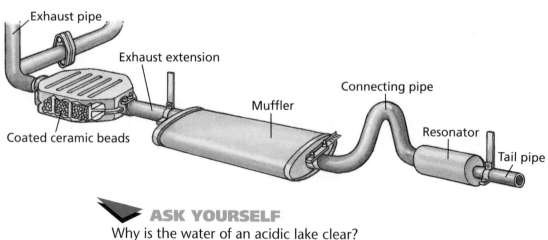

ASK YOURSELF

Why is the water of an acidic lake clear?

Nuclear Energy Is NOT Waste Free

A typical 1000-megawatt (MW) nuclear energy plant creates about 25 tons of spent, or used, fuel each year. This fuel is highly radioactive and can have a half-life of hundreds or even thousands of years. That means that the material will be dangerous for extremely long periods of time. Spent fuel and other radioactive products from nuclear reactions are called **nuclear waste.** These wastes are extremely dangerous to most life forms and, if not disposed of properly, can cause great environmental damage.

Even though nuclear energy has been used for decades, no permanent storage facility for nuclear wastes currently exists. The wastes from the first experiments in nuclear energy, which took place in 1919, are still in temporary storage. The nuclear energy industry identifies categories of nuclear wastes. The most critical storage problem relates to spent nuclear fuel.

Currently, spent fuel is stored temporarily in cooling ponds on the grounds of the nuclear power plants that produce it. In this case, "temporary" seems to mean several decades. Some engineers and scientists hypothesize that the spent fuel rods will eventually corrode and crack open, releasing radioactive waste into the environment. Also, the storage ponds at some older nuclear power plants are reaching their capacity. It is vitally important that we find new and improved methods of storing nuclear wastes.

A number of ways of dealing with spent fuel rods have been proposed. One way is to reprocess the rods to extract reusable uranium and plutonium. This reprocessing would lessen, but not eliminate, the radioactivity of the wastes. Disposing of the reprocessing wastes would still be a problem.

Another plan, which is receiving serious consideration, is to store wastes in underground caverns. The wastes would first be incorporated into solid cylinders of glass or ceramic. These cylinders would then be covered with stainless steel shells and

Figure 7–9. Spent fuel from nuclear reactions is currently stored in cooling ponds, as shown here. These ponds are located on the same site as the nuclear reactor.

Figure 7–10. This diagram shows a planned underground facility for permanently storing nuclear waste.

placed along the hallways of underground caverns. When full, each hallway would be sealed off with rock and stone.

This plan may not be foolproof. Critics of the plan maintain that certain circumstances, such as earthquakes, could cause the release of radioactive materials into the environment. The critics also point out that the security of the storage area would have to be maintained for hundreds or thousands of years. Supporters of the plan claim that the chances of wastes escaping from a well-designed storage facility are very low compared with the chances of wastes escaping from temporary storage. As you can see, there is disagreement about how nuclear waste should be handled. Competent scientists and engineers are found on both sides of the technical arguments.

 ASK YOURSELF

Where and in what form is spent nuclear fuel currently stored?

SECTION 1 *REVIEW AND APPLICATION*

Reading Critically

1. What problems are caused by high concentrations of nitrogen oxides, sulfur oxides, carbon dioxide, and carbon monoxide in the air?

2. What are some ways acid rain damages the environment?

Thinking Critically

3. Provide background information that supports this statement: There are no easy solutions to the pollution problem, only difficult choices.

4. If the United States benefits from the processes that produce acid rain, why should the United States be concerned about forests in another country being damaged by that acid rain?

INVESTIGATION

Determining the Effect of Acid Rain on Seeds

▶ MATERIALS

- safety goggles
- laboratory apron ● gloves
- petri dishes with lids ● paper towels
- solution to simulate acid rain ● seeds ● tape
- aluminum foil ● labels ● litmus paper

▼ PROCEDURE

CAUTION: Put on safety goggles, a laboratory apron, and gloves.

1. Fold a paper towel to cover the bottom of a petri dish. Then dampen the towel with "acid rain" solution.

2. Count out 25 seeds, and place them on the towel.

3. Cover the seeds with another folded paper towel that has been dampened with "acid rain" solution. Put the top on the petri dish, tape it shut, and wrap the dish with foil to protect the seeds from light.

4. Design a control for your experiment. Have your teacher OK your ideas before you proceed. Then set up the second petri dish.

5. Use litmus paper to determine the pH of the "acid rain" solution and the regular tap water, and record it in your journal.

6. Allow the petri dishes to sit undisturbed for the period of time recommended by your teacher. The precise time period will be determined by the kind of seeds you are using.

7. Unwrap and open the petri dishes. Count the seeds that have germinated and the seeds that have not in each dish.

▶ ANALYSES AND CONCLUSIONS

1. Calculate the germination rate for seeds under both conditions. To calculate the germination rate, divide the number of seeds that have germinated by the total number used. How do they compare?

2. Collect data from others in the class who used the same kinds of seeds you did. Average the results for plants germinated in "acid rain" and plants germinated in tap water. How do the average results compare to your results?

▶ APPLICATION

A farmer in your area complains to the county commission that her lettuce crop has suffered ever since the new coal-fired power plant was built. Based on your results from the germination experiment and your knowledge of pollution from coal-fired power plants, how likely is it that she is right?

Discover More

Collect some weed seeds and test them as well as seeds purchased at a seed store. Based on your new research, are food plants or weeds more likely to survive in an area of acid rainfall?

Energy Conservation

Objectives

State three ways new technology can help people conserve energy, and give an example of each.

Evaluate the positive and negative effects of methods of conserving energy.

Fossil fuels are nonrenewable energy resources. Because coal and oil are also raw materials for drugs, plastics, and many other useful compounds, it would be unwise to burn them up completely. How can we use fossil fuels wisely? What are the options?

Spending Energy Wisely

Using a resource efficiently is called **conservation.** Energy conservation has several advantages. By conserving nonrenewable energy resources, existing supplies will last longer. Because any energy transformation creates waste heat, using less energy will produce less pollution. Overall, reducing the amount of energy used will reduce, but not eliminate, the negative impact of energy use on our environment.

There are only three ways to conserve energy—decrease activities that require energy, perform these activities more efficiently, or use different methods of producing and transforming energy that are more efficient. These options can involve new technology, new processes and procedures, and new behavior.

Figure 7–11. This chart shows the sources of energy in the United States.

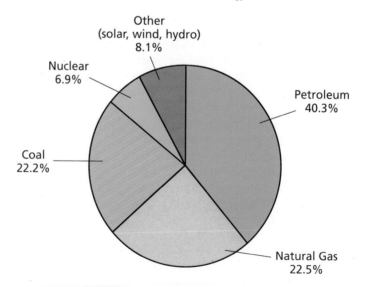

The Sources of Energy
Source: U.S. Dept. of Energy

Other (solar, wind, hydro) 8.1%

Nuclear 6.9%

Petroleum 40.3%

Coal 22.2%

Natural Gas 22.5%

Conservation in the Workplace In industrial settings, using new technology is a common way to increase efficiency. New technology is often applied in two ways—using new materials and using new processes.

Chemists and engineers are constantly trying to produce new materials with useful properties. Some of these materials help to conserve energy by increasing efficiency. For example, synthetic materials have replaced metal panels in some automobile bodies. These panels are much lighter than comparable metal panels. Automobile bodies made with these panels weigh less and can be equipped with smaller, more energy-efficient engines with no loss of performance.

Using new processes can also result in energy savings. A simple example of this can be found in the trucking industry. Many large trucks today are equipped with wind deflectors on their cabs, as shown in the photograph on this page. These simple wind deflectors reduce resistance to air flow around the trucks' trailers. This reduction in resistance increases the fuel mileage of the trucks.

Figure 7–12. Wind deflectors on trucks (left) save considerable amounts of fuel. Why do you think some drivers have substituted this netting device (right) for the metal tailgate?

Only a few decades ago, no one had even thought of putting wind deflectors on trucks. Simple ideas that can make a difference are often overlooked. The idea of the wind deflector, it is rumored, was thought up by a high school student for a science project. This idea resulted in greater fuel efficiency in the trucking industry.

Another example of a simple process that increases efficiency is something called *cogeneration*. This refers to the use of the heat produced in manufacturing to boil water to use in a steam generator. Electricity is generated along with the manufactured product, so it is *co*generated. By producing two useful forms of energy from one fuel source, cogeneration results in fuel savings. You can explore co-generation by doing the next activity.

ACTIVITY

How does a steam turbine work?

MATERIALS

safety goggles; laboratory apron; ring stand (2); utility clamp (2); ring with wire gauze; flask, 250 mL; boiling chip; two-hole stopper to fit flask with bent glass nozzle and thin glass tube (safety valve); Bunsen burner; toy pinwheel

CAUTION: Put on safety goggles and a laboratory apron. Keep them on during this activity.

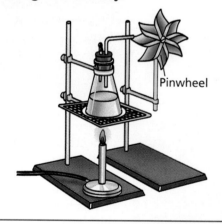
Pinwheel

PROCEDURE

1. Assemble the equipment as shown. Fill the flask about half full of water, and add the boiling chip.
2. Insert the stopper with the nozzle and valve assembly into the flask so that the nozzle points toward your pinwheel.
3. Light the burner and heat the water.
4. Once the water starts boiling, adjust the location of the nozzle so that the turbine (pinwheel) starts to spin.
5. Observe the interaction of steam and turbine as you make the turbine spin. Carefully feel the air near the turbine on the side away from the nozzle.

APPLICATION

How is your steam turbine similar to the steam turbines used in electric-energy plants? How is it different? What could you do to make your steam turbine more efficient?

Robots   An example of a technology that involves both new materials and new processes is robotics. Robots are computer-controlled tools. Today, most robots are used in factories to assemble products. Robots can do jobs that are dangerous to humans because they are unaffected by the heat, fumes, and blinding arcs produced in welding. Also, robots do not need

Figure 7–13. Robots and other computer-controlled devices can improve the efficiency of manufacturing. Here, robots are being used to weld car frames.

good lighting or a comfortable environment. The use of robots results in energy saved on lighting, heating, and air conditioning. A disadvantage of using robots is that people who once did the jobs of the robots have to find work elsewhere. The introduction of new technologies often produces a shifting of jobs.

 ASK YOURSELF

In what ways can technology help conserve energy?

Conservation at Home

As stated earlier, the only ways to conserve energy are to limit activities that require energy, to do these activities more efficiently, or to use different methods of producing and transforming energy that are more efficient. In industry, increasing efficiency is the goal, and this is most often done with new processes and technology. In the home, even small changes in behavior can result in large increases in efficiency. For example, an average dishwasher uses about 25 to 30 L of hot water per load. It has been estimated that if every family that owned a

Figure 7–14. If all Americans turned down their heat by six degrees, we would save 500 000 barrels of oil each day! What benefits could this have to the environment?

dishwasher did one less load of dishes a week, the energy savings would equal the energy necessary to heat about 150 000 homes during the winter.

It is sometimes a struggle to change your behavior. Many people don't even realize what their behavior is. To become aware of your behavior, try the next activity.

DISCOVER BY *Writing*

Write in your journal all the applications of energy you make during the day. Beside each use, identify the energy conversion. For example: Electric iron—electric energy to heat energy. Make a chart of your energy use. Next to each use, mark whether the use is necessary or a convenience. Next to that, suggest at least one way you could change your behavior to decrease the energy used for that activity.

Table 7–1 lists some of the ways that you and your family can decrease your personal energy use and help conserve energy resources. Many suggestions for conserving energy are based on common sense. For example, turning off the television when you are not watching it is a matter of common sense. So is turning out the light when you leave the room, or keeping the doors and windows closed during cold weather. Remember, energy costs money. The more energy you save, the more money you save, and the environment benefits as well.

Table 7-1 — Some Ways to Conserve Energy

1. In the winter, turn down the heat and put on a sweater to help keep warm. In the summer, use the air conditioner less often.
2. Take short, warm showers instead of long, hot showers. Remember, heating that water takes a lot of energy.
3. Turn off lights and appliances when you are not using them.
4. Avoid letting the ice build up in the freezer compartment of your refrigerator, and keep the heat exchanger coils on the back of the refrigerator clean and free of dust. Both ice and dust act as insulators and make it harder for the refrigerator to transfer heat.
5. Use washers, dryers, and dishwashers only when you have a full load. They use about the same amount of energy whether they are full or not. Hang clothes out to dry on a clothesline when the weather permits.
6. Walk or ride your bike when you need to go somewhere. Remember that cars are among the major energy users in the United States.

ASK YOURSELF

What are several ways of conserving energy?

SECTION 2   REVIEW AND APPLICATION

Reading Critically

1. How can energy conservation help reduce pollution?
2. List three ways new technology can help decrease the amount of energy used in industry, and give a short example of each.
3. How do robots help conserve energy?

Thinking Critically

4. How would putting plastic over the windows and using curtains properly help to decrease home energy consumption during the winter months?
5. Table 7–1 lists six behaviors around the home that will help your family conserve energy. List two others things you could do in your home, and explain how they will help conserve energy.

Energy Decision Making

Objectives

Evaluate the economic factors to be considered in energy decision making.

Discuss three issues in international politics that are related to energy policy.

How much does energy cost? This is a very important question, but it is not a simple one to answer. You have probably heard about new energy systems that are designed to reduce energy use. In many situations, these new systems do use less energy and do save money spent on fuel. However, the benefits are not the total story.

Energy Systems

Jazz, Rock, Country, Rap, the Blues, and Classical—these are all different types of music. Whenever you use a stereo system to listen to your favorite music, you are using electric energy. However, this energy is the end product of many processes that also require energy. In other words, energy was required to make energy available to your stereo.

Energy in the form of coal may have been used at the electric generating plant to drive the turbine generators that produced the electricity. Energy was used to mine the coal and transport it to the energy plant. Energy was also used in manufacturing the equipment necessary to harness and control all the other energy. For example, energy was used to create equipment such as coal-mining machinery, turbine generators, steam boilers, and electric lines. The total energy system includes much more energy than may be obvious when you use the electricity that is finally produced.

Figure 7–15. Digging machine at the Peabody coal strip mine in Kankakee, Illinois

By thinking of energy use as a system, you can compare the energy output of a particular system to its input and calculate the net energy. For example, to determine the net energy produced by petroleum, you must first determine its energy input. This includes the energy used to find the petroleum, drill for it, pump it into tankers, transport it to refineries, refine it, transport it to users, and deal with its wastes. If a particular energy system receives support from taxes, that must also be considered an energy input, since money also requires energy to be produced. This combined energy input must then be subtracted from the total energy produced by burning the petroleum. In most energy systems, net output is less than 20 percent of the combined inputs.

Any thought about using or saving energy can be much more complex than it may first appear. Much information must be considered before sometimes tough choices must be made.

ASK YOURSELF

Why is energy use considered a system?

Energy Economics

Before you can determine the benefits of a new energy system, you must figure the total price of a new system and compare the energy savings with the amount of energy used by the old system.

DISCOVER BY *Calculating*

Suppose your family is considering spending $2,000 on a new solar water heater. The solar hot-water system is designed to cut your previous fuel charge for hot water in half. If your family spends $40 per month on hot water, the new system will reduce this bill to $20 per month. Is this system a wise investment? Explain your answer. How long will it take for the system to pay for itself? ✎

Figure 7–16. A solar hot-water system can help cut energy costs. However, it is also an expensive investment.

Another way of looking at the investment is from a break-even standpoint. At a savings of $240 a year, it will take a little more than eight years for the system to pay for itself, or for you to break even. If your family is planning to live in the house for only a couple of years, the new system will not be worth the investment, unless it adds significant value to the house.

The economics of energy also include some hidden costs. Fossil fuels are relatively inexpensive fuels for producing energy. However, the acid rain caused by fossil fuels has damaged food crops and killed fish. The reduced food supply has caused increases in the prices of certain foods. In addition, many people have required additional medical care as a result of pollution from fossil fuels, driving up the cost of insurance. Paying for this care and the increased insurance premiums are also hidden costs of using fossil fuels.

Figure 7–17. Fossil fuels have hidden costs. The pollution from using fossil fuels can cause health problems that require expensive medical attention.

DISCOVER BY *Writing*

Imagine that you have the authority to select the type of energy plant to be built in your community. What would be your choice and why? Look in newspapers and magazines for examples of effective advertisements. Then write three advertisements that would help convince the citizens in your community that your choice is the best one. ✎

 **ASK YOURSELF**

Name two hidden costs of the use of fossil fuels.

Energy Politics

Energy plays an important role in domestic and international politics. One political and economic alliance, called the *Organization for Petroleum Exporting Countries,* or *OPEC,* was formed to help member countries obtain control of their oil resources and to control oil prices. From their point of view, petroleum is their most valuable natural resource. The countries feel that they need to control the production and price of their oil so they can adequately prepare for the future, when their petroleum will be used up. Countries that import oil, on the other hand, feel they are overcharged and captive consumers.

Petroleum is not the only energy resource that affects international politics. Other international issues involve the control of nuclear energy and acid rain. Yet another international issue is the control of carbon dioxide emissions, which may alter climates and produce the greenhouse effect and global warming. These issues can only be confronted with international cooperation.

A step toward the goal of international cooperation was taken in 1992, with the first Earth Summit held in Rio de Janeiro, Brazil. That meeting, designed to address global environmental problems, was attended by a collection of world leaders. Among the issues discussed at the first Earth Summit was the control of carbon dioxide emissions to decrease global warming.

Figure 7–18. The United Nations Conference on Environment and Development—the Earth Summit—was held in June 1992 in Rio de Janeiro.

 ASK YOURSELF

Why are acid rain and the control of carbon dioxide emissions considered international issues?

SECTION 3   *REVIEW AND APPLICATION*

Reading Critically

1. Why was OPEC formed?

2. Describe how the medical costs of treating illnesses resulting from pollution can be considered hidden costs of energy use.

Thinking Critically

3. Develop a plan for diversifying United States energy sources so that the United States no longer has to depend on imported petroleum.

4. Present two arguments in favor of and two arguments against energy independence for the United States.

SKILL Collecting and Organizing Information

▶ **MATERIALS**
- paper ● pencil

▼ **PROCEDURE**

1. Make a chart to collect data on your home electric-energy use. Organize the chart by appliances found in your home. Your chart should also have at least three columns: one column for data, one column for calculations, and one column for the products of your calculations.

2. For a week, survey how often and how long your family uses lights and appliances.

3. From the results of your survey and from the table, calculate how many watts of power are used by each appliance in the course of a year. Show your calculations in your chart.

4. Add the individual values to determine the total power in watts your family uses in a year.

5. Convert your energy total to kilowatt-hours. A kilowatt-hour is 1000 watts, or 1 kilowatt, of energy used for 1 hour. To convert kilowatts to kilowatt-hours, divide the number of kilowatts by 3600 (the number of seconds in an hour).

TABLE 1: ENERGY CONSUMPTION FOR COMMON APPLIANCES	
Appliance	**Energy Consumption**
Automatic dishwasher	4.3 kWh/use
Clothes dryer	3.0 kWh/use
Coffee maker	8.9 kWh/month
Dishwasher with no dry cycle	3.8 kWh/use
Electric clock	1.4 kWh/month
Electric range	
Manually cleaned oven	85 kWh/month
Self-cleaning oven	100 kWh/month
Freezer	
Chest-type	110 kWh/month
Upright, frost-free	210 kWh/month
Hand iron	0.83 kWh/hr
Hot shower (5 minutes)	1.5 kWh/use
Portable electric heater	1.3 kWh/hr
Radio (3 hours/day)	0.08 kWh/hr
Refrigerator	150 kWh/month
Room air conditioner	0.9 kWh/hr
Stereo (3 hours/day)	0.12 kWh/hr
Television (6 hours/day)	
Color	0.21 kWh/hr
Black-and-white	0.05 kWh/hr
Toaster	3.3 kWh/month
Vacuum cleaner	3.9 kWh/month
Warm bath (tub half-full)	1.3 kWh/use
Washing machine	
Hot wash, warm rinse	7.8 kWh/use
Hot wash, cold rinse	6.3 kWh/use
Warm wash, warm rinse	5.5 kWh/use
Warm wash, cold rinse	2.8 kWh/use

▶ **APPLICATION**

Using your survey, propose ways your family could decrease its energy consumption. In each case, calculate the amount of energy and money that would be saved in a year.

✳ *Using What You Have Learned*

1. Multiply the number of kilowatt-hours in your estimate by the cost of a kilowatt-hour of energy. The result is the estimated cost of the electric energy your family uses in a year.

2. Which appliance in your home uses the most energy in the course of a year? the least?

The Big Idea

Energy may be the most important topic you will ever study. Everything you do requires energy, and energy is central to civilization as we know it. Energy production and consumption produces pollution. Even if we solve problems like the greenhouse effect and acid rain, we may never escape creating pollution with our energy activities. The best we can do is to reduce the pollution to a minimum by using alternative types of energy, more efficient technology, and changing our behavior. Solving energy problems will always be politically and economically difficult because energy production and consumption make up a very complex system of interactions.

For Your Journal

Review the questions you answered in your journal before you read the chapter. Then add an entry that shows what you have learned from the chapter. Include ideas about the relationship between energy and the environment.

Connecting Ideas

Below are a series of illustrations relating to this chapter. In your journal list and explain the concepts you can find illustrated.

REVIEW

Understanding Vocabulary

1. Explain how the terms in each set are related.
   a) acid rain (182), smog (178), nuclear waste (185)
   b) greenhouse effect (179), global warming (180), conservation (188)

Understanding Concepts

MULTIPLE CHOICE

2. Which of the following is not a way to conserve energy?
   a) Wash and dry small loads of laundry more often.
   b) Walk or ride your bike when you need to go someplace.
   c) Take short, warm showers instead of long, hot showers.
   d) Turn off lights and appliances when you are not using them.

3. The ozone layer is important because it
   a) increases global warming.
   b) protects Earth from ultraviolet radiation.
   c) protects Earth from smog.
   d) increases the greenhouse effect.

4. An air pollutant that is not the direct result of burning fossil fuels is
   a) carbon dioxide.
   b) sulfur dioxide.
   c) nitrogen oxides.
   d) chlorofluorocarbons (CFCs).

5. Which of the following is not the result of burning fossil fuels?
   a) greenhouse effect
   b) acid rain
   c) nuclear waste
   d) smog

SHORT ANSWER

6. Using petroleum production as an example, explain how the net output of energy is generally much less than the combined energy input.

7. An energy-efficient air conditioner costs $120 more than another model, but is $30 a year cheaper to run. How many years would it take before your savings equaled the additional cost?

Interpreting Graphics

8. This graph shows energy consumption by major sources. What patterns of energy use do you observe? What predictions about future energy use can you make? Explain your predictions.

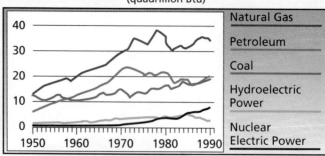

Energy Consumption* 1950–1990
(quadrillion Btu)

Natural Gas
Petroleum
Coal
Hydroelectric Power
Nuclear Electric Power

*One British thermal unit (Btu) = 1055 joules

Reviewing Themes

9. Energy
Earth radiates heat energy into space as a way of getting rid of waste heat. Explain how the greenhouse effect interferes with this process.

10. Environmental Interactions
Chlorofluorocarbons have been linked to the destruction of Earth's ozone layer. Use the release of Freon into the atmosphere as an example of how this interaction is harmful.

Thinking Critically

11. Energy has become increasingly important in global politics. Which topics related to energy might you expect to be most important during future international negotiations? Explain your answer.

12. If Earth's average temperature rose a few degrees, less energy would be needed to heat homes. Why, then, is global warming not considered beneficial to the energy problem?

13. A "fuel-saver" thermostat can be programmed to set back the heat or air conditioning for periods of time a homeowner chooses. To be most effective, the time periods must be longer than two hours. Based on this information, when would it be most practical to set back the heat or air conditioning?

14. The problems of the environment are global in aspect. Explain why international cooperation is necessary to solve these problems.

15. For many industrial tasks, robots are reportedly more efficient than humans in assembling a simple product. Offer pros and cons that should be considered when introducing more robots into industry.

Discovery Through Reading

Pack, Janet. *Fueling the Future*. Children's Press, 1992. This illustrated book presents the relationships between issues pertaining to energy and the environment.

Science

PARADE

Physics in the Circus

Acrobats, high wire artists, and trapeze performers all use physics in their acts.

The crowd grows quiet as Carla Hernandez mentally prepares herself for the quadruple somersault attempt. Her brother, Raul, is already in position to catch her. She jumps from the platform and begins to swing, back and forth, building up height and speed. Suddenly, she lets go! Her body is a blur as she flies through the air, spinning like a cannonball. She extends her arms, and Raul is there. He grabs her wrists, and the crowd goes wild.

Trapeze Tricks

As if by some kind of sixth sense, trapeze artists always seem to know just where the trapeze will be as they fly through the air. A physicist, of course, could calculate where the trapeze would be using the laws of motion. However, by the time the calculations were complete, the trapeze would have moved and the physicist would have fallen into the net below.

When they perform, trapeze artists appear to be flying free, making their bodies move at will. What is really happening is that they are pulling against the force of gravity to do their tricks. At the same time, gravity is pulling on the center of their bodies. They practice pushing and pulling their bodies against gravity. The different parts of their bodies also push and pull on each other using muscle power. Whether they realize it or not, trapeze artists take advantage of the laws of physics.

It's Not the Fall . . .

Trapeze artists know that even if they fall into a net, they probably won't be hurt. The physics of this is not difficult to understand. When a performer falls into a net, the net sags. This slows the fall gradually, instead of stopping it quickly. If the stop were sudden, the force applied to the artist's body would be tremendous, and he or she could be seriously hurt. But with a long, drawn out stop, the force applied to the artist is small. As long as the net is properly placed below the performers and they land on the broad parts of their bodies, falls are not especially dangerous. It is important to know, however, that trapeze artists must stay in superb physical condition. Strains on joints and muscles due to the rapidly changing accelerations of the act could hurt a performer much more than any fall.

Obey the Laws

Once you learn the physics of speed, acceleration, momentum, inertia, work, and mechanical advantage, understanding circus tricks becomes much easier. Figuring out how circus performers take advantage of the laws of physics makes watching them even more fun. Knowing the science behind a trick may even give you insight into just how rigorously circus performers must train in order to accomplish their amazing feats successfully. ◆

Physics helps the "human cannonball" determine where he will land.

Even clowns use physics in their stunts.

You Are There:

RIDING A ROLLER COASTER

Thrills, Chills, and Fun Physics
by Hugh Westrup from *Current Science*

How much of a daredevil are you? Daredevil enough to ride the Magnum XL-200, the world's tallest roller coaster? I did, and now that my hands have stopped shaking, I can tell you all about it.

I rode the Magnum to learn about the science behind the sensations–the exhilaration and the extreme fear–of a roller coaster ride. I wanted to find out how roller coasters achieve some of their heart-stopping effects and still remain safe.

One of the first things I learned was that we are now living in a golden age of roller coasters. Every year several new coasters spring up across the United States, each one attempting to break the previous year's records for the tallest, fastest, and steepest coaster. With names like King Cobra, Great American Scream Machine, and Shock Wave, these new megacoasters are also breaking records for thrills and chills.

Getting to the Point

Surely no roller coaster is scarier than the Magnum XL-200, a hump-backed monster of crosshatched tubular steel that looms over Cedar Point Park in Sandusky, Ohio, on the shore of Lake Erie. So wicked is the Magnum that its designer, Ron Toomer, president of Arrow Dynamics, Inc., refuses to ride on it!

Having braved the Magnum, I'm not sure which was more frightening: the ride itself or standing in line to reach it! The Magnum is so popular that waiting for a ride takes up to an hour. All the while, the blood-curdling shrieks of riders on the Magnum mingle in the air with a continuous recorded message warning away anyone with a heart, back, or blood-pressure problem. Some fun!

But all too soon the wait was over and I found myself quickly ushered into a seat on the 36-passenger train. I took a long, deep breath as a ride operator lowered a padded steel bar over my head and into my lap. An instant later, the train pulled out as the crowd behind me burst into a round of applause. And away we went!

On the Fright Track

Some people who have ridden the Magnum say that the first 30 seconds of the ride may be the most agonizing ones of your life. As the little train leaves the station, it latches on to a moving chain that pulls you slowly up the first hill. On my trip, my riding companion, a Cedar Park tour guide, motioned toward the magnificent view of Lake Erie unfolding before us. She said that on a clear day you can see the skyscrapers of Cleveland, Ohio, 60 miles (96 kilometers) away.

But who could be in a mood for sight-seeing at a time like this? The only sight I could focus on was the alarming gap growing between the train and the ground as we reached 12 stories, then 13, 14, 15 . . . . Was it too late to turn back?

As I learned later, most roller coasters begin with their tallest hill so as to supply each train with a large reserve of potential energy. *Potential energy* is energy that is stored in an object. For the train to achieve this potential energy, a motorized chain pulls the vehicle uphill. After the train reaches the top and starts downward, the potential energy changes to *kinetic energy*, or energy of motion. This kinetic energy provides much of the fuel for the effects that send chills up and down the spine.

As our train rounded the crest of the hill at a height of 20 stories (205 feet, or 62 meters), for a half-second we seemed to hang in mid-air, not going up or down. A small weather-vane at the top turned gently in the breeze.

But then, with a gutwrenching whoosh, the train's potential energy abruptly changed to kinetic energy and we barreled down the back of the hill at a speed of 10 . . . 20 . . . 30 . . . 40 . . . 50 . . . 60 . . . 70 miles per hour! Eeeyyyooowiii!!!

Because the first drop is so long (195 feet, or 60 meters) and the angle of the track so steep (60 degrees), the Magnum accelerates to a speed that broke the record for fastest roller coaster in the world.

As we hurtled downward, the track beneath us seemed to disappear. Like Wile E. Coyote plunging off a desert cliff, we appeared to be in a free-fall, with the ground rushing upward to smash us to bits.

G Whiz

With only a few feet to spare, our fall was suddenly broken as the train bottomed out on the coaster's first curve. Entering the curve, I found myself overcome by a strange, sinking sensation, as if I had just swallowed a cow. Various factors had combined to hold me in my seat with a force of 3.5 g's.

A *g* refers to the combination of forces acting on a moving vehicle and the people in it. The number of g's that a person or roller coaster is subjected to depends upon such factors as the acceleration of the roller coaster, the pull of gravity, and the curvature of the roller coaster tracks.

Earth's gravity pulls all things down with a force of 1 g. As you sit in class, Earth's gravitational attraction is exerting a force of 1 g on you. As I waited in line for a ride or sat in the train at the top of the first hill, the g force on me was 1.

Riders hurtle down the back side of the Magnum XL-200's first hill. As they move into the first curve, they experience large g forces that make them feel much heavier than normal.

However, the number of g's quickly changed during my first plunge. As the roller coaster accelerated downhill, the increasing speed offset somewhat the effect of gravity. The g force became less than 1 but not quite zero.

As the roller coaster tracks began to curve, I felt an increasing pull on my body. The force pressing on me jumped to 3.5 g's, making me feel as if I were 3.5 times heavier than normal.

A view from the top of the Magnum XL-200, 205 feet in the air. Riders accelerate to more than 70 miles (112 kilometers) an hour on the downhill slide.

Later I learned that 3.5 g's is no small force, considering that space shuttle astronauts experience 4 g's while blasting off from Cape Canaveral. At 5 g's, human eyesight dims and a person loses consciousness.

As our train pulled out of the first curve, the momentum gained from our precipitous drop shot us to the top of the second hump, 157 feet (45 meters) in the air. As the car zoomed toward the top, I found myself on a sudden crash diet, losing all that "weight" I had just gained—and more. At the top of the hump, the number of g's holding me in my seat dropped to below zero (*negative g's*). The acceleration of the train overcame the pull of gravity and I was lifted off my seat. I was weightless!

Had it not been for the lap bar holding me down, I would have taken off into space. The train would have taken off too if not for its two sets of wheels, one riding above the track, the other riding below.

The National Aeronautics and Space Administration trains astronauts for the weightlessness of space by giving them rides in jet aircraft that take roller coaster rides in the sky. Each time a jet flies over the top of a "hill," its passengers become weightless and float around the cabin of the plane.

After the Magnum's second hill, what followed was a pretzel-shaped series of turns and more out-of-seat experiences, including numerous speed bumps; more breathless plunges; a hairpin corner that seemed about to dump us into the lake; a sharp dive into a tunnel that threatened to cut off our heads; and a fantastic show of smoke, strobe lights, and sound effects inside the tunnel.

Thrills but No Spills

Every jolt and whipcrack corner on the Magnum reminded me of what designer

An Exciting Journey of Great Gravity

This diagram shows how the number of g's varies during the first part of the Corkscrew, another roller coaster at Cedar Point. Unlike the Magnum XL-200, the Corkscrew has a loop-the-loop turn. During the ride, you feel as if you're lifting out of your seat when the force on your body is only one-half normal gravity. You become weightless when your acceleration totally offsets the pull of gravity (zero g's) at the top of a hill. At the top of the loop, you feel a force of 1 g that is directed straight up into the air because the car is upside down.

Ron Toomer once said about making roller coasters: "The aim is to build in every fright imaginable." But as I learned at Cedar Point, a roller coaster is one of the safest forms of transportation. The Magnum XL-200's tracks get a careful inspection every day. And once a year, a team of engineers uses X-ray equipment to examine those parts of the coaster where riders experience excessive positive and negative g forces.

Look at it this way. The odds of getting killed on a roller coaster are about 1 in 140 million, compared with 1 in 60 million on a domestic airliner. Statistically speaking, roller coasters are safer than merry-go-rounds. The few deaths that happen on roller coasters usually result from the antics of riders who disregarded the safety rules.

But surely as roller coasters grow taller and steeper, they must test the outer limits of steel and concrete technology. Not so, say roller coaster designers. Roller coasters could climb much higher with no increase in danger to riders.

The limit to a roller coaster's height is not in the engineering; it's in the economics. Roller coasters are expensive; Cedar Park spent $8 million on the Magnum.

Passengers ride upside down in a double helix turn on the Typhoon roller coaster at Busch Gardens in Tampa, Florida.

The other limit is human endurance. Twenty-story structures like the Magnum may be about as high as roller coaster riders will dare to go. How many thrills can a person take anyway?

Well, speaking for myself, I can only say that when the train came to a final stop, my throat was hoarse from screaming and my knees were like jelly. I was breathing hard. Stepping off the train, I couldn't help but turn to the tour guide and exclaim: "Let's do it again!" ◆

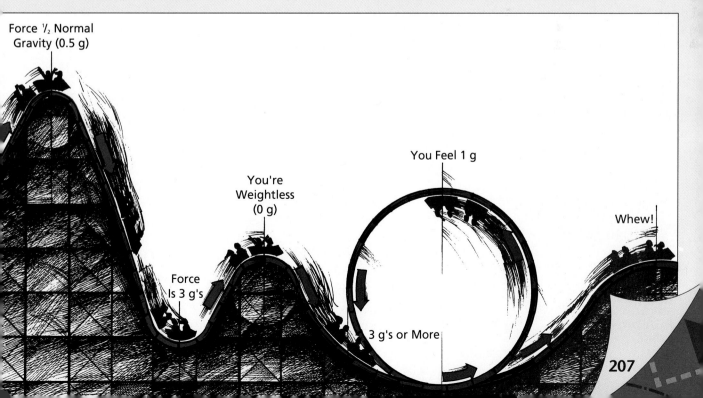

Force ½ Normal Gravity (0.5 g)

You're Weightless (0 g)

You Feel 1 g

Whew!

Force Is 3 g's

3 g's or More

Ellen Richards (1842-1911)

Ellen Richards was probably one of the first women to be concerned with water pollution and energy conservation. In the 1870s, she studied the quality of water in Massachusetts rivers and wrote several books about energy conservation in the home.

Richards was born in Boston in 1842. She grew up on a farm in the nearby town of Westford. Although it was unusual for a young woman to receive a formal education, Richards attended Vassar College and Massachusetts Institute of Technology (MIT). In fact, she was the first woman to be accepted at MIT. She received a B.S. degree in chemistry from MIT and an M.A. from Vassar.

Richards was dedicated to improving the home environment. She started the Home Economics Association and also helped to improve opportunities for women in science through the Association of University Women. As a result of her efforts, the Women's Laboratory was established at MIT. The laboratory offered women training in chemistry, biology, and mineralogy. ◆

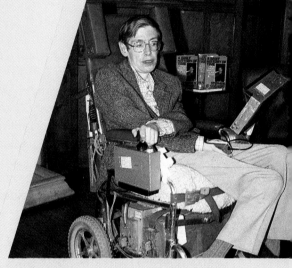

Stephen Hawking (1942-)

While a student in graduate school, Stephen Hawking was diagnosed with amyotrophic lateral sclerosis (ALS), a disease of the nervous system. Also called *Lou Gehrig's disease*, ALS slowly destroys the nerves that control muscle activity. Since then he has been a wheelchair user and needs help with most physical activities.

ALS has not kept Hawking from scientific research, however. He has made some of the most important discoveries about gravity since Albert Einstein. His work supports the theory that the universe started with a big bang.

Hawking was born in Oxford, England, in 1942. At the age of 20, he received a B.A. degree from University College, part of Oxford University, and went on to receive a Ph.D. from Cambridge University. Hawking is perhaps best known for his work in applying Einstein's general theory of relativity to conditions that might lead to the formation of black holes in space. His book, *A Brief History of Time*, introduces people to his theories. ◆

CONNIE ENSING
CART Race Official

As the sleek race car downshifts into pit row, Connie Ensing springs into action. During the few seconds that the car is in the pit, she carefully observes the activity of the pit crew that services and fuels the racer. She makes sure that everything is done safely and according to the rules. Ensing is a technical official for CART, **C**hampionship **A**uto **R**acing **T**eams, Inc.

Officials of CART travel to races all over the United States and even to Canada and Australia. At each event, Ensing recruits and trains as many as 25 volunteers who will assist her before, during, and after the race.

Before a race, her team examines each car to certify that it meets all regulations. Designs are checked to make sure they satisfy CART rules. Measurements are taken of each car's length, wheel width, and suspension system. Eight-foot tape measures are used for large measurements; calipers for small measurements. Other tools include levels and plumb bobs. These help the team to determine whether the angle of the rear wing on each race car is correct.

Team members report their findings to Ensing, who then determines whether a car has an unfair advantage or a safety problem. This information is shared with the car's pit crew, so that corrections can be made.

During a race, Ensing and her team watch for oil leaks, mechanical problems, and even driver fatigue. If a pit crew violates safety procedures or maintenance rules, Ensing calls the control tower. If necessary, crews or drivers are warned on the spot. Imagine stepping out in front of a race car to signal its driver! Sometimes Ensing does exactly that.

After a race, Ensing and her team may recheck some of the cars. They do this if they think any cars have been altered during the race. Cars can still be disqualified at this point. The findings of the inspectors sometimes change the outcome of a race.

Ensing enjoys her work as a CART technical official. If you are interested in car racing and like using math and science skills, you, too, might enjoy this kind of work. ◆

Discover More

For more information about a career as a CART official, write to the

Championship Auto Racing Teams, Inc.
390 Enterprise Court
Bloomfield Hills, MI 48013

Free Fall and Space Travel

Remember the last movie you saw about travel in space? People probably moved around inside the spacecraft pretty much as they do on Earth. Most movies don't show one common feature of space travel—weightlessness.

FREE FALL

Since most space travel is done relatively close to Earth, astronauts are not truly weightless. What makes space travel unique is that astronauts are actually falling freely. Gravity is not making them push on anything as it would on Earth. Nothing under them provides support, because their spacecraft or space station is also falling freely.

Even though astronauts are not totally weightless, they and everything around them behave in rather strange ways. For example, if an astronaut placed a lead weight in a bucket of water, the weight would not sink. Both the lead weight and the bucket of water would be falling toward Earth at the same speed. The lead weight would float. If an astronaut placed a drop of water in the center of a spacecraft, it would stay there, because the spacecraft and the water drop would fall at the same rate.

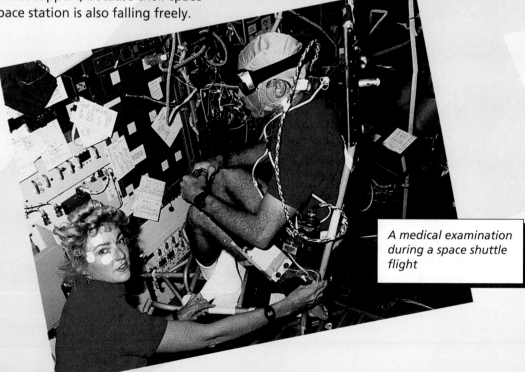

A medical examination during a space shuttle flight

In free fall, astronauts can lift and move heavy objects with ease.

PLUSES AND MINUSES

Free fall has some advantages and some disadvantages. One big disadvantage is that it causes space sickness. Imagine how your stomach would feel as you dropped down the first hill of the Magnum XL-200 roller coaster you read about earlier. Now imagine how your stomach would feel if, instead of reaching the bottom of the hill and starting back up again, you just kept falling. This is what space sickness is like.

The advantages of free fall are many. Weight is a problem for many manufacturing processes on Earth. For example, many metals cannot be mixed because they have different densities. When mixed, these metals separate as oil and water do. In space, metals of different densities stay mixed. This allows for many new kinds of metals to be produced.

MADE IN SPACE

Free fall has other advantages, too. Many kinds of semiconductor crystals that cannot be grown on Earth can be grown in space. Also, very strong but lightweight ball bearings can be made in space by inflating drops of steel like balloons, then allowing them to cool. Because the molten steel and the air bubble are falling at the same rate, the air does not rise through the steel and pop the bubble as it would on Earth.

New metals, semiconductor crystals, and lightweight ball bearings are only three products that could be made in space. There will undoubtedly be many more manufacturing processes that will benefit from a free-fall environment. ◆

211

UNIT 3

WAVES — Energy in Motion

CHAPTERS

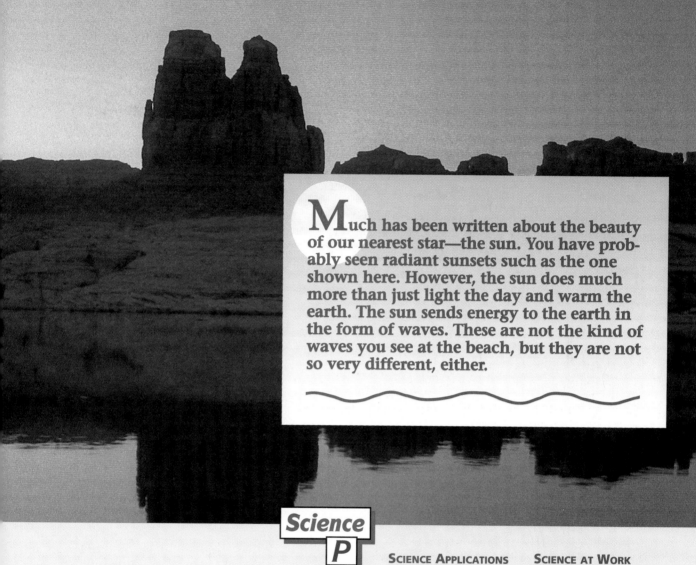

Much has been written about the beauty of our nearest star—the sun. You have probably seen radiant sunsets such as the one shown here. However, the sun does much more than just light the day and warm the earth. The sun sends energy to the earth in the form of waves. These are not the kind of waves you see at the beach, but they are not so very different, either.

Science PARADE

WAVES

The sea moves with such regular and ceaseless rhythms that humans mark time by it. But the wind can suddenly whip it into such a fury that the water's awesome force smashes ships and houses as if they were made of matchsticks.

Down on a long lonely stretch of beach, an hour or so before high tide, the surf was rushing in, splashing and foaming over stones. A pair of sandpipers flew over the water. It is amazing how the birds can fly so fast and low, so near the breaking waves, without getting splashed. Perhaps they do! I'm sure they felt the ocean's spray. I could feel it where I was standing.

After a few hours painting the incoming surf, you begin to sense the rhythm of the waves—first a series of small waves, then others slightly larger, followed by the roaring forms of great big waves, each in its turn breaking on the shore.

Watch a great wave taking shape beyond the surf. A moving mass of energy raises a ridge of water surface that swells as it gains momentum, building to a white-capped wedge, then cresting in a great curving arc of water, luminous, sparkling, hurling itself toward the shore. As the roaring wall of water advances, one end of the arc curls forward and rolls over, forming a funneling, moving mouth of water that would swallow the entire wave if the whole shebang did not break and come crashing splashing down.

from **Near the Sea**
by Jim Arnosky

Copyright © Jim Arnosky, from NEAR THE SEA: A PORTFOLIO OF PAINTINGS by Jim Arnosky.
By permission of Lothrop, Lee & Shepard, a division of William Morrow & Co., Inc.

For Your Journal

- Observe water in motion—in a downspout, water fountain, or faucet—and describe its movements.
- How are water waves like other kinds of waves?
- How can waves carry energy?

Characteristics of Waves

Distinguish between transverse and longitudinal waves.

Recognize the following properties of waves: amplitude, wavelength, frequency, and speed.

Calculate the wavelength, frequency, and speed of a wave.

Summer is a favorite time of year for many people. Along coastlines, people walk along the beaches or ride the ocean waves. At freshwater lakes, people swim or water-ski. Sailboats skim along the open water, from wave to wave, as the wind catches their sails.

Imagine that you are spending a month at camp. Today is your first day, and camp counselors are introducing you to one another. The sun dances in the blue July sky as you enjoy a swim in the nearby lake. The sights and sounds of the city, which you left just a few hours ago, seem very far away.

Waves of Energy

You and other campers go to the beach to meet Ms. Evans, the counselor in charge of boating. You find a stone on the beach and toss it in a pool of water trapped in the sand. You've created a **wave**—a disturbance that travels through matter or space.

As Ms. Evans approaches, she says, "A wave expert, I see. I know that you've all seen waves, but did you ever stop to think that this wave was created as energy from the stone was transferred to the water? The wave then carried the energy outward, away from where the stone hit the water. The energy of this wave traveled in water. The material a wave travels through is called the *medium*. Ocean waves travel on the surface of water; seismic waves travel through the earth."

Everyone looks at one another, and you roll your eyes in exasperation. What is this? Aren't you at camp to have a good time? One of the campers is bold enough to ask, "What's the big deal about waves anyway?"

"The big deal about waves," Ms. Evans responds, "is that they carry energy. Some waves carry energy in a form that's important to you—for example, sound and light are waves. No waves, no stereo! No waves, no TV! By understanding and studying water waves, which you can see, you can understand the behavior of other kinds of waves."

Everyone is silent, not quite sure how to react. Finally, someone asks, "Are the waves in the ocean like the waves caused by a stone thrown into a pool?" Ms. Evans tells the campers that water waves formed in a bathtub or out in the middle of the ocean are alike. In water, waves of energy consist of high points called **crests** and depressions called **troughs.**

She uses a stick to draw a picture of a wave on the beach, and it looks like a snake moving along the sand. Many people think that a wave at the beach is caused by water moving toward the shore. But it isn't. The water simply moves up and down as energy passes through it. Some waves do deliver their energy to the shore in the form of breakers. Only then does water rush to the shore. The water immediately returns to the sea.

ASK YOURSELF

What is a wave?

Figure 8–1. The waves of oceans and lakes can carry enormous amounts of energy. Most of this energy comes from the wind.

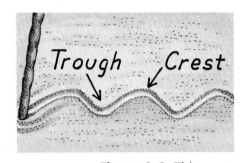

Figure 8–2. This wave has crests and troughs.

Types of Waves

Later that day, you pass by the cabin where Ms. Evans lives. She's sitting on the porch, reading a book. She smiles as you pass by. You say, "I've decided that I'd like to know more about waves, especially since I'll be spending the summer around so many of them."

Up and Down Ms. Evans shows you a book about waves that she's using in her university studies. "Look at the picture of a person creating a rope wave with crests and troughs. Through what medium is this wave traveling?"

To study waves in more detail, look at Figure 8–3. When the person moves her hand up and down, she sets the rope in motion. As she raises and lowers her hand, she creates a wave in the rope that travels down the rope to the other end. The wave consists of a series of crests and troughs that follow each other down the rope. Ms. Evans explains that the wave carries energy through the rope because each point on the rope passes its energy on to the next point.

Figure 8–3. A person is making a transverse wave with a rope to operate a bicycle pump. As the wave moves to the right, the points of the rope move up and down. As a crest reaches the pump, the handle is raised. As a trough reaches the pump, the handle is lowered.

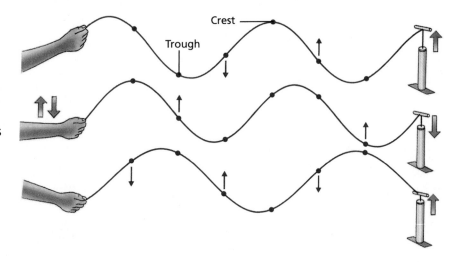

"In a water wave, as in a rope wave, particles move perpendicular to the path of the wave. That means the water moves up and down. This kind of wave is called a **transverse wave**." Then she shows you a picture of a bucket brigade from her book and has you try to solve an interesting problem.

DISCOVER BY *Problem Solving*

Study the picture that shows how a bucket brigade gets buckets of water to the fire. How is this similar to the way energy travels as a wave? Meet with some other students in a small group to discuss this question. ✎

Figure 8–4. The way a bucket brigade gets water to a fire is similar to the way energy travels as a wave.

Side to Side   Ms. Evans turns the page of her book and shows you another kind of wave, a **longitudinal wave.** In this kind of wave, the particles move parallel to the path of the wave. That is, they move back and forth rather than up and down.

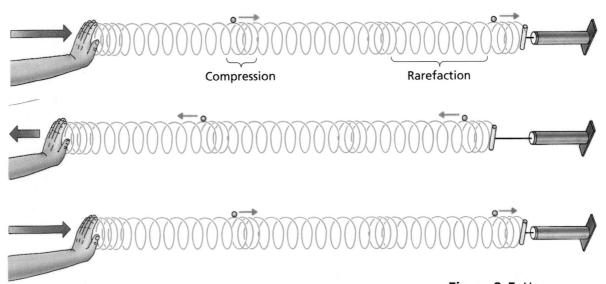

Compression   Rarefaction

Notice that you don't see crests or troughs in this kind of wave. The particles in a longitudinal wave move differently from those in a transverse wave. Every time the person in Figure 8–5 pushes the spring forward, the coils get closer together. The areas where the coils are closer together are areas where the molecules are closer together. These areas are called **compressions.**

When the person pulls her hand back, she separates the coils. In the areas where the coils separate, the molecules are farther apart. These areas are called **rarefactions.** A longitudinal wave consists of a series of compressions and rarefactions following each other down the spring. The compressions and rarefactions move the pump handle in and out.

Remember that in both types of waves, energy is transferred from one place to another. The particles may move up and down perpendicular to the wave's direction as in a transverse wave, or the particles may move back and forth as in a longitudinal wave. The particles don't move down the rope or spring, but the energy does.

Figure 8–5. Here a longitudinal wave created by compressing a coiled spring is used to operate a bicycle pump.

▼ **ASK YOURSELF**

What do transverse and longitudinal waves have in common?

Properties of Waves

You go back to your cabin and tell your bunkmates what you've just learned about two different kinds of waves. They groan. Geez, what a nerd! However, you convince another person to go back with you to Ms. Evans's cabin to find out even more about waves.

"I'm glad you came back—and with a friend, I see," Ms. Evans says, pleased by your appearance on her porch. You watch as she draws a diagram.

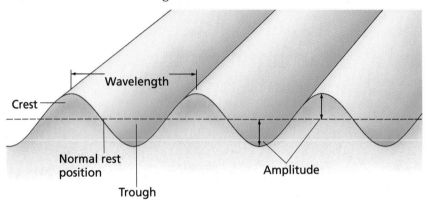

Figure 8–6. These are the parts of a transverse wave.

How High Is the Wave? "There are properties of waves you can measure: amplitude, wavelength, frequency, and speed," she says. "The height of a wave, or the distance a wave moves from a rest position, is called **amplitude.**" She points out amplitude on the diagram and explains that on a calm day, waves with small amplitude allow a boat to rise and fall gently a few centimeters. In stormy weather, however, waves with large amplitude can toss a boat several meters high. Ms. Evans pulls out a fresh sheet of paper and asks you to draw two pictures.

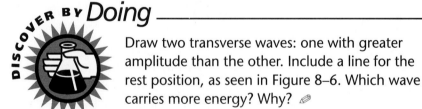

DISCOVER BY *Doing*

Draw two transverse waves: one with greater amplitude than the other. Include a line for the rest position, as seen in Figure 8–6. Which wave carries more energy? Why? ✎

The Wave Is How Long? Waves are characterized not only by amplitude but also by wavelength. Ms. Evans draws two more diagrams: one of a longitudinal wave and one of a transverse wave.

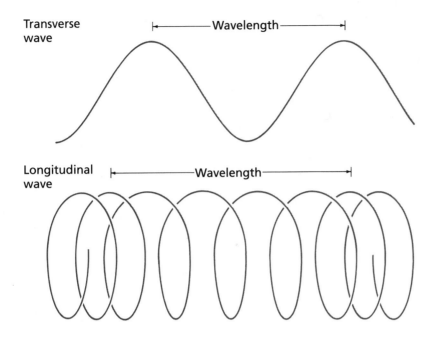

Transverse wave

Wavelength

Longitudinal wave

Wavelength

Figure 8–7. The top diagram shows the wavelength of a transverse wave. The bottom diagram shows the wavelength of a longitudinal wave.

You can compare waves by measuring their **wavelength,** the distance between one point on one wave and the identical point on the next wave. The easiest way to find the wavelength of a transverse wave is to measure the distance from one crest or trough to the next. The easiest way to find the wavelength of a longitudinal wave is to measure the distance between compressions. This tells you the length of one complete wave cycle.

Good Vibrations

Now Ms. Evans draws two more diagrams. She says, "If you shook one end of a rope slowly, you would not produce many vibrations in the rope in a given amount of time. But if you shook the end of the rope quickly, you would produce many vibrations in that same amount of time. **Frequency** measures the number of vibrations a wave produces each second. In other words, it tells you how frequently a wave is produced."

Figure 8–8. The two boats shown here are one wavelength apart. They will move up and down in the water at exactly the same time if they maintain this position.

Figure 8–9. The rope on the bottom is being shaken twice as fast as the rope on the top. What happens to the wavelength of a wave as the frequency increases?

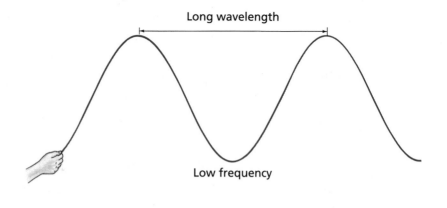

Long wavelength

Low frequency

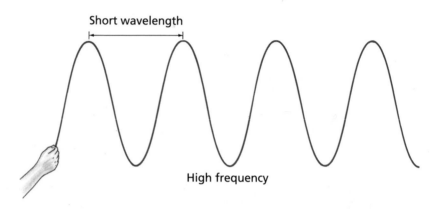

Short wavelength

High frequency

She explains that scientists measure frequency in a unit called *hertz (Hz)*; one hertz is equal to one vibration per second. Then she gives you and your friend a sheet of paper and asks you to draw diagrams.

DISCOVER BY *Writing*

In your journal, make a diagram of two transverse waves, one with a high frequency and one with a low frequency. Label all the parts and properties of the waves. Then diagram two longitudinal waves showing the same parts and properties. ✏

Faster Than a Speeding Bullet   "There's one more property of waves you can measure—speed," Ms. Evans says. "How could you find out how fast a wave is moving?" Your friend says, "Maybe I could swim beside the crest of a wave and time the trip." The counselor replies, "You could measure the speed of a wave by following the crest of a transverse wave or the compression of a longitudinal wave and timing the trip, but you'd have to move pretty fast. Here's an easier method."

Ms. Evans then explains that scientists measure the speed of a wave mathematically in meters per second (m/s). In any wave, the speed of the wave is equal to the product of the wavelength and the frequency. The counselor then writes an equation.

$$v = \lambda f$$

OR

speed = wavelength $\times$ frequency

The symbol for wavelength is the Greek letter lambda (λ). In the next activity you'll have a chance to calculate the speed, wavelength, and frequency of a wave.

ACTIVITY How can you find the relationship among the properties of a wave?

MATERIALS
metal spring about 3 m long, meter stick, stopwatch, chalk

PROCEDURE
1. Place a spring on the floor in a straight line. Measure and record its length in meters.
2. Have one group member shake the spring on the floor from side to side two or three times to produce a 2 Hz or 3 Hz frequency.

3. With a stopwatch, time how long it takes the crest of one wave to travel the entire length of the spring.
4. Determine the speed of the wave. Divide the length of the spring by the time it took a single crest to travel from one end of the spring to the other.

5. With a stopwatch, time how long it takes to produce 10 complete cycles in the spring. Use the formula again to determine frequency.

$$f = \frac{v}{\lambda}$$

Now to find this same quantity by measurement, divide 10 into the total time. Do the two quantities match?

6. Find the wavelength. Use chalk to mark the floor at the top of the first crest. Then mark the top of the second crest. Measure the distance between the marks.
7. Repeat the procedure using a frequency of 4 Hz to 5 Hz.

APPLICATION
1. How does the speed of the wave with the lower frequency compare with the speed of the wave with the higher frequency?
2. What happens to the wavelength when the frequency is increased?

ASK YOURSELF

How does the changing frequency of waves affect their wavelength?

What's It All About?

Amplitude, wavelength, frequency, speed—the properties of waves are used to describe waves. But Ms. Evans said waves carry energy. How do they do that?

The frequency of a wave can tell you how much energy a wave has. The higher the frequency, the more energy in the wave. So if everything else is the same, a wave that has a frequency of 100 Hz has more energy than a wave that has a frequency of 50 Hz.

What if two waves have the same frequency? How can you tell which one has more energy? Simple. If two waves with the same frequency have different amplitudes, the one with the greater amplitude has more energy. If they have the same amplitude, they have the same energy. This principle is especially important with sound waves, which you will learn about later.

ASK YOURSELF

If you wanted to create a wave along a rope carrying the maximum energy, how would you move the rope?

SECTION 1 *REVIEW AND APPLICATION*

Reading Critically

1. What causes areas of compression and rarefaction in a longitudinal wave?

2. How could a person producing a wave in a rope increase the frequency of the wave?

3. A wave in a rope is traveling at 6 m/s and at a frequency of 2 Hz. What is the wavelength of the wave produced? Be sure to show all your calculations.

Thinking Critically

4. What would happen to the frequency and the wavelength of the transverse wave in Figure 8–3 if the student shook her hand faster or shook the rope from side to side? How do changes affect the energy carried by the wave? Explain your answers.

5. A small boat is bobbing up and down in the water as the crests move under it. Are these waves transverse or longitudinal? Explain your answer.

Changing the Direction of Waves

The next afternoon you and the other campers who already know how to swim meet with Ms. Evans. Before you go boating, she wants to observe your swimming abilities. She has you swim to a raft, one by one. You are told to climb the ladder up to the raft and wait for the other campers. You are the first to go; the water feels cool and refreshing. You watch the others swim to the raft and notice how the waves they create make a disturbance in the water. The waves of water lap against the side of the raft, creating a relaxing rhythm. This is the life, you think. Waves and summer skies!

Objectives

Describe reflection and refraction of waves.

Compare the angle of incidence with the angle of reflection.

Explain why the direction of a wave changes when it enters a new medium.

Waves That Bounce

You observe that as the waves created by the swimmers hit the side of the raft, the waves seem to bounce back. As the others climb aboard the raft, you point this out. This is interesting stuff, you think to yourself, and decide that you'll talk to Ms. Evans about how waves change direction.

Figure 8–10. Why has this wave changed direction?

Later you see Ms. Evans in the dining hall and tell her what you observed about waves against the raft. "Good observation," she says. "Waves move in a straight line until they strike a barrier. The 'bouncing back' of a wave after striking a barrier is called **reflection.** You can see this in a basin of water."

DISCOVER BY Doing

Get a basin of water. Let the surface of the water become very calm, and then create a wave with your hand. Watch the wave travel and hit the side of the basin. In your journal, record what happens.

There is a way to predict how a wave will bounce off a barrier. Try the next activity to see if you can figure it out.

DISCOVER BY Doing

Find a hall where you can safely use a small rubber ball. Stand about a meter away from the wall, and then roll the rubber ball to the wall from a variety of angles, such as 25, 45, 75, and 90 degrees. Predict the angle at which it will bounce back. Use a protractor to measure the angle at which you will roll the ball to the wall and the predicted angle of the reflection. Mark the angle with masking tape. How close were your predictions?

Look at the diagram on this page. Notice both the green and the red lines. The green line has been drawn perpendicular to the barrier. A line perpendicular to a barrier is called the *normal*. The red line shows the direction in which the wave is traveling as it approaches and leaves the barrier. The angle between the normal and the direction in which the wave is traveling is called the *angle of incidence*. Think of it as the angle of the wave as it runs into the barrier. The angle is labeled *i*. The angle between the normal and the direction in which the reflected wave is traveling is called *angle of reflection*. This angle is marked *r*. The angle of incidence always equals the angle of reflection.

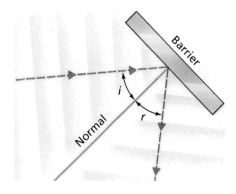

Figure 8–11. When a wave strikes a barrier, it is reflected, as shown here.

 ASK YOURSELF

How does the angle of incidence compare with the angle of reflection when a wave reflects from a barrier?

Waves That Bend

The next day before boating class, Ms. Evans tells you that in addition to reflection, there is another way waves change direction. She asks you to observe what happens when a wave created by a motor boat approaches the shore.

You can see that the crests of the wave get closer together as the wave approaches the shore. "Why do you think that is?" she asks. You don't have an answer, so she continues. "The waves get closer together because the lake bottom becomes more shallow. The more shallow the water, the slower the wave travels. When one crest line slows down, the crest behind it, still in slightly deeper water, catches up a little."

She goes on to explain that ocean waves behave the same way. As an ocean wave enters shallow water, the crest lines bend and become nearly parallel to the shore. The direction at which the wave approaches the shore becomes nearly perpendicular.

Figure 8–12. This motorboat creates waves that will eventually approach the shore.

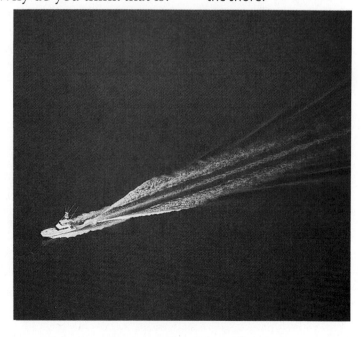

You might wonder what happens to waves if they travel from shallow to deeper water. The opposite happens. They speed up. Why? Because the speed of waves in water depends only on the depth of the water. The speed of other types of waves depends on the medium in which you find them. The change of direction when a wave enters a different medium is called **refraction**. Refraction can occur whenever a wave either speeds up or slows down.

ASK YOURSELF

Why does the speed of a wave change as it enters water of different depths?

Figure 8–13. As an ocean wave comes to shore, it bends. This happens because of friction between the wave and the shallow ocean bottom.

SECTION 2   *REVIEW AND APPLICATION*

Reading Critically

1. If a wave approaches a barrier, making an angle of incidence of 30°, what is the angle of reflection? What is the angle between the direction of travel of the reflected wave and the barrier? Draw a diagram showing the angle of incidence and the angle of reflection.

2. Explain why the crests of waves coming toward a shore tend to become parallel to the shoreline.

Thinking Critically

3. A fast boat is moving parallel to the shore of a lake with a gently sloping bottom. Why is the shape of the wake not a perfect V?

4. As this worker rolls a barrel across different surfaces, what happens? Explain your answer.

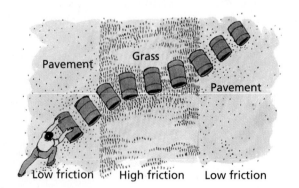

SKILL Diagramming

▶ **MATERIALS**
 • pencil • metric ruler • sheet of unlined paper • protractor

▼ **PROCEDURE**

1. Draw a straight line across the paper at the top of the page. This represents the barrier the wave will strike.
2. At the center of the line, draw a line perpendicular to the barrier. This is the normal. It should run vertically down the center of the page.
3. Now draw a line to represent the direction of travel of an incident wave. Draw the line at an angle so it strikes the barrier at the intersection of the normal. Measure this angle of incidence. Draw arrowheads on the line to show that the wave is traveling toward the barrier.
4. Draw a set of six lines representing the crest lines of the wave. The crest lines should be perpendicular to the direction in which the wave travels.
5. Now draw a line showing the direction of travel of the reflected wave. The line should start at the center of the barrier at the normal and have the same angle as the first one. Put arrowheads on the line to show which way the reflected wave is traveling.
6. Draw in six crest lines of the reflected wave. You will find that these lines cross the crest lines of the incident wave.

▶ **APPLICATION**

Diagram a wave that has a wavelength of 2 cm. Make each wave line 10 cm long. Draw six wave lines to represent the incident wave and six wave lines to represent the reflected wave. The angle of incidence should equal 25 degrees. Make your diagram as clear as possible by using labels and notes to identify key points and features.

✳ ***Using What You Have Learned***

1. If the angle of incidence is 25°, what is the angle of reflection? Measure the angle of reflection in your diagram. Is it correct? Explain.
2. What is the wavelength of the reflected wave?
3. What would the relationship between the incident and reflected waves be if the angle of incidence were changed to 0°?

Waves and Vibrations

Objectives

Describe *what happens when two waves arrive at a point at the same time.*

Analyze *the parts and properties of a standing wave.*

In the evening, all the campers and counselors gather around a campfire to talk and sing songs. The moon shines overhead and is reflected in the water. One counselor takes out a guitar and sings a folk song about a ship called the *Edmund Fitzgerald*. This ship sank in huge waves on Lake Superior. One camper says that waves more than 30 m have been recorded on the Great Lakes, where he comes from. Another camper knows about tsunamis—giant sea waves caused by earthquakes or volcanos. She says that in the Ryukyu (ree OO kyoo) Islands of Japan, where her family originally came from, a powerful tsunami once produced breakers 85 m high! That's about the same height as a 25-story building.

Hokusai, 36 Views of Fuji, Great Wave of Kanagawa. Private Collection.

Figure 8–14. This drawing is an artist's idea of what a tsunami looks like.

Interference

After hearing these stories about great waves, you ask Ms. Evans how waves can become so large and powerful. She explains that waves do not often occur in isolation. One wave usually follows another, and not all waves travel at the same speed with the same frequency. If the crests of two waves meet, their energy combines to form a more powerful wave. This process is called *constructive interference.*

After the campfire, you and your friends join Ms. Evans in the dining hall to continue your discussion about waves. She

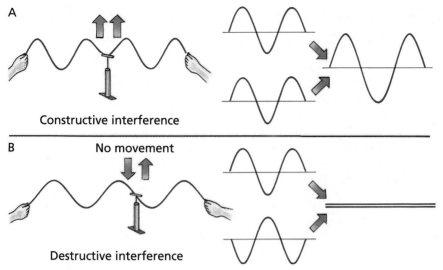

A

Constructive interference

B

No movement

Destructive interference

Figure 8–15. In the top drawing, two waves undergo constructive interference and produce a single, stronger wave. In the bottom drawing, the waves undergo destructive interference and cancel each other out.

has you look at a diagram in her book. Ms. Evans points out that in Part A of the diagram, which shows constructive interference, the bicycle pump handle rises twice as high when the crests of two waves meet as it does when the crest of one wave raised the handle.

Then she points out what happens in Part B of the diagram when the trough of one wave and the crest of another wave meet at the pump handle: the energy of one cancels out the energy of the other. Because the waves cancel each other out, the handle does not move. When two or more waves come together in such a way that a crest meets a trough, the waves undergo *destructive interference.*

Figure 8–16 shows waves created in a ripple tank. In the picture you can see that as two waves meet, constructive interference takes place in some areas while destructive interference takes place in other areas.

Figure 8–16. Two pointers are vibrating to produce two waves that interfere with one another. The green lines show where the interference is constructive. The red lines show where the interference is destructive.

"What happens to something floating on top of an area where interference is taking place?" you ask. Ms. Evans replies, "Remember the other day when we were all floating on rafts in the lake? Some of us were bobbing up and down in one place, and some of us were not moving at all. Tell me what was happening."

"Let me think," you reply. "The people who were bobbing were riding the waves. But the rest of us weren't moving because the waves under our rafts were undergoing destructive interference!"

Figure 8–17. What type of interference are these waves experiencing?

 **ASK YOURSELF**

What is the difference between constructive and destructive interference?

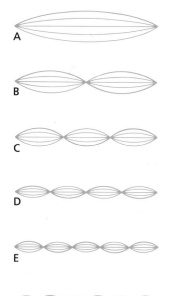

Figure 8–18. These diagrams show the vibrations of a guitar string. In each example, a standing wave has been created.

Standing Waves

Ms. Evans has one more wave concept to share with you. She says, "Imagine that you send a wave through a rope and the wave is reflected at a barrier. If you send one crest to the barrier, one crest will be reflected toward you. The crest will encounter no interference. If you keep making waves with a rope, the crests traveling toward the barrier will eventually meet crests reflected from the barrier. Now interference occurs. If the opposite waves have the same amplitude, frequency, and speed, standing waves will be formed. **Standing waves** are waves that form a pattern in which portions of the waves do not move and other portions move with increased amplitude."

The points on standing waves that have no vibration are called *nodes*. These nodes are caused by destructive interference between waves moving in opposite directions. The points on standing waves that vibrate with the greatest amplitude are called *antinodes*. Antinodes are the result of constructive interference between waves moving in opposite directions.

Standing waves look very different from traveling waves. Standing waves do not appear to move through the medium. Instead, the waves cause the medium to vibrate in a series of loops. Figure 8–18 shows standing waves in a vibrating guitar string. Each loop is separated from the next by a node, or point of no vibration. The distance from one node to the next is always one-half of a wavelength. Only certain frequencies of waves can form standing waves in a string. Standing waves and vibrations are responsible for almost all the sounds you hear.

Figure 8–19. The people in this raft are experiencing standing waves firsthand.

DISCOVER BY *Problem Solving*

Look at the diagram on the previous page. Each loop is separated from the next by a node, or point of no vibration. The distance from one node to the next is always one-half of a wave-length. If Figure 8–18A is one-half of a wavelength, what are the wavelengths for the standing waves in Figures 8–18B through 8–18F? ✎

"Excuse me, Ms. Evans," says one of the campers. "I went rafting with my family a while ago. As we were going down the river, we came upon some waves about a meter high. The water in the waves just seemed to be stuck in the air. Were those standing waves?" "Yes," says Ms. Evans. "That kind of standing wave occurs because the movement of water creates the same kind of pattern that you see in the diagram."

► ASK YOURSELF

How are standing waves produced?

SECTION 3 *REVIEW AND APPLICATION*

Reading Critically

1. What happens when two waves arrive at a point in such a way that the crest of one coincides with the trough of the other?
2. How does constructive interference make it possible to increase the energy of a wave?

Thinking Critically

3. The tuning fork shown on the right is struck to put it into free vibration. Where is the node? Where are the antinodes?
4. Sometimes large waves suddenly appear and disappear on the surface of what looks like a calm ocean. How is this possible?

INVESTIGATION

Creating and Measuring Standing Waves

► MATERIALS
- rope clothesline, about 3 m ● stopwatch ● meter stick

▼ PROCEDURE

1. Prepare a data table like the one shown.

TABLE 1: WAVE CHARACTERISTICS			
Number of Antinodes	Wavelength (m)	Frequency (Hz)	Speed (m/s)
1			
2			
3			

2. Tie one end of the rope to a fixed support, such as a doorknob. Measure the length of the rope.

3. Shake the other end of the rope up and down. Be sure not to swing the rope around. Adjust the frequency until the entire rope is vibrating with just one loop. There is now a node at each end of the rope and an antinode in the middle.

4. Remember that nodes are always one-half of a wavelength apart. Therefore, with one loop in the rope, the rope is one-half wavelength long. What is the wavelength of the standing wave in the rope? Once you know the frequency and the wavelength, you can calculate the speed of the wave.

5. Determine the frequency of the wave. Using a stopwatch, time how long it takes to shake the rope up and down 10 times. This is equal to 10 cycles. Calculate the frequency by dividing 10 into the total time. Record your answer in the data table.

6. Increase the frequency of the wave until there is a node in the middle of the rope. The rope will vibrate with two loops. The rope is now one wavelength long. Determine the frequency and the speed of the wave in the rope. Enter this data in your table.

7. Continue increasing the frequency to get a vibration with three loops in the rope. Now the rope is one-and-one-half wavelengths long. Make the calculations, and fill in your table.

► ANALYSES AND CONCLUSIONS
1. What have you found out about the speed of the wave?

2. What have you found out about the frequencies that produce standing waves? Explain your answer.

► APPLICATION
Standing waves also occur in rivers, especially in fast-flowing areas called *rapids*. What conditions do you think cause standing waves in rivers?

✳ Discover More
Repeat the procedure with a rope that is much thicker and/or heavier in weight. Make sure it has the same length as the first rope you used. What are your results? If they differ, why do you think this is so?

HIGHLIGHTS

The Big Idea

Waves carry energy; they deliver their energy from one point to another as they travel. They can travel through various media as either transverse or longitudinal waves. All waves have properties, such as amplitude, wavelength, frequency, and speed, that tell us how much energy they possess.

Waves react with one another predictably to form patterns. Waves may bounce as reflections or bend as refractions. They may combine either to increase their energy or to cancel it. The energy carried by waves may be heard as sound or seen as light. It can even be used to see through objects. The energy of waves is a powerful force in nature.

Copyright © Jim Arnosky, from NEAR THE SEA: A PORTFOLIO OF PAINTINGS by Jim Arnosky. By permission of Lothrop, Lee & Shepard, a division of William Morrow & Co., Inc.

For Your Journal

Look back at the ideas you wrote in your journal before you began this chapter. Revise your journal entry to reflect what you have learned about waves. Be sure to include information that tells how waves carry energy.

Connecting Ideas

This chart shows how the big ideas of this chapter are related to both transverse and longitudinal waves. Copy the chart into your journal and complete it.

PROPERTIES OF WAVES					
Type of wave	Zone of most dense molecules	Zone of least dense molecules	Properties that can be measured	Kinds of possible interference	Ways in which direction changes
Transverse	crest	_____	speed _____ _____ _____	constructive _____	reflection _____
Longitudinal	_____	rarefaction	_____ _____ _____ amplitude	_____	_____

Understanding Vocabulary

1. For each set of terms, explain the similarities and differences.
   a) transverse wave (218), longitudinal wave (219)
   b) amplitude (220), frequency (221), wavelength (221)
   c) reflection (226), refraction (228)
   d) crests (217), troughs (217)
   e) standing wave (232), wave (216)

Understanding Concepts

MULTIPLE CHOICE

2. Refraction occurs when a wave
   a) enters a different medium.
   b) bounces off a fixed barrier.
   c) interferes with another wave.
   d) strikes a mirror.

3. When a wave reflects from a barrier, it changes
   a) amplitude.
   b) wavelength.
   c) frequency.
   d) direction.

4. Amplitude is a measure of the distance
   a) between a wave crest and a trough.
   b) between a crest and the rest position of a wave.
   c) between the crests of adjacent waves.
   d) a wave travels in a second.

5. In a longitudinal wave, the movement of energy can be seen through a series of
   a) crests and troughs.
   b) frequencies and wavelengths.
   c) compressions and rarefactions.
   d) angles of incidence and reflection.

SHORT ANSWER

6. Explain how a rope attached to a bicycle pump can demonstrate how waves carry energy.

7. If two strings are identical except that one is longer, why does the shorter string vibrate with higher frequency?

8. If a wave has a frequency of 120 Hz and a wavelength of 1 cm, what is its speed?

Interpreting Graphics

9. How do these wavelengths compare?

10. Which wave has more energy? Explain why.

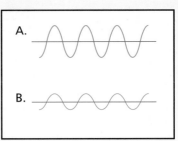

Reviewing Themes

11. *Energy*

Energy is defined as the ability to do work. All waves possess energy. Describe some ways in which waves are harnessed to produce usable energy.

12. *Energy*

Name three machines you use in your home and describe the kinds of waves they use.

Thinking Critically

13. You are outside during a thunderstorm and see a flash of lightning. Then you hear thunder 8 seconds later. Assume that sound waves travel 2 km in 6 seconds; light waves travel 900 000 times as fast as sound waves. In kilometers, how far away is the lightning?

14. How could an aerial photograph enable you to detect the presence of an underwater sand bar near the shore?

15. Professional surfers travel around the world to ride the best waves. Describe in detail the type of waves you think surfers prefer in terms of wavelength, amplitude, and speed.

16. On December 23, 1986, pilots Dick Rutan and Jeana Yeager completed a nonstop flight around the world. The flight lasted for nine days, three minutes, and forty-four seconds, and set an aviation endurance record. Rutan and Yeager accomplished this feat in the aircraft *Voyager*. In making the flight, Rutan and Yeager wore specially-designed headsets that protected their hearing from the monotonous drone of the aircraft engines. The headsets contained an electronic system that reduced noise by creating antinoise. From your knowledge of waves, explain the principle by which these headsets work. In what other applications would such headsets be useful?

Discovery Through Reading

Mayersohn, Norman. "Hear No Evil." *Popular Science* 240 (April 1992): 84–88+. This extensive article discusses active noise cancellation.

CHAPTER 9

Sound

*F*ew sounds are as familiar or unique as the human voice. The human voice is so distinctive that police can identify criminal suspects by voiceprint (the electronic record of an individual voice). Telephone company computers may soon identify long-distance callers by their voiceprint instead of a credit card number.

Wynton Marsalis

The sound of an individual voice is a blend of the other voices we have heard all our lives. Family and friends teach us the peculiar ways of speaking that make for regional accents. A simple word like "nice" sounds much different in an Alabama drawl than in the clipped tones of the Midwest. Regional accents may themselves be blends of different ways of speaking, as in the American South, where the speech of African slaves affected the speaking style of whites.

Louis Armstrong

Charles Mingus

A musician seeking an individual voice must also find a balance between a unique sound and the remembered voices of past masters. Trumpeter Wynton Marsalis has studied the giants of jazz music since he began playing at age twelve. Marsalis grew up in the birthplace of jazz, New Orleans, and learned from his pianist father. Marsalis has studied the pioneering sounds of Louis Armstrong on trumpet, Art Tatum on piano, and Charles Mingus on bass. He has transcribed their music recordings into sheet music and learned to play their styles by ear, seeking the spirit of their music. Yet when Marsalis performs, his trumpet must sound like his own voice, not an imitation of his teachers. His audiences judge him by his unique contribution to the sounds of jazz.

When asked whether he had anything to say to young musicians, Marsalis said, "Work on your sound. Understand that the control and the production of expressive sound is the highest aspect of music."

Art Tatum

For Your Journal

🖍 How do you think sound travels?

🖍 How can sound be harmful to your health?

🖍 What are some of the uses of sound?

Properties of Sound

Relate *the speed of sound to the medium through which it travels.*

Recognize *that volume indicates the amount of energy in a sound wave.*

Identify *the cause of noise pollution and its effect on hearing.*

Imagine that your science class has been chosen to sit in on a rehearsal of a jazz combo called the Five Notes. The night before the rehearsal, you listen to a tape of their music. The piano seems like a rich, complex net of sound that catches the clear melody of the trumpet. The drum and string bass keep pace while the singer's voice leaps up and down the scale, from high notes to low, like an acrobat. The songs are sometimes humorous, sometimes melancholy.

Sending Out Sound Waves

When your class arrives at the city auditorium the next morning, the musicians are warming up on stage. The director of the science museum, Ms. Mendoza, will be your "sound guide" for the day. She ushers your group into a room adjacent to the box office in the lobby. This room will serve as your "sound classroom."

Ms. Mendoza asks for your ideas about how sound travels. There's a lot of whispering and murmuring, but no one volunteers an answer. She explains that the sounds you just heard from the Five Notes are caused by vibrations. These vibrations travel in the form of invisible sound waves to your ears. She motions everyone over to a table where she has set up a demonstration in which you can actually see the pattern of sound waves as they travel. You can try this sound demonstration for yourself.

Figure 9–1. Sound travels to your ears in the form of waves.

DISCOVER BY Doing

 CAUTION: Use care around the lit candle. Place a lit candle near the surface of a flat piece of glass until the glass is blackened with soot. Attach a piece of thin, stiff wire with tape to one prong of a tuning fork. Strike the tuning fork so that it vibrates. Let the tip of the wire just touch the blackened area of the glass. What does this show about the motion of the tuning fork? ✏

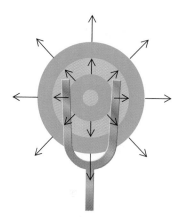

Figure 9–2. As this tuning fork vibrates, it creates waves.

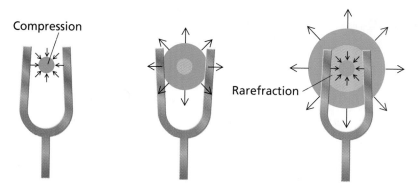

Compression

Rarefraction

How did the pattern of sound waves form when the tuning fork prongs vibrated? As the tuning fork prongs got closer together, they squeezed the air between them to form a dense zone of molecules called a *compression*. As the tuning fork prongs moved apart, they formed a zone in which molecules were less densely packed, called a *rarefaction*. A series of compressions and rarefactions made up the pattern of the sound waves. You may recall reading about compression and rarefaction in the previous chapter.

▶ ASK YOURSELF

How does a tuning fork produce sound waves?

Faster Than the Speed of Sound

As you and your classmates are walking toward the auditorium, one student wants to know how fast sound travels. Before the question can be answered, you see a bolt of lightning flash outside in the sky. Seconds later, you hear the crash of thunder. Ms. Mendoza speaks loudly over the thunder. "Nature just gave us a classic demonstration of the speed of sound," she says. "The lightning and thunder occur at the same time, but it takes longer to hear the thunder because the speed of light is faster than the speed of sound."

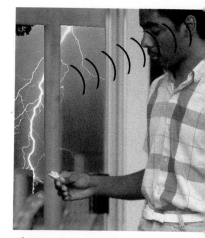

Figure 9–3. How far away is the lightning? Count the number of seconds between the lightning and the thunder. Divide by three to determine the approximate distance of the lightning in kilometers.

She then explains that sound travels at different speeds through different media. "You know that sound can travel through gases, such as air. It also travels through solids and liquids," she says. Do you think sounds travel more easily through solids or gases? Make a prediction. You can test your prediction by doing the following activity.

DISCOVER BY Doing

You and a partner should stand at opposite ends of a table. Ask your partner to face away from you. Drop an eraser and then a paper clip onto the table. Through what material does your partner hear the sound traveling? Ask your partner to put one ear against the table and cover the other ear with a hand. Drop the eraser and paper clip onto the table again. Through what material does your partner hear the sound traveling? Now trade places with your partner and repeat the activity. In your journal, write a description of each sound. Does sound travel more easily through a gas or a solid?

Most solids transmit sound better than gases do. The reason for this has to do with the action of molecules. When a sound is made, molecules vibrate. As you may recall, molecules in a solid are very close together. A vibrating molecule in a solid will quickly bump into another molecule. Vibrations pass from molecule to molecule faster through solids than they do through liquids, where the molecules are not so close together.

Figure 9–4. How did Westerns demonstrate that solids can conduct sound?

The molecules in a gas are farther apart than the molecules in a liquid. A vibrating molecule must travel farther before it bumps into another molecule, so vibrations pass more slowly through gases than through solids or liquids. In water, sound travels about 1500 m/s. This is because the molecules of a liquid are closer together than those of a gas.

Temperature also affects the speed at which sound waves travel. You may recall that molecules move faster in warmer temperatures. When the molecules are moving faster, the sound waves are transmitted more quickly. Ms. Mendoza tells your group that when the air temperature is 0°C, sound travels at about 331 m/s. But, since this is a hot day, the thunder outside could be traveling at a speed as great as 350 m/s.

In Figure 9–3, you learned to use the speed of sound to figure out how far away a bolt of lightning is. How can you measure the speed of sound itself? Try the next activity to find out.

Figure 9–5. Traveling at approximately 350 m/s, it takes a sound wave about 0.08 s to travel across this room. How far is it across the room?

ACTIVITY

How can you measure the speed of sound?

MATERIALS
large building with a flat surface in an open space, hammer, wooden board, stopwatch, meter stick

PROCEDURE
1. One person should stand in front of the building about 50 m away, strike the board with the hammer, and listen for the echo.
2. Repeat step 1 until each blow on the board coincides with a returning echo from the previous blow.
3. With the stopwatch, another person should measure the time interval for 10 round trips of the sound. Start the stopwatch when the hammer strikes and say "zero." Count the blows, and record the time at the count of 10.
4. Divide the time by 10 to get the time for a single round trip.
5. Measure the distance to the building with the meter stick.
6. Double this distance to find the distance the sound wave travels to the building and then back as an echo.
7. Divide the distance by the time to find the speed of sound.

APPLICATION
1. What is the speed of the sound wave?
2. Why is it better to measure 10 round trips of the sound wave than just one?

 ASK YOURSELF
What affects the speed at which sound travels?

Volume

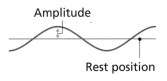

(a) small amplitude

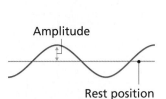

(b) medium amplitude

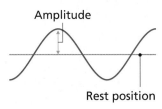

(c) great amplitude

Figure 9–6. Which sound wave has more volume? How do you know?

Your group is now sitting in the auditorium as the drummer and trumpeter rehearse a passage. Sometimes the sounds seem soft and gentle like a spring breeze. At other times, the sounds seem harsh and angry. In fact, you feel as though the loudness of the sound has so much energy it will hurt your ears. Are you right about this?

The greater the vibrations, the more energy will be carried in the sound wave. The more energy carried in the sound wave, the louder the sound. So when the trumpeter blows hard and creates greater vibrations in the trumpet, the result is a sound wave with greater energy. When the drummer softly taps the snare drum, the result is a sound wave with less energy. Both softness and loudness of sound are referred to as **volume.** The amplitude—that is, the distance a sound wave moves from a rest position—indicates the intensity of volume. The greater the amplitude, the greater the volume.

You saw the pattern sound waves made in an earlier activity. How does a change in volume affect the pattern of sound waves? Look at the Figure 9–6. It shows three different volumes of sound.

 ASK YOURSELF

How does the amount of energy in a sound wave affect volume?

Pitch

Later in the morning, the pianist rehearses a solo. She stops and asks the singer if she is playing in a high enough pitch. The vocalist suggests she play in a higher pitch. What does that mean, you wonder? As you listen to the pianist adjust the sound, you infer that the **pitch** of a sound means its highness or lowness. Pitch is not affected by the volume—the softness or loudness of the sound.

During a rehearsal break, your group is invited on stage. You ask the pianist how she increased the pitch on the piano. She invites you and the rest of the group to look under the lid of the grand piano. The pianist points out that shorter strings are attached to keys that play higher notes; longer strings are attached to keys that play lower notes. She tells you that 2000 years ago ancient Greek philosophers knew that when a string was shortened to half its length, the note it produced was twice as high in pitch. The Italian astronomer and physicist Galileo (1564–1642) suggested that the reason this happens is because the frequency doubles.

Figure 9–7. Higher pitched notes are created when keys attached to shorter strings are struck. The pitch produced by each of these piano keys is determined by the frequency.

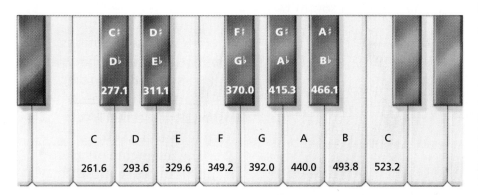

You now understand that piano keys attached to shorter strings have higher pitch. And, according to Galileo, the shorter strings create a higher pitch because they have a higher frequency when struck. This means they vibrate more times per second than longer strings. Try the next activity to find out how pitch affects the pattern of sound waves.

DISCOVER BY *Observing* _____

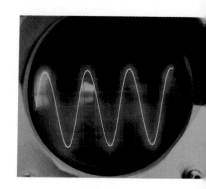

Observe the measurements of an oscilloscope that is connected to a microphone. An oscilloscope is shown here. Observe the wave patterns of five different sounds. How does the pitch change as frequency changes? How does the oscilloscope pattern show the amplitude of the wave? 🖉

▼ **ASK YOURSELF**
What determines the pitch of a sound?

Hazardous Noise

A classmate sitting next to you fidgets and complains that this jazz music is nothing more than "noise pollution." Your classmate might not like the music of the Five Notes, but it isn't **noise pollution**—noise that is harmful to human health. Although the musicians do occasionally play their instruments loudly, they do not do so for extended periods of time.

Ms. Mendoza has noticed the whispering and joins in on the discussion. She asks, "Can someone give me an example of noise pollution?" You mention the sound of jackhammers and other tools involved in construction work outside of the auditorium. She agrees with your example and remarks that listening to loud noises for extended periods of time can be harmful to your health.

The energy in sound waves can injure the human ear by damaging delicate tissues through intense vibrations. Scientists measure volume in units called **decibels (dB).** The chart on this page shows examples of decibel levels from 0 dB to 120 dB.

The jet engine of an airplane produces sound with a volume of approximately 130 dB. This is why ground crews at airports must wear ear protectors. Listening to sound at a high volume over a long period of time can gradually make portions of the inner ear less sensitive and less capable of converting sound waves into nerve impulses. For example, listening to loud music at concerts or on headphones can cause permanent hearing damage.

Figure 9–8. A sound with a volume of 90 dB is loud enough to cause permanent damage to your eardrums.

Decibels (dB)	
120	
110	RESULTS IN HEARING LOSS
100	
90	VERY LOUD
80	
70	LOUD
60	
50	MODERATE
40	
30	FAINT
20	
10	VERY FAINT

Jack hammer
Firing range
Explosion

Truck unmuffled
Noisy machine shop
Rock concert

Noisy office
Loud radio
Orchestra fortissimo

Noisy home
Loud conversation
Movie theater

Private office
Quiet home
Quiet conversation

Rustle of leaves
Whisper
Human breathing

ASK YOURSELF

How can noise be harmful to your health?

Ultrasonic Waves

In the auditorium, the discussion about sound continues. Someone asks if there are sounds people can't hear. Ms. Mendoza tells you that yes, there are pitches far higher than those produced by striking the highest piano keys. She explains that the human ear of someone your age can hear frequencies between about 18 Hz and 20 000 Hz. Although the human ear cannot perceive higher frequencies, humans have found uses for these *ultrasonic waves*—sound waves of very high frequency with short wavelengths.

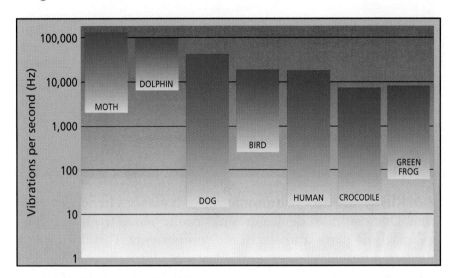

Figure 9–9. This graph shows the range of sound waves that can be heard by humans and various animals.

Sending Out a Signal Many ships have a navigation system that uses echoes of ultrasonic waves to find the depth of water. This system is called **sonar,** which is an acronym for **so**und **na**vigation **r**anging. The sonar device sends short pulses of ultrasonic waves through the water. When the sound waves hit the sea floor, some of the waves are reflected back to the ship as an echo. The echo is then detected by a receiver.

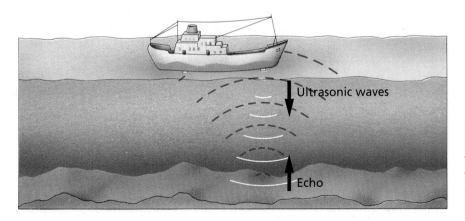

Figure 9–10. The amount of time it takes the ultrasonic waves to return is used to calculate the depth of the ocean floor.

Sonar is even used in some robots. The sonar enables the robots to "sense" obstacles or to detect moving objects. Computers process the signal from the sonar, allowing the robots to react to the location or movement of objects. In this manner, some robots are able to respond to unexpected situations.

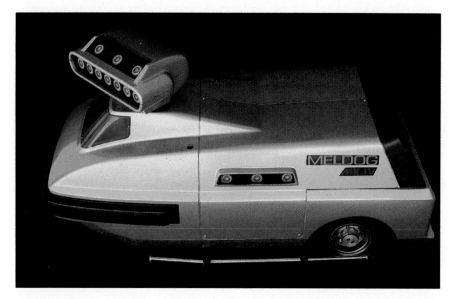

Figure 9–11. Some robots are equipped with sonar to detect obstacles in their path. The robot shown here was designed to replace guide dogs used by the visually impaired.

Sparkling Clean   Ultrasonic waves have uses in addition to sonar. One common device uses ultrasonic waves to clean jewelry, machine parts, and electronic components. This device, called an *ultrasonic cleaner,* consists of a container that holds a bath of water and a mild detergent. Sound waves of about 50 kHz are then sent through the bath. These sound waves create such intense vibrations in the water that they remove dirt from the items placed in the bath. The major advantage of ultrasonic cleaners is that they are nonabrasive, so they do not scratch the items as they are cleaned.

Figure 9–12. An ultrasonic cleaner such as the one shown here uses high-frequency sound waves to clean jewelry.

Using Sound Waves in Medicine   Ultrasonic waves used for medical applications are referred to as *ultrasound.* In some cases, ultrasound is used to remove kidney stones without

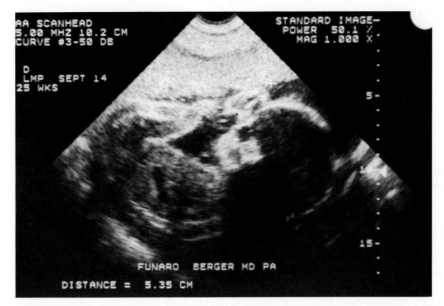

Figure 9–13. This sonogram shows a 25-week-old fetus. Sonograms are used to determine the size, health, and position of a fetus.

surgery. Kidney stones are crystals that form in the kidneys. Ultrasound is used to shatter these crystals without damaging the soft tissues nearby. The tiny fragments can then pass out of the body in the urine.

Ultrasound also provides a way to "see" inside the body. The ultrasonic waves bounce off high-density tissue, such as a tumor. The reflected waves are converted into electrical signals and fed into a computer. The computer uses these signals to create an actual picture, called a *sonogram*. Using this technique, physicians can locate tumors and gallstones. They can also examine developing fetuses inside their mothers.

 ASK YOURSELF

Explain two uses of ultrasonic waves.

SECTION 1 *REVIEW AND APPLICATION*

Reading Critically

1. Explain how sound waves travel through air.
2. What two measurements would you make to determine accurately how far away a lightning bolt struck in a thunderstorm?

Thinking Critically

3. Does the volume of a sound increase or decrease as sound waves move further and further from their source? Explain your answer.
4. Tell which of the following sounds is best described as noise pollution. Then explain your answer.
   a. Three people singing different songs at the same time in high pitches.
   b. The sound of an airplane taking off from a runway.

SKILL *Graphing Musical Pitch*

▶ **MATERIALS**
 ● graph paper ● pencil

Study the table. It shows the frequency of the sound produced each time a guitar string was plucked. The string was plucked eight times, each time shortened 10 cm by the placement of the musician's finger. The frequency was measured in Hz by an audio oscillator—an electronic device used to identify the pitch of sounds accurately.

TABLE 1: FREQUENCY OF GUITAR STRING	
Length (cm)	Frequency (Hz)
80.0	124
70.0	142
60.0	165
50.0	198
40.0	248
30.0	331
20.0	496
10.0	992

▼ **PROCEDURE**

1. Use graph paper to create a graph, and mark the horizontal axis in centimeters. Make the scale large enough so that the final reading is near the right side of the paper.
2. Now mark the vertical scale of the graph in Hz. The 1000-Hz mark should be near the top of the page.
3. Plot each of the eight measurements, or data points, on the graph, and connect them with a smooth curve.

▶ **APPLICATION**
 1. What happens to the frequency produced by the string as the string gets shorter?
 2. Middle C has a frequency of 261.6 Hz. Use your graph to determine how long the string would have to be to sound a middle C.

3. A♭ and G♯ correspond to 25.3 cm. What frequency is that?

✳ ***Using What You Have Learned***
Considering what you know about a guitar, what besides the length of the string determines its pitch?

Hearing Sound

During lunch break, your group sits on benches outside the auditorium. Instead of the rhythms of the Five Notes weaving together, you hear the daily sounds of a busy city street. Ms. Mendoza asks you to use your imagination to discover a rhythm in these street sounds. You close your eyes and listen: a mourning dove repeats soft coos from a tree branch above; car engines hum steadily in a chorus; the traffic light lets out rhythmic clicks as the colors change from green to yellow then red and back to green again. Your ear has learned to hear a new kind of music!

The Doppler Effect

Just as you step back inside the building, the intense scream of a siren signals a firetruck's approach. The pitch of the siren gets higher and higher as the truck gets closer to where you are standing. Then, when the truck passes by, the pitch of the siren suddenly drops. Why did that happen?

Look at the illustration of the firetruck. You can see sound waves in front of the truck and behind it. The sound waves in front of the truck are packed tightly together as the truck moves closer to the last wave it produced. They have a shorter wavelength and a higher frequency. That's why you hear

Figure 9–14. The sound waves from a siren change pitch due to the Doppler effect.

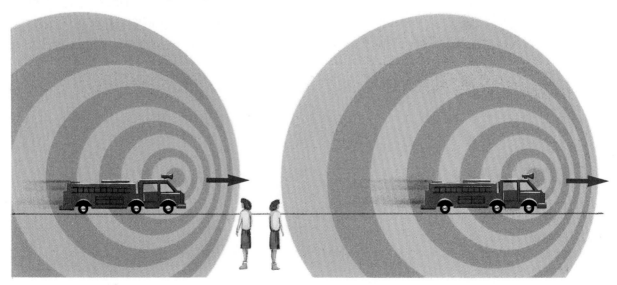

the pitch of the siren get higher as the firetruck approaches. Then look at what happens when the truck passes. The sound waves are farther apart from one another because the truck is now moving away from the last wave it sent toward you. That's the reason the pitch of the siren drops as the firetruck passes. A change in the frequency of waves caused by a moving wave source or a moving observer is called the **Doppler effect.**

 ASK YOURSELF

Give an example of the Doppler effect, and explain how it occurs.

Acoustics

Everyone noisily reassembles in the auditorium. The voices, laughs, and coughs seem to echo all around you. In fact, the design of an auditorium affects the quality of the sound produced by musicians or speakers on stage. Ms. Mendoza tells you that the city auditorium will soon be renovated to improve its acoustics. **Acoustics**, she explains, is the branch of physics that deals with the transmission of sound. She points out that lighting fixtures, seats, and other decorative details in an auditorium affect the acoustics.

Ms. Mendoza hands out a diagram of an auditorium, showing the way sound waves travel to four different seats. She wants you to determine whether or not there are acoustical problems with reverberation (rih vuhr buh RAY shuhn). **Reverberation** is the combination of many small echoes, occurring very close together. It is caused when reflected sound waves meet at a single point, one right after another. Some reverberation gives music or speech a richness or fullness. Too much reverberation, though, makes speech hard to understand and music fuzzy.

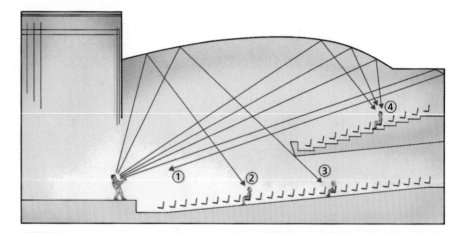

Figure 9–15. This diagram shows how sound reflects from the ceilings and walls of an auditorium.

DISCOVER BY Writing

Look at the diagram of the auditorium in Figure 9–15. Follow the sound pathways from the speaker to the people in the audience marked "2" through "4". Is this auditorium well designed or poorly designed? In your journal, write a paragraph that supports your answer. Make sure that you explain whether or not reverberation would be a problem and why. ✎

► ASK YOURSELF

How can reverberation affect the way you hear voices or music in an auditorium?

How the Human Ear Hears

As you think about the variety of sounds you've heard so far today, you can't help but wonder how the human ear hears. You ask Ms. Mendoza and she says, "Actually, your ears don't hear sounds. Rather, they convert energy from sound waves into nerve impulses. These impulses travel from your ears through nerves that lead to your brain, where they are interpreted as sound."

She passes out an illustration that shows the ear and describes how it works. Then she suggests an activity that simulates a vibrating eardrum. After you look at the illustration of the human ear, try the activity to see how the eardrum receives sound.

Figure 9–16.
(1) The outer ear collects sound waves and passes them through the outer ear until they strike the eardrum. (2) The eardrum vibrates when sound waves hit it. The hammer, a small bone of the middle ear, is connected to the eardrum. When it vibrates, the anvil and stirrup amplify the vibrations from the hammer. (3) Energy from the vibrations pass to the cochlea in the inner ear. This snail-shaped tube is filled with liquid and lined with nerve endings. (4) Vibrations in this liquid move small hairs that stimulate nerve cells, which send messages to the brain. The brain interprets these messages as sound.

The Human Ear

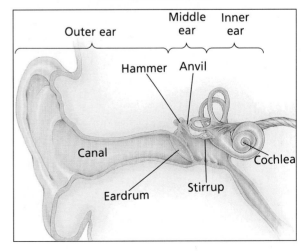

Discover by Doing

Cut a large square from some plastic wrap and then stretch it tightly over the top of a tin can. Next put a rubber band over the plastic to hold it in place. Then sprinkle some sugar on the plastic. Now hit a metal tray hard near the tin can. What happens to the sugar? How is the plastic wrap like your eardrum? What inferences about the eardrum can you make based on this activity? ✐

I Hear a Voice   After completing a song, the singer for the Five Notes steps down from the stage and joins Ms. Mendoza. She explains that singers and nonsingers alike have organs that enable them to make sounds. She suggests the following demonstration.

Discover by Doing

Place two fingers from one hand at the top of your larynx—the hard spot on your throat that moves when you swallow—and two fingers from the other hand at the bottom. Hum a low note, and then hum a high note. Where is the vibration greatest for each of these notes? Then say each of these consonant sounds: kkk, ddd, shhh, hhh. What do you feel? Say each of these vowel sounds: ee, aah. What do you feel? How do you think each of these sounds is produced? ✐

Figure 9–17. This diagram shows the organs that help create the human voice.

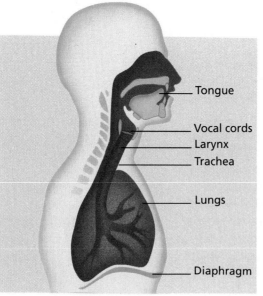

- Tongue
- Vocal cords
- Larynx
- Trachea
- Lungs
- Diaphragm

The organ that produces your voice is called the **larynx** (LAIR ingks), or voice box. This is a small box made of cartilage that is located in the front of your neck. You can feel it easily with your fingers. Two folds of tissue are stretched across the top of the larynx. These folds are called the *vocal cords*. The sound of your voice is produced when air from your lungs flows through the larynx, making the vocal cords vibrate.

As you speak or sing, you constantly adjust the pitch of your voice by stretching and relaxing your vocal cords to produce the correct frequency. Notice that whatever you say, you also change the tone quality of your voice by changing the shape of your mouth and the placement of your tongue to achieve a different sound. Some of these sounds are "voiceless."

Cool Jazz   At the end of the day, you listen to the Five Notes play a new and unusual composition called "Too Much to Do." Sometimes you can detect a melody in the music. At other times the trumpet seems to pop out sounds; the pianist runs her fingers frantically up and down the keyboard; the bass player plucks out low, bellowing notes; and the drummer dazzles you with an improvised solo. The singer uses her voice to produce unusual sounds as if it were an instrument. Like the rhythms you discovered in the street sounds earlier this day, you've learned to use your musical imagination and your knowledge about sound to appreciate the music of this innovative jazz group.

 ASK YOURSELF

How do the vocal cords produce high-pitched sounds and low-pitched sounds?

SECTION 2   *REVIEW AND APPLICATION*

Reading Critically

1. Why does the pitch of a sound change as the source of the sound moves past you?
2. Why is it difficult to distinguish the words of a speaker in some auditoriums?
3. How is sound transmitted from the eardrum to the inner ear?

Thinking Critically

4. A train sounding its horn starts to move toward you. As it speeds up, what happens to (a) the wavelength, (b) the frequency, and (c) the speed of the sound waves reaching your ears?
5. Some animals, such as rabbits, have very large outer ears that they can move in different directions. How does this adaptation help their ability to hear?

INVESTIGATION

Calculating the Speed of Sound

▶ MATERIALS
- standard water-column apparatus ● tuning fork, G = 392 Hz
- thermometer

▼ PROCEDURE

1. Put enough water in the metal reservoir of the water column so that the length of the air column can be adjusted.
2. Adjust the height of the water in the reservoir until the air column is about 20 cm long.
3. Strike the tuning fork, and hold it over the open end of the tube.
4. Lower the water in the reservoir until you hear a sudden increase in the volume.
5. Continue to move the reservoir up and down until you are sure you have found the place where the volume is at the maximum.

6. Measure the length of the air column (in meters), and multiply it by four to get the wavelength.
7. Multiply the wavelength by the frequency of the tuning fork to find the speed of the sound. The frequency of a tuning fork is stamped on its handle.
8. Measure the air temperature, and double-check the speed you just calculated using the following procedure. At 0°C, sound travels though air at 331.5 m/s. For every 1°C increase in temperature, the speed of sound increases by 0.6 m/s. Calculate the speed of sound at your room's air temperature.

▶ ANALYSES AND CONCLUSIONS

1. According to your measurements of the air column, what is the speed of sound?
2. How does the speed compare with the speed calculated from the temperature?
3. What would happen to the length of the air column on a warm day? How would you use the information above to support your response?
4. Describe the results of the same experiment performed in a much colder room.

▶ APPLICATION

If you were in the desert with a friend, would your voice travel faster during the day or during the night? Why?

✳ *Discover More*

If the air temperature outside is substantially different from the air temperature inside, repeat this investigation outdoors. How do the results change if the temperature outdoors is much warmer than it is indoors? much colder? What do you think accounts for the different results?

The Big Idea

Sound is vibrations—the rapid back-and-forth movement of matter. Sound, like other forms of energy, can travel through matter. The amount of energy in a sound wave determines the intensity of volume. The frequency of a sound wave determines the pitch. Sound travels through some forms of matter faster than through others. The medium through which the sound wave moves and the temperature of the medium affects the speed at which a sound wave travels.

Properties of sound are understood not only by characteristics of the source that produces the sound wave but also by the object that receives the sound wave. The human ear receives sound waves; the human voice produces sound waves.

For Your Journal

Look back at the ideas you wrote in your journal at the beginning of the chapter. How have your ideas changed? Now that you know that sound can be harmful to your health, what will you do to protect your hearing? Add your new ideas about the uses of sound in medicine and industry.

Connecting Ideas

These illustrations show examples of some of the concepts presented in this chapter. In your journal, write a caption that describes each of the three scenes.

Understanding Vocabulary

1. For each set of terms, explain the similarities and differences in their meanings.
   a) volume (244), pitch (245)
   b) noise pollution (246), decibels (246)
   c) ultrasonic waves (247), sonar (247), ultrasound (248)
   d) acoustics (252), reverberation (252)

Understanding Concepts

MULTIPLE CHOICE

2. A sonar device can use the echoes of ultrasonic waves to find the
   a) speed of sound.
   b) temperature of water.
   c) height of any waves.
   d) depth of water.

3. Ultrasonic waves are *not* used to
   a) locate tumors.
   b) improve acoustics.
   c) clean jewelry.
   d) shatter kidney stones.

4. During a thunderstorm you see lightning before you hear thunder because
   a) thunder occurs after lightning.
   b) thunder is farther away than lightning.
   c) light travels faster than sound.
   d) sound travels faster than light.

5. A person with his or her ear to the rail hears an approaching train sooner than someone standing up next to the tracks because
   a) sound travels faster closer to the ground.
   b) the train is within sight.
   c) the first person has better hearing.
   d) sound travels faster through a solid than through a gas.

SHORT ANSWER

6. You complain of the noise pollution caused by chain saws being used to remove trees brought down by a storm. Your sister complains that while the school band does not play excessively loud, it causes noise pollution because it plays too many marching songs. Which of you is correct?

7. Describe how ultrasonic waves are used to "see" inside the body.

Interpreting Graphics

8. Which part or parts of the ear
   a) vibrate when sound waves hit them?
   b) collect sound waves and pass them to the eardrum?

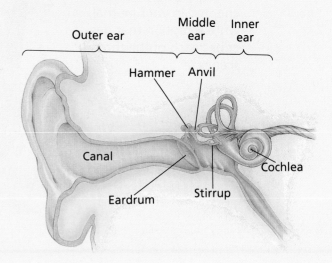

Reviewing Themes

9. *Environmental Interactions*
Explain how each of the following contributes to the sounds you make when you speak: lungs, larynx, vocal cords, mouth, tongue, lips.

10. *Environmental Interactions*
If a tree falls in a forest and no one is there to hear it, does the tree make a sound as it falls? Explain your answer.

Thinking Critically

11. An acoustic engineer might place drapes on the walls of an auditorium. How might the drapes improve the acoustics in the auditorium?

12. A popular science fiction film used this statement in its advertisements: "In space, no one can hear you scream. . . ." Explain the statement in terms of what you have learned about sound waves.

13. How might a robot equipped with sonar be more useful than a guide dog? How might the robot be less useful?

14. Your friend has heard about something called a sonogram, which physicians use to examine a fetus inside the womb. Your friend says, "If it has to do with sound, it must hear and measure the heartbeat." How would you answer your friend?

15. An orchestra is playing in a huge outdoor amphitheater. Thousands of listeners sit on a hillside far from the stage. The amphitheater has an amplification system to increase the volume produced by the orchestra. A special computer slows down the amplified sound of the orchestra by a fraction of a second. Why is this computer used? What would happen if the speed of the amplified sound was not slightly decreased?

16. Police often use radar guns to help them determine the speed of approaching vehicles. These guns send out high-frequency waves that are reflected back from moving objects. How does the Doppler effect relate to the operation of these radar guns?

Discovery Through Reading

"New Devices Wipe Out Noise." 77 *Current Science* (March 13, 1992): 11. This article discusses the development of a "noise-cancellation device."

LIGHT AND LENSES

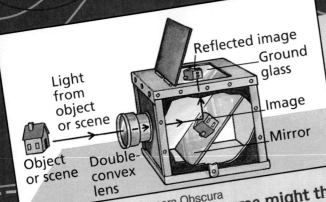

Light from object or scene

Object or scene

Double-convex lens

Reflected image

Ground glass

Image

Mirror

Camera Obscura

Telescopes fitted with the first glass lenses brought new worlds closer to Europe in the 1600s. With the telescope, scientists discovered moons and stars never before seen by human eyes. Yet the glass lens also brought the familiar world closer. In the hands of a Dutch master artist, the lens revealed an inner world of marvelous light.

Jan Vermeer is admired as one of the greatest painters of the 1600s. However, some might think Vermeer was something of a cheat. He probably composed his vivid interior scenes using an early camera-like device called the *camera obscura.*

Camera obscura is Latin for "darkened room." For centuries scientists had known that a tiny shaft of light entering a darkened chamber projected an image on the back wall. By Vermeer's time, inventors had added glass lenses to focus the image and made the camera obscura into a portable box.

How did the images created by the camera obscura affect Vermeer's paintings? Historians find several clues that Vermeer used the camera obscura to trace scenes such as the geographer gazing out a window. Notice that the geographer's table and woven tablecloth seem overly large. The camera devices of Vermeer's day enlarged objects near

Jan Vermeer's *The Geographer*. Frankfurt am Main, Staedelsches Kunstinstitut.

the lens, a distortion noticeable in many Vermeer paintings. Another clue is that our eye views the geographer on a level with the windowsill, about the height of a portable camera obscura. And, finally, note the white blobs of light reflected in the painting. Vermeer faithfully reproduced these distorted points of light just as they appeared through the crude glass lenses of his time.

If Jan Vermeer had possessed a modern camera, he might have just snapped a picture of this geographer in his study. Some might say that Vermeer's pictures are simply the record of light projected through a glass and wood device. Others would claim that no photograph made by glass lens and film could possess the quiet brilliance of Vermeer's painted masterpieces. To the light of science, Vermeer added the artist's vision.

For Your Journal

- How does light travel?
- How does the way in which light travels affect how we see things?
- How can lenses affect the way we view objects?

What Is Light?

Recognize *that the brightness of light depends upon the distance from the light source.*

Diagram *how light travels using the ray model.*

Summarize *the theories that have been proposed to explain the nature of light.*

Imagine that you've joined the school photography club. For your first meeting, the club members go downtown to a sporting goods store where an athletic apparel company is having a photography session for some new advertisements. Two teenage models, wearing T-shirts and caps with the company logo, are taking directions from the photographer. The photographer sets up her camera and lights. She places framed white sheets behind the models.

Even though it is a cloudy and cool day, the models look warm under the hot lights. The scene itself looks bright and sunny because light reflects off the white material. The models pose over and over again, shot after shot, until the photographer captures the exact image she wants.

The Brightness of Light

Before each shot, the photographer checks to make sure that the camera setting matches the reading from a light meter. She allows a photography club member to look at an extra light meter on the set. You can try the same thing.

Figure 10–1. To adjust a camera for proper exposure, photographers use a light meter. The light meter measures the brightness of the light in an area.

Discover by Doing

Using the light meter of a camera or a separate light meter, measure the brightness at 10 different locations. Make a table of your measurements, and compare it with the tables of your classmates. Which areas are the brightest? ✎

The photography club's advisor, Mr. Tallchief, explains that photographers are not the only people who measure the brightness of light. Scientists measure brightness also. They use a unit of measure called the **lux.** Sunlight may be up to 50 000 lux. A lamp by which you might read a book is about 1000 lux. The dim light in a hall or theater lobby is about 1 lux. The closer you are to a light source, the brighter it is, so the higher the lux.

▶ ASK YOURSELF

How does the lux measurement change as the brightness of light increases?

The Ray Model of Light

You know that the closer you are to a light source, the brighter it is. Why? The reason is that light moves in straight lines outward from a source. What does that have to do with brightness? As light leaves the source, the rays spread apart. The farther the light travels, the more the rays spread apart.

Consider the models at the photo shoot. The closer they stand to the lights, the more rays actually strike them. Therefore, the closer they stand, the brighter the light. As they move away from the lights, fewer rays strike them. So the light is less bright.

A diagram of how light travels, such as the one shown here, can help you understand brightness. Light rays, just like the other waves you have studied, move in straight lines until something happens to make them bend. You can experiment with light rays by doing the next activity.

Figure 10–2. A ray model can help you understand how light behaves.

You ask the photographer's assistant setting up the photo area if he knows how fast the light travels. He looks at you and chuckles. "It gets there as soon as I turn the lights on," he says.

Mr. Tallchief explains that the speed of light is not easy to calculate. In fact, it has been a long-time problem for scientists. Early scientists tried to measure the speed of light, but it couldn't be done by ordinary means. It wasn't until 1676 that the Danish astronomer Olaus Roemer (ROH muhr) successfully calculated the speed of light. By studying the eclipses of the moons of Jupiter, he was able to figure out how long it takes for light to travel to Earth. Mr. Tallchief shows you a diagram that explains how Roemer made his measurements.

Figure 10–3. Roemer was able to calculate the speed of light by comparing the distance between Jupiter's moon, Io, and Earth in two different positions in Earth's orbit.

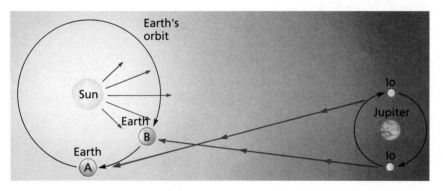

Mr. Tallchief amazes you when he says, "In a vacuum, light travels at a speed of about 300 million m/s. That means that a light beam could get from Los Angeles to Atlanta in less than a hundredth of a second!"

 ASK YOURSELF

Draw a diagram that shows how light rays travel from a reading lamp to the page of a book a person is reading.

Other Models of Light

At school the next day, you look up the subject *light* in an encyclopedia and several other library books. You discover that, through history, some scientists have believed that light behaves like waves while other scientists have believed that light behaves like streams of particles. You make a poster showing what you learned.

Summary of the Theories of Light

In 1672 the British scientist Sir Isaac Newton (1642–1727) wrote a paper about light. In this paper, Newton hypothesized that light consists of streams of tiny particles. Around the same time, the Dutch physicist Christiaan Huygens (HOY guhnz) (1629–1695) hypothesized that light behaves like waves, since light beams can pass through each other undisturbed—something waves can do, but particles cannot.

In 1804 the British scientist Thomas Young (1773–1829) performed an experiment that showed how light, like waves, has the property of interference. He passed a beam of light through two narrow slits onto a screen. The light produced a pattern of bands that proved light underwent interference.

A century later, Albert Einstein (1879–1955) explained how the energy of light comes out in tiny packages called photons. Photons are given off by hot objects. Einstein's photons are similar to Newton's light particles.

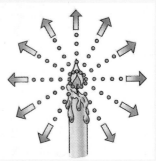

The particle theory of light

The wave theory of light

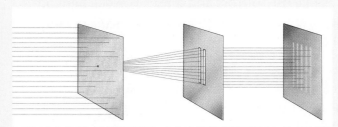

Thomas Young's experiment: On the screen, bright bands result from constructive interference and dark bands result from destructive interference.

 ASK YOURSELF

In what ways does light behave like waves?

SECTION 1   *REVIEW AND APPLICATION*

Reading Critically

1. Why does moving closer to a light source increase its brightness?
2. In what ways does light behave like a ray?
3. What are two models of the nature of light?

Thinking Critically

4. Astronomers measure distances in light-years. One light-year is the distance light travels in a year. How many kilometers are there in a light-year?
5. Make a sketch showing light from a lamp striking an object. Using the ray model, explain why the light source looks dimmer when viewed from a greater distance.

Using the Ray Model

Describe *the processes by which the direction of a light ray can be changed.*

Compare *the light rays that emerge from a lens when the light source is at various distances from the lens.*

Explain *what is meant by the focal length of a lens.*

Y ou arrive at the next photography club meeting. On a table you find an exhibit of optical devices, including mirrors, lenses of different shapes, and a 35 mm camera. You pick up the camera and look through it at a tree outside the window. The tree resembles something from a bad dream. The branches look as wide as the trunk, and the whole image looks fuzzy. Then you focus the lens and continue to look at the tree. Now it seems as though the tree is moving toward you. As it does, the image becomes clearer and clearer—no longer a dream, but a reality.

Bending Light Rays

Mr. Tallchief says, "Before you begin to use a camera, you should know what happens when waves of light energy reach optical devices like the mirrors and lenses you see here on display. Remember that light rays are really waves of energy, and they behave just as all waves do."

Reflection   When light hits the smooth, flat surface of a mirror it acts as any wave does when it strikes a barrier—it reflects. Light rays obey the rule that the angle of incidence equals the angle of reflection.

Of course, most objects in the world do not have smooth, flat surfaces; they have rough and irregular surfaces. What happens when light from different directions strikes these irregular surfaces? It bounces off in many directions. This scattering is called *diffuse reflection*.

Figure 10–4. When a light beam reflects from a mirror, the angle of incidence equals the angle of reflection.

Refraction   There's another way in which light changes direction. Mr. Tallchief sets up a simple demonstration to show what happens when light moves from one medium to another. You can do the same thing.

Discover by Doing

Take a pencil, and view it through a piece of glass held at an angle to your line of sight. What happens to the image of the pencil? Stand a spoon in a glass half filled with water. What happens to the image of the spoon? Explain why you think the spoon looks as it does. ✎

When light travels from air to glass, it moves from one medium to another medium. Light waves, like other waves, travel at different speeds in different media. When these waves of light energy change speeds, they also change direction, or refract. The part of the pencil behind the glass appears to move to one side, and the spoon seems to bend when it hits the water because the light waves refract.

▶ ASK YOURSELF

Choose an object close to you, and explain how light reacts to its surface.

Lenses and Light Rays

You are looking at the lenses on the table at the front of the room. Mr. Tallchief notices your interest and asks if you would like to know how the lenses work. "Of course," you respond, so he begins to tell you about the two basic types of lenses.

Mr. Tallchief explains that the shape of a lens determines how it focuses the light rays that enter it. He asks, "What can a prism do that a plain piece of glass can't do?"

"Oh, I learned in science class that a prism causes refraction as the light comes into it and as the light goes out of it too," you respond.

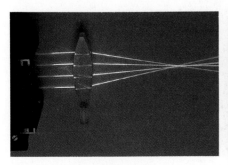

Figure 10–5. A convex lens causes parallel light rays to converge. Six prisms combined in a circle act just as a convex lens does.

"That's right," he says. He goes on to explain that a lens acts as if it is made up of a series of prisms arranged in a circle. Sometimes the arrangement produces a **convex lens,** which is thicker in the middle than at the edges. Sometimes the arrangement produces a **concave lens,** which is thinner in the middle than at the edges.

Here's what the photography club members found out about how the convex lens of a camera focuses.

Convex Lenses
A light ray coming into a convex lens is bent toward the center of the lens. When light rays come together in a convex lens, they are said to converge. The focus is the point at which all the light rays entering the lens converge. Convex lenses make objects look larger than they really are. You can find out more about how convex lenses work by trying the next activity.

DISCOVER BY *Observing*

Get several convex lenses. Focus sunlight or a distant light source onto a sheet of paper so that all rays converge into a bright pinpoint of light. Measure the distance from the paper to the lens. This distance is the focal length of the lens. Do all the lenses have the same focal length? What is the relationship between the focal length of the lens and its power to magnify images? ✎

Look at the markings on a camera lens. It may be marked "f = 45 mm." The distance from the lens to the focal point is called the **focal length.** This means the incoming beam, if it consists of parallel rays, will be brought into focus 45 mm from the center of the lens.

Figure 10–6 shows what happens as a source of light gets closer. Notice when the rays are parallel, when they converge,

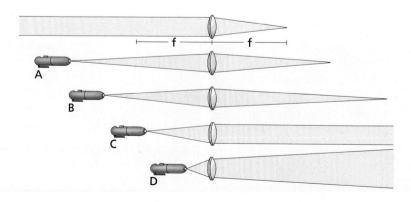

Figure 10–6. As a light source gets closer to a convex lens, the emerging light converges less and finally diverges.

and when they diverge. Explain what happens as the light source gets closer.

Concave Lenses   How would you find the focal length of a concave lens? A concave lens might have a focal length marked "f = –60 mm." Why would the focal length be a minus? Unlike a convex lens, a concave lens causes light rays to diverge. Thus, you can use a concave lens to enlarge an area of light, but it cannot focus light.

When a beam of light leaves the concave lens, it is more divergent than when it entered. You can find the point of convergence by drawing dotted lines that extend the rays back through the lens. The point at which all the dotted lines meet is called the *virtual focus*. A focal length of –60 mm means that a parallel beam coming into the lens will diverge as though coming from a virtual focus 60 mm behind the lens.

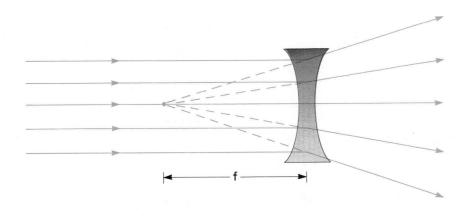

Figure 10–7. Parallel rays of light entering a concave lens diverge from the virtual focus.

 ASK YOURSELF

Describe several differences between convex and concave lenses.

SECTION 2   *REVIEW AND APPLICATION*

Reading Critically

1. What is the angle of reflection if a beam of light strikes a mirror at 47 degrees? Explain your answer.

2. What does it mean when the lens of a camera is marked "f = 35 mm"?

Thinking Critically

3. Suppose you are at a swimming pool. You stand in shallow water—your legs appear to bend sharply! Explain why this happens.

4. Suppose you had a block of glass and a bucket of water. Using a lamp that produces a narrow beam of light, how could you find out whether light travels faster in the glass or in the water?

5. How could you find the focal length of a convex lens? a concave lens?

SKILL *Communicating Using a Diagram*

▶ MATERIALS
- object • small mirrors (2)

▼ PROCEDURE

1. Get two small mirrors, and sit under a table as shown in the picture.
2. Ask a classmate to put an object you have not seen onto the table.
3. Hold one mirror above the table so light from the object can hit the mirror. That mirror will reflect the light downward to the second mirror, below the table. Then the second mirror will reflect the light to your eyes. You should be able to see the object on the table.
4. Draw and label a diagram showing the path of reflected light from the object to your eyes.
5. Use what you know about the ray model of light to explain why you can see the object on the table.

▶ APPLICATION
How could you use lenses to magnify the object you saw with the mirrors? What use would you have for an optical instrument that would enlarge an image? Name as many instruments of this type as you can.

✳ *Using What You Have Learned*
A periscope is a device consisting of a tube that holds a system of lenses, mirrors, or prisms. A periscope is used to see objects not in your normal range of vision. One use for a periscope is on a submarine—when the submarine is submerged the periscope allows the sailors to see above the water. Use your diagram to construct a periscope.

Images

Objectives

Distinguish between a real image and a virtual image.

Describe how the human eye focuses on objects.

Explain the properties of images formed in a plane mirror.

At the beginning of the next meeting of the photography club, you leaf through books of photographs. You admire the nature photographs, in particular. One photograph of a ship on a stormy sea is so real looking that you can feel your knees buckle as you stand on the tilted deck along with the other sailors. In another photograph you feel as though you are actually breathing fresh pine-scented air and touching craggy, mineral-encrusted rocks on the steep slope of a stately mountain.

You are so impressed with these photographs that you decide you'd like to take photographs of nature, from simple wildflowers to the vast grassy plains that stretch out in every direction from your town. Before you embark on your first photographic adventure, you should know more about images and the way the human eye sees.

Figure 10–8. El Capitan

Real Images from Lenses

A camera lens can focus on objects that are very close and those that are far away. "Remember what you just learned about the f-setting for a camera lens?" Mr. Tallchief asks. "Well, imagine that you are visiting Yosemite National Park, taking pictures of the mountain El Capitan. Tiny amounts of the reflected sunlight from the mountain enter the lens of the camera. The lens uses the light to form a picture on the film. That picture is an image. Now think of El Capitan as composed of tiny points. Light reflects off each one of those points and scatters in all directions. The camera lens focuses the light from each point onto the film so that each point is represented by a point of light on the film. The image created when the collection of points that make up El Capitan is focused by the lens is a **real image.**"

Remember that the distance from the center of the lens to the film is the focal length of the lens. This distance equals the focal length when the camera is set at "infinity." When you take a close-up photograph, the lens can still focus, but it has to be farther away from the film.

 ASK YOURSELF

How can you focus your camera to take a photo of a distant mountain and to take a photo of a flower at the base of the mountain?

Figure 10–9. A lens is positioned further from the film for a close-up. Light passes through the lens and converges on the film to form an enlarged real image.

Not So Real Images

Have you ever used a hand lens to look closely at an object? A hand lens is a convex lens. It creates an image larger than the real object. When you use a hand lens, the object is placed close to the lens—at a distance less than the focal length. The rays from a point on the object do not come to a focus. The rays coming out of the lens are divergent. When some of these rays enter your eyes, they seem to be diverging from some point beyond the object. What you see is an enlarged image of the object. The image that forms does so only because the light seems to diverge from it. This type of image is called a **virtual image**.

A concave lens, also called a diverging lens, causes light to spread. Therefore, it can never focus light to form a real image.

Figure 10–10. With a convex lens, the light rays you see seem to diverge from points beyond the real image.

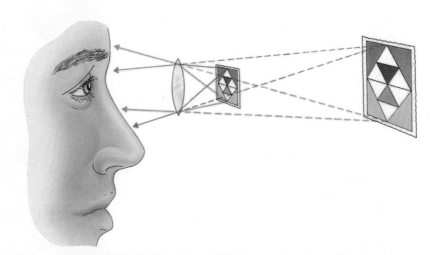

It can, however, form a virtual image. This type of image is always smaller than the actual object. See if you can find a virtual image by doing the next activity.

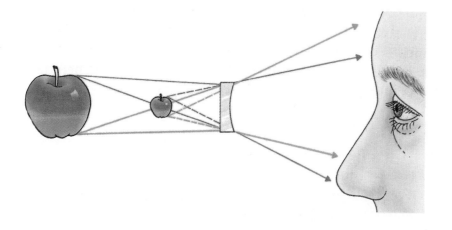

Figure 10–11. With a diverging lens, every point on the object is represented by a point on the virtual image. A concave lens always forms images reduced in size.

ACTIVITY

How can you find a virtual image?

MATERIALS
colored glass; corrugated cardboard, 10 cm × 10 cm; straight pins (6); metric ruler; protractor

PROCEDURE
1. Mount the glass on the cardboard with straight pins, as shown.

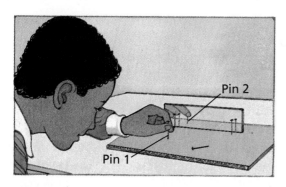

2. Place *Pin 1* upright in the cardboard, 10 cm in front of the glass.

3. Look into the glass at the image of *Pin 1*.
4. Place *Pin 2* behind the glass in the same place as *Pin 1*. Move your head to different positions to make sure *Pin 2* stays in the same place as the image of *Pin 1*.
5. Measure and record the distance from *Pin 1* to the glass and from *Pin 2* to the glass.

APPLICATION
1. What happens to the image if you change the point from which you look at the pin?
2. If you draw a line from *Pin 1* to the image, what angle does it make with the glass? Relate this to what you know about the angle of incidence and the angle of reflection.

 ASK YOURSELF
What is the difference between a real image and a virtual image?

The Human Eye

Mr. Tallchief explains that a camera is like the human eye in some ways. "Like a camera lens, your eyes use their own convex lenses to produce real images. At the back of your eye is a sensitive layer of nerve cells called the *retina*. This layer of cells, which is similar in function to the film in your camera, receives the image and codes it into nerve impulses. You see when the nerve impulses reach your brain and are decoded."

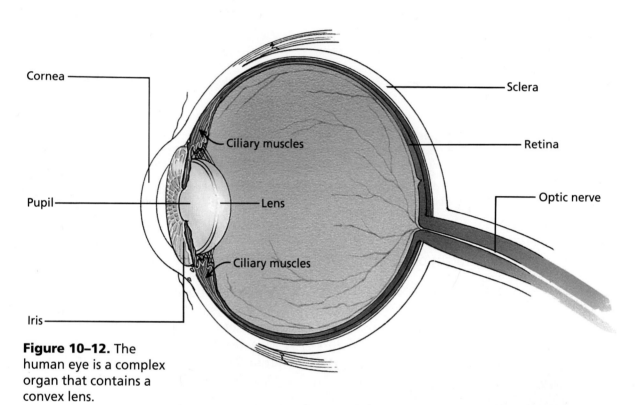

Cornea

Sclera

Ciliary muscles

Retina

Pupil

Lens

Optic nerve

Ciliary muscles

Iris

Figure 10–12. The human eye is a complex organ that contains a convex lens.

Adjusting Vision   "Obviously," says Mr. Tallchief, "your eye can't focus the same way a camera does. You can't move your eyeballs in and out the way you do a camera lens."

"What happens is this," he explains. "The lens in your eye can change its focal length by changing its thickness. The result is the same as when you change the distance between the film and the lens in your camera. The thickness of the lens is controlled by the ciliary muscles attached to the edges of the lens. When you look at something far away, the ciliary muscles relax, thus thinning each lens. To see something close up, the ciliary muscles contract, allowing each lens to become a little thicker. This increases the converging ability of the lens so that the light rays focus on the retina."

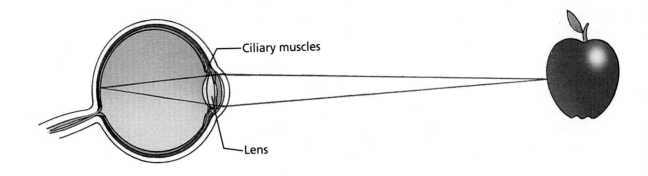

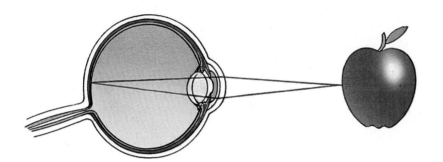

Figure 10–13. Action of the ciliary muscles change the shape of the lens. A thinner, flatter lens focuses on objects far away. The lens becomes thicker and more rounded to focus on objects close by.

Correcting Vision When one club member wants to know how close up or far away an object has to be for the eye to focus, Mr. Tallchief answers, "On average, a young person can focus objects at 7 cm, while 60-year-olds can't focus closer than 200 cm. Why? The lenses, like the rest of the human body, age and become less flexible. That is why many people, like myself, wear eyeglasses when they get to be around 40 years old. The glasses provide an additional convex lens for their eyes."

Another club member wants to know why, then, some young people wear eyeglasses. Mr. Tallchief explains, "If a person's eyeballs are a little too short from front to back, images of near objects are formed by their lenses behind the retina. This condition is called *farsightedness.* Eyeglasses containing convex lenses that allow extra convergence enable the image to focus on the retina."

"If a person's eyeballs are too long from front to back, images of distant objects are formed in front of the retina. This condition is called *nearsightedness,* and eyeglasses containing concave lenses that reduce the amount of convergence enable the image to focus on the retina."

Figure 10–14.
Farsightedness is the result of an eye shape that places the retina in front of the focus. Nearsightedness is the result of an eye shape that places the retina behind the focus.

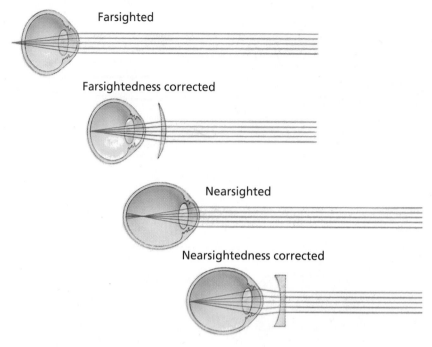

Farsighted

Farsightedness corrected

Nearsighted

Nearsightedness corrected

 ASK YOURSELF

How does the lens of the human eye focus on objects close up?

Flat Mirrors

"Now it's my turn to ask a question," says Mr. Tallchief. "What do you see when you look into a flat mirror?" No one says a thing, so Mr. Tallchief suggests you all try the following activity.

Look at your face in a mirror. Is the image you see exactly like the one facing the mirror? Hold up a sheet of paper with handwriting on it. What do you see? Now hold up a transparent piece of plastic with handwriting on it. What do you see? Write a hypothesis that explains your observations. ✐

Some club members think that the mirror image is reversed right to left. Mr. Tallchief responds: "That's a strange idea, if you think about it. If you lie on your side, will the mirror interchange your head and your feet? Of course not! The fact is that a mirror cannot reverse right and left. Did the mirror reverse the handwriting on the sheet of paper? No. The images you see in a mirror are reversed front to back. So if you look into the mirror

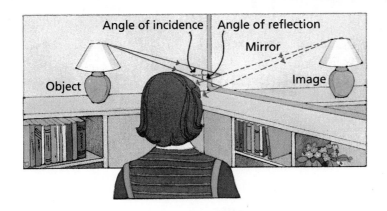

Figure 10–15. This figure shows how all virtual images appear in flat mirrors. The virtual image point is just as far behind the mirror as the lamp is in front of it.

once more and wave the hand on the western side of your body, the image from the mirror will wave back with the hand from the western side."

What you see in a mirror is a virtual image. How do you know? Here's one important clue. With mirrors, as with lenses, virtual images are right side up and real images are upside-down. Think about it. You know the image cannot be real, since light can't get through the mirror.

Now you know about images in flat mirrors. Remember, light waves are reflected from the surface of the mirror so that the angle of incidence is equal to the angle of reflection. In Figure 10–15, the virtual image appears to be as far behind the mirror as the object is in front of it. Extend the reflected rays beyond the mirror with dotted lines. Your eye tells you that the rays are coming from the point where the dotted lines intersect, or the virtual image point.

 ASK YOURSELF

How does an image form in a flat mirror?

SECTION 3 *REVIEW AND APPLICATION*

Reading Critically

1. What is the difference between a real image and a virtual image?
2. Why is it necessary for the lens of the eye to change shape?
3. As you walk toward a flat mirror, what happens to the size and position of your image?

Thinking Critically

4. Why is it impossible for a concave lens to form a real image?
5. Why do older people often need eyeglasses that correct for both nearsightedness and farsightedness?
6. Make a table comparing and contrasting the uses and effects of concave and convex lenses.

INVESTIGATION

*F*inding the Properties of Images Formed by a Convex Lens

▶ MATERIALS
- white cardboard ● convex lens ● meter stick ● lamp

▼ PROCEDURE

1. Place your cardboard screen facing the window of the room as shown. Move the lens until a sharp picture appears on the screen.
2. Measure and record the distance from the lens to the screen. Is the image real or virtual? right side up or upside-down? smaller or larger than the object? Explain the reasons for your results.

3. Turn on your lamp, and use it as the object. Place the lamp several meters away from the lens, and focus the image of the lamp on the screen. With an object distance of several meters, what is the image distance?
4. Move the lamp closer to the lens. As the object distance decreases, what happens to the image distance? As the image distance increases, what happens to the size of the image? Explain why.

5. Keep moving the lamp closer until you cannot see an image. What is the object distance at that point?
6. Now use the lens as a magnifying glass to read the letters on this page. Start with the lens about 1 cm from the page. Is the image real or virtual? right side up or upside-down? larger or smaller than the object?
7. Increase the object distance. What happens to the image? Why?
8. Keep increasing the object distance until you can no longer find an image. What is the object distance at that point?

▶ ANALYSES AND CONCLUSIONS
1. Describe the situations in this experiment that allowed you to find the focal length of the lens.
2. In what ways would you expect the results to be different if you used a lens with a smaller focal length?
3. Describe two ways in which you can tell whether an image is real or virtual.
4. Why is the image on the screen always upside-down?

▶ APPLICATION
A concave lens works like a convex mirror. This type of mirror is often used on the passenger side of cars so that the driver can see cars on the right and behind. Why does a convex mirror allow the driver to do this? Draw a diagram to support your answer.

Discover More
Repeat this procedure using a concave lens. How do the results differ?

Jan Vermeer, The Geographer. Frankfurt am Main, Staedelsches Kunstinstitut.

The Big Idea

Light may act and react as waves do or as particles do. Lenses and mirrors change the direction in which light moves. One way to understand the behavior of light is to study it using the ray model. The brightness of light rays depends upon an object's distance from the light source.

Light energy in the form of waves may reflect or refract. When light rays enter a lens, they refract. Through a convex lens, parallel light rays converge at the focal point and diverge after passing through it. Through a concave lens, parallel light rays diverge. When light is focused by a lens, a real image is produced. An image formed by a divergent lens is a virtual image.

For Your Journal

Look back at the ideas you wrote in your journal at the beginning of the chapter. How have your ideas about light changed? Revise your journal entry to show what you have learned. Be sure to include information on how lenses can affect light.

Connecting Ideas

This concept map shows how the big ideas of this chapter are related. Copy the concept map in your journal. Extend the concept map to show how light and lenses interact to form images.

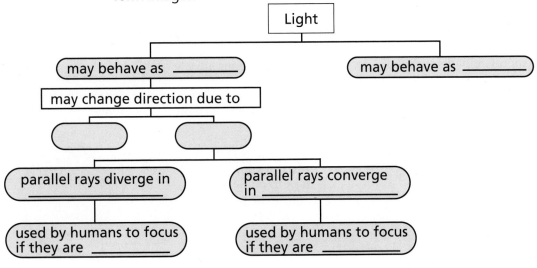

CHAPTER 10

REVIEW

Understanding Vocabulary

1. For each set of terms, explain the similarities and differences in their meanings.
   a) convex lens (268), concave lens (268)
   b) real image (271), virtual image (272)
   c) lux (263), light rays (263)
   d) focus (266), focal length (268)

Understanding Concepts

MULTIPLE CHOICE

2. When you see an image, it's because your brain decodes nerve impulses that have been registered on your
   a) sclera.   b) retina.
   c) lens.   d) iris.

3. An important assumption of the ray model of light is that light
   a) always travels at a certain speed.
   b) acts like waves.
   c) acts like particles.
   d) travels in straight lines.

4. In seeing things, your eyes make use of light beams that are
   a) divergent.
   b) convergent.
   c) virtual.
   d) focused.

5. The Danish astronomer Olaus Roemer successfully determined
   a) the colors found in light.
   b) the angle of incidence.
   c) the speed of light.
   d) the degree of refraction.

6. Who was responsible for proving the theory that light acts like a wave?
   a) Isaac Newton
   b) Christiaan Huygens
   c) Albert Einstein
   d) Thomas Young

7. Convex lenses make objects look
   a) smaller than they really are.
   b) brighter than they really are.
   c) larger than they really are.
   d) darker than they really are.

SHORT ANSWER

8. Describe the function of the ciliary muscles.

9. What causes farsightedness in people as they grow older?

Interpreting Graphics

Answer each question below. Then write the letter of the part of the diagram that identifies your answer.

10. Which part enables your eye to focus light rays?

11. Which part changes the thickness of the lens to allow it to adjust the focus?

12. Which part captures the image that you eventually see?

13. Which part carries nerve impulses to the brain?

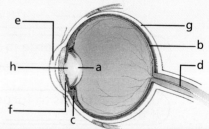

Reviewing Themes

14. Systems and Structures
Explain why a camera lens must produce a real image rather than a virtual image.

15. Environmental Interactions
Many light bulbs used in reading lamps are made with frosted glass. So are the glass doors in shower stalls and some inside walls in buildings. How does frosted glass affect the intensity of light passing through it?

Thinking Critically

16. Stars have different absolute magnitudes. In other words, each star has a degree of brightness determined by how much energy it produces. Astronomers assign each star a degree of *apparent* magnitude or brightness—how bright the star *seems* to viewers on Earth. What factor would affect this measurement?

17. A light beam passing from benzene into an alcohol solution at a high angle of incidence does not bend. What can you conclude? Explain your answer.

18. In the novel *Lord of the Flies,* a very nearsighted character starts a fire with his glasses. Explain why this could not happen.

19. One ballerina, but many images. Each drop of water is a tiny lens, producing an image of the dancing ballerina. Explain what it is about the water that makes it different from the surrounding air. Why does each drop of water act just like a tiny lens? Why does each droplet form an image in a different place?

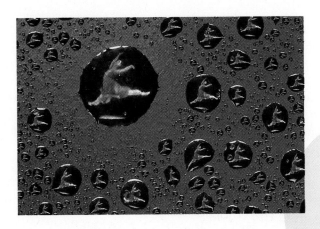

Discovery Through Reading

Gloeckner, Carolyn. "The Human Eye: A Powerful Little Package." *Current Health* 18 (October 1991): 4–9. This article includes fascinating facts about the human eye, including how we see in three dimensions and differentiate the colors of light.

Dear Theo:

In my picture of the "Night Café" I have tried to express the terrible passions of humanity by means of red and green. The room is blood-red and dark yellow, with a green billiard table in the middle; there are four lemon-yellow lamps with a glow of orange and green. Everywhere there is a clash and contrast of the most alien reds and greens in the figures of little sleeping hooligans in the empty dreary room, in violet and blue. The white coat of the patron [landlord], on vigil in a corner, turns lemon-yellow, or pale luminous green.

Vincent

From *Dear Theo: The Autobiography of Vincent van Gogh* edited by Irving Stone

Scientifically speaking, colors are created by light of different wavelengths striking the eye. Traditionally, the spectrum of colors are also associated with a spectrum of human emotions. The color yellow suggests gaiety to most people, while blue is associated with sadness. That's why those smiling happy faces are colored yellow, but mournful singers croon the "blues."

The artist Vincent van Gogh was fascinated by colors. He was also gripped by powerful emotions, often suffering bouts of depression. Influenced by Japanese art, van Gogh freed himself from the traditional color spectrum in order to use colors more imaginatively. In one self-portrait he painted the shadows on his face in green. In another, painted after he cut off part of his ear, the background is red and orange.

In letters to his brother Theo, van Gogh explained the colors of his picture "The Night-Café." To van Gogh, the unusual colors expressed a wide spectrum of emotions, from darkness to "Japanese gaiety." The cafe was "a place where one can ruin oneself" or "go mad," Vincent wrote to Theo. By choosing such unusual colors for his paintings, van Gogh tried to capture the whole spectrum of human feeling. The human eye can distinguish thousands of different colors. In van Gogh's paintings, they might all be matched to a different feeling.

Vincent van Gogh, *The Night Cafe*, oil on canvas, 1888, 28½" x 36¼".
Yale University Art Gallery, Bequest of Stephen Carlton Clark, B.A., 1903.

For Your Journal

- ✏ *What kinds of waves make up the colors you can see?*
- ✏ *Why do you see colors?*
- ✏ *How do electromagnetic waves, similar to those you can see, affect your everyday life?*

Introducing the Spectrum

Objectives

Compare the characteristics of electromagnetic waves to their position in the electromagnetic spectrum.

Relate frequency to wavelength.

Demonstrate the separation of white light into colors of the visible spectrum.

It's career day. Everywhere you turn, you find posters advertising careers for this afternoon's festival. There's no particular career that especially interests you. You do know, though, that you want excitement on the job and you like to be involved with people. Sometimes you imagine yourself as a famous surgeon, performing complicated operations and saving lives. While watching the evening news, you've often daydreamed about becoming the anchorperson on a national network, informing the nation of late-breaking, global events. Then again, you also find adventure in quiet contemplation, like when you draw pastels of magnificent sunsets over the river or abstract compositions that include mysterious images from your dreams. There are so many interesting career possibilities you want to investigate.

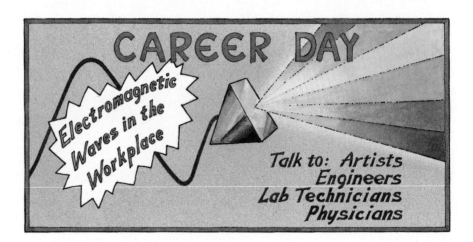

The Electromagnetic Spectrum

You are still thinking about careers as you walk into your first class. Ms. Oyama, your science teacher, is talking about **electromagnetic waves**—the invisible transverse waves that carry both electric and magnetic energy. She asks, "How do electromagnetic waves relate to the career choices you might someday make?" Everyone looks puzzled, but then she explains.

"Say you want to become a radio disc jockey. What kinds of waves would you use?" One student confidently answers, "Sound waves." Ms. Oyama agrees that we hear music from a radio because of sound waves. "But," she goes on, "before radio signals are turned into sound waves, they're transmitted as radio waves to the receiver on your radio. Radio waves are just one type of electromagnetic wave. All the types of electromagnetic waves together are the **electromagnetic spectrum.**"

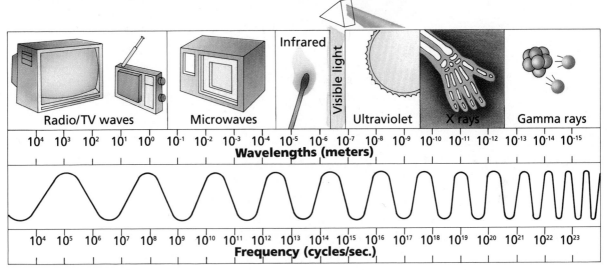

Wavelengths (meters)

| 10⁴ 10³ 10² 10¹ 10⁰ | 10⁻¹ 10⁻² 10⁻³ 10⁻⁴ | 10⁻⁵ 10⁻⁶ 10⁻⁷ 10⁻⁸ 10⁻⁹ 10⁻¹⁰ 10⁻¹¹ 10⁻¹² 10⁻¹³ 10⁻¹⁴ 10⁻¹⁵ |

Wavelengths (meters): 10^4 10^3 10^2 10^1 10^0 10^{-1} 10^{-2} 10^{-3} 10^{-4} 10^{-5} 10^{-6} 10^{-7} 10^{-8} 10^{-9} 10^{-10} 10^{-11} 10^{-12} 10^{-13} 10^{-14} 10^{-15}

Frequency (cycles/sec.): 10^4 10^5 10^6 10^7 10^8 10^9 10^{10} 10^{11} 10^{12} 10^{13} 10^{14} 10^{15} 10^{16} 10^{17} 10^{18} 10^{19} 10^{20} 10^{21} 10^{22} 10^{23}

Figure 11–1. The electromagnetic waves in this diagram of the electromagnetic spectrum are presented in order, according to their wavelength and frequency. The visible spectrum takes up only a small portion of the spectrum.

You wonder what electromagnetic waves have in common and what other kinds there are. You raise your hand and ask Ms. Oyama. She tells you that the frequencies of these waves are high, but each type differs in frequency and wavelength from the rest. The entire class looks confused. "OK," she says. "Electromagnetic waves have a certain range of frequency, from less than 100 cycles per second to more than 10^{23} cycles per second. Most technologies, however, use electromagnetic waves in the range of 10 000 Hz to 10^{23} Hz. Compare that to the highest sound you can hear, which has a frequency of 20 000 cycles per second. That's what I mean when I say that electromagnetic waves have a high frequency."

Look at the diagram of the electromagnetic spectrum. You can see that radio and TV waves have lower frequencies than X-rays and gamma rays.

Someone asks how fast these different waves travel. "Good question," she answers. "The speed of electromagnetic waves is another common property. They all travel at the speed of light—300 000 000 m/s. What's more, the speed of a wave, its frequency, and its wavelength are all related mathematically. The wavelength of an electromagnetic wave multiplied by its frequency equals the speed of light."

Ms. Oyama asks you to use this fact to clarify some relationships among these quantities. Try the next activity.

DISCOVER BY *Problem Solving* _____

How would you describe the wavelength of an electromagnetic wave with very high frequency? the frequency of a wave with a very long wavelength? Use the diagram in Figure 11–1 to help you. Explain your answer. ✏

▶ ASK YOURSELF

What characteristics do different kinds of electromagnetic waves have in common?

The Visible Spectrum

Ms. Oyama tells you that visible light—the light you can see—takes up only a small part of the electromagnetic spectrum. She explains that nearly 300 years ago, the British physicist Sir Isaac Newton passed a ray of sunlight through a prism. The ray refracted and produced the colors of the rainbow. You can do the same experiment.

ACTIVITY

How does a prism form colors from white light?

MATERIALS
tape, white poster board, light source, prism

PROCEDURE
1. Tape the white poster board on a wall or against a stand.
2. Darken the classroom, and turn on the light source. Hold the prism between the light source and the poster board until the colors of the rainbow appear on the poster board.
3. Move and turn the prism to see whether the colors change. Record your results.

APPLICATION
1. What colors could you see clearly? In what order did the colors appear?
2. Were there shades of colors in between the most noticeable colors? Why do you think this is so?
3. Rainbows in the sky are formed by refraction from water drops in the atmosphere. How do water drops act like prisms?

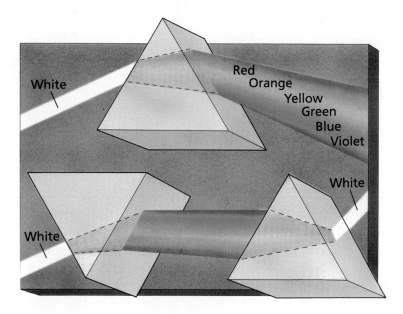

Figure 11–2. A diamond sparkles with color because it acts as a prism. A prism can separate white light into the colors of the visible spectrum.

The order of the colors in Sir Isaac Newton's experiment appeared from the color with the longest wavelength (red) to the color with the shortest wavelength (violet). This is the same order of colors in the rainbow—longest wavelength to shortest, or red to violet. The colors produced—red, orange, yellow, green, blue, violet—are the colors of the *visible spectrum*.

You ask Ms. Oyama, "Then how do you make white light?" Another student correctly deduces, "White light is the combination of the colors in the rainbow."

At the end of class Ms. Oyama asks everyone to stay alert during the afternoon career festival to find out the ways in which electromagnetic waves relate to many different careers.

 ASK YOURSELF

In what order do the colors of the visible spectrum appear? How does this order relate to wavelength?

SECTION 1 *REVIEW AND APPLICATION*

Reading Critically

1. What is the electromagnetic spectrum?
2. What is the relationship between wavelength and frequency?
3. What is the relationship between wavelength and color in the visible spectrum?

Thinking Critically

4. Why is it possible to separate the colors of the visible spectrum with a prism and then recombine the light waves to form white light?
5. Longer wavelengths of light bend less than the shorter ones. Why do you think this is true?

SKILL Communicating Using Scientific Notation

▶ MATERIALS
- paper • pencil

▼ PROCEDURE

When numbers are very large or very small, scientists and engineers write them using a method called scientific notation.

1. In scientific notation, a number is stated by writing the number 10 raised to some exponential power. For example:

$$\text{a hundred} = 100 = 10^2$$
$$\text{a hundred thousand} = 100\,000 = 10^5$$

In each of these cases, the number of zeros after the 1 is the same as the exponent of the 10 in scientific notation.

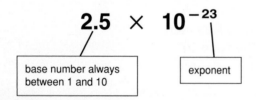

$$\mathbf{2.5} \times \mathbf{10}^{-23}$$

base number always between 1 and 10 | exponent

2. You can also write numbers that are not simply powers of ten when you use scientific notation. For example:

$$\text{four hundred and fifty thousand} = 450\,000 = 4.5 \times 10^5$$

The trick is to place the decimal point after the first digit of the number. This makes the base number always less than 10 and greater than or equal to one. Then count the number of places from the decimal to the end of the number to get the exponent number.

17 100 000 000 000 000 cycles per second is a frequency of ultraviolet light. To put this number into scientific notation, put the decimal after the 1. This gives you a base of 1.71. From the decimal point there are 16 digits to the end of the number. Therefore, this frequency of ultraviolet light is about 1.71×10^{16} cycles/second.

3. For numbers less than one, the exponent is negative. Place the decimal point after the first nonzero digit. Then count the number of places to the right of the decimal to get the negative exponent of 10.

0.000 000 0176 m is the wavelength of the ultraviolet light above. To change this number to scientific notation, put the decimal after the 1. This gives you a base of 1.76. From the decimal point there are 8 digits to the front of the number. Therefore, the wavelength of this ultraviolet light is about 1.76×10^{-8} m.

▶ APPLICATION

Write the following in scientific notation.
1. The speed of light in a vacuum is 300 000 000 m/s.
2. The thickness of a piece of gold foil is 0.000 32 m.
3. The mass of a dust particle is 0.000 000 000 753 kg.
4. A terameter is 1 000 000 000 000 m.

✳ Using What You Have Learned
1. What purpose does using scientific notation serve?
2. Name three quantities, not discussed here, that would be easier to express in scientific notation.

The Nonvisible Spectrum

Distinguish between ultraviolet radiation and infrared radiation and explain the uses of each.

Identify the major uses of X-rays, radio waves, and microwaves.

At the career festival, the first booth you visit is called Medicine and Technology. Physicians and medical technicians from a nearby hospital are there to help you learn about their profession. You're amazed by what you see: X-rays that show a person's skull; infrared photographs that show the outline of the human body filled in with different colors; photographs of an operating room stocked with more equipment, you think, than a spaceship. Now you realize that this profession would satisfy the mechanic and engineer in you as well as your desire to help people.

You imagine a day in the future when over a hospital loudspeaker you hear: "Dr. Rey, your expertise is requested in the OR (as a famous brain surgeon, you know OR means "operating room," of course!). We have the virtual reality scanners ready and waiting for you."

X-Rays

At the Medicine and Technology booth, a medical technician tells students that X-rays were an important, early innovation in modern medical technology. The German physicist Wilhelm Roentgen (1845–1923) discovered X-rays in 1895 and in 1901 received the Nobel Prize for his work. Then the technician hands out old dental X-rays for students to study.

DISCOVER BY Observing

Ask your dentist for an old dental X-ray. Light spots may be fillings or other metals. Are there any dark spots on the X-ray? What do you think the spots are? Why are the spots dark? How can a dentist find cavities on an X-ray?

X-rays are very high-frequency electromagnetic waves. They are usually produced when electrons traveling at high speeds strike a solid object. The technician explains that when

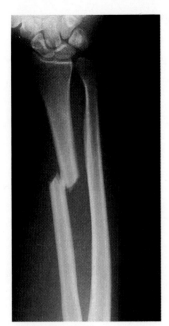

Figure 11–3. X-rays that pass through an object darken film to create an image.

part of your body gets X-rayed, electricity is passed through a tube and the electrons that are released hit a target. The X-rays produced shoot out of the tube through a small hole, pass through the object being X-rayed, and strike the film. The X-rays pass through matter that has low density, such as skin and muscle, but are absorbed by high-density matter, such as bones and teeth. Where X-rays are absorbed, a light area appears on the film.

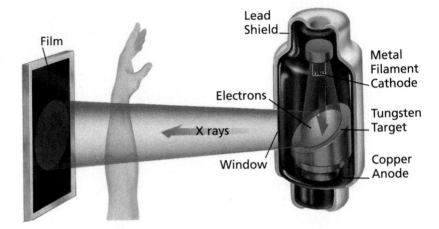

The technician explains that X-rays can be harmful because they can kill healthy body cells. X-rays can also kill cancer cells and are, therefore, used to treat certain types of cancer. Because X-rays are potentially harmful, the number of them you receive should be carefully monitored. Always ask for a protective lead apron to wear during an X-ray if you are not offered one.

 ASK YOURSELF

If an X-ray was taken of your foot, what parts of your foot would show? Why?

Ultraviolet Waves

You ask the medical technician about the photograph in the booth that interests you most—one that shows the interior of a hospital operating room. The technician points out the special lights and tells you they are UV lights. UV stands for ultraviolet. These lights produce *ultraviolet radiation*—that is, waves with a frequency just higher than visible light. UV waves are a part of the electromagnetic spectrum.

UV can kill bacteria; therefore, UV lights help keep an operating room sterile. The technician adds that UV radiation is invisible, but it is present in sunlight. In fact, it is the UV rays in sunlight that cause skin to tan or burn. She then tells everyone that while overexposure to UV can cause skin cancer, a certain amount of UV helps skin cells produce vitamin D—a substance necessary for healthy bones and teeth. She passes out copies of the following chart to students who visit her booth.

Table 11-1	Sunscreen Protection
No SPF	You may get a fast tan with these products, but you can also get a bad burn. They offer no protection.
SPF 4 to 6	These are good if you tan easily and rarely burn. It's the least you should start with on your face.
SPF 7 to 8	These provide extra protection that still permits tanning.
SPF 9 to 10	Excellent protection is provided by these products if you burn easily but would still like a minimal tan.
SPF 15+	These products provide the best protection against burns. They also prevent tanning.

The students who stop at the booth study the chart and discover that the higher the SPF (sun protection factor) number, the more completely the chemical blocks UV rays from your skin. You learn that you should cover your body with clothing, wear a hat, and use a sunscreen when you go outdoors in the sun.

 ASK YOURSELF

What are some ways in which ultraviolet waves help people?

Infrared Waves

The technician next shows photographs developed from film sensitive to infrared waves. *Infrared radiation,* or IR, is part of the electromagnetic spectrum. It has wavelengths slightly longer than those of visible light.

First the technician shows a photo that looks like an outline of a person, but it has many different colors. "This is a thermogram," the technician explains. "When an object receives IR waves, it absorbs the waves and becomes warmer. Then the object radiates the IR back into the environment. The different

Figure 11–4. Infrared sensitive film can be used to photograph the IR radiation given off by objects.

 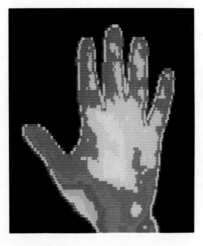

colors of a thermogram represent different amounts of IR waves, and therefore different temperatures, given off by the object."

The technician explains that infrared rays are often used in cafeterias to keep food hot. When special viewers are used that can "see" IR, the wearer can locate objects or people in foggy or dark areas because of the IR waves they give off. IR even helps people with night blindness see at night through special glasses. Then she shows you two photographs that show IR waves as they are given off by two hands.

DISCOVER BY *Observing*

Study the thermograms on this page. What colors do you think correspond to hot areas? to cold areas? Explain the reasons for your answers.

ASK YOURSELF

How are infrared waves used to help people?

Radio Waves

The idea of seeing yourself nightly on television as a television journalist or hearing your voice make announcements on the radio lures you to the Communications booth at the career festival. As you arrive, a communications engineer begins a presentation on *radio waves*—electromagnetic waves with frequencies at the lower range of the electromagnetic spectrum. You're surprised to find out that the first voice radio broadcast took place in 1910. The program featured the famous opera singer Enrico Caruso, and it was broadcast live from the Metropolitan Opera House in New York City.

You have been listening to radio broadcasts for years. Sometimes you tune in to FM 105.9 to listen to rock music, and sometimes you like to listen to AM 680 for the new country sound. What do those numbers on a radio dial really mean?

Observing

Turn on a radio to AM, and move the tuning dial from station to station. How do the numbers of the radio stations relate to frequency? Now do the same thing with FM. How do these numbers relate to frequency? Which have longer wavelengths, radio stations that broadcast on AM or those that broadcast on FM? How do you know?

You find out from the communications engineer that AM radio stations broadcast on frequencies from 535 kilohertz (kHz) to 1605 kHz. FM stations use much higher frequencies than AM stations, from 88.1 megahertz (MHz) to 107.9 MHz. When you tune in a radio station, you are really adjusting it to receive a particular frequency of radio waves. Then the engineer shows you an illustration.

Amplitude modulation (AM)

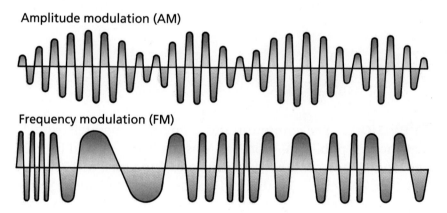

Frequency modulation (FM)

Figure 11–5. Radio waves are either amplitude modulated for AM broadcasts or frequency modulated for FM broadcasts. The frequency remains constant in an AM wave; the amplitude remains constant in an FM wave.

In the AM wave, the sound information is encoded by varying the amplitude of the wave. When the wave is picked up by an AM radio receiver, the sound, or audio, information is converted into an electrical audio signal. This signal is then amplified and sent to a loudspeaker, where it is converted into a sound wave.

In FM, the audio information is encoded by varying the frequency of the radio wave. The amplitude of the wave is kept constant. When the radio wave is picked up by an FM radio receiver, its frequency variations are converted into an electrical audio signal. This signal is then converted into a sound wave in the same manner as in an AM radio.

You ask the communications engineer why it is possible to hear an AM station farther away from its source than an FM station. He explains that AM waves reflect from the ionosphere, an electrically charged layer of atmosphere, just as ultraviolet light is reflected from a windowpane. Due to this reflection, many AM broadcasts can be received over distances of thousands of kilometers or more.

FM waves, with their shorter wavelengths, pass through the ionosphere into space, just as visible light passes through a windowpane. They do not reflect back to Earth and, therefore, have a limited range. Shortwave radio broadcasts, with frequencies just above the AM broadcast band, reflect from the ionosphere even better than AM waves. This is why shortwave transmissions are used for international broadcasts.

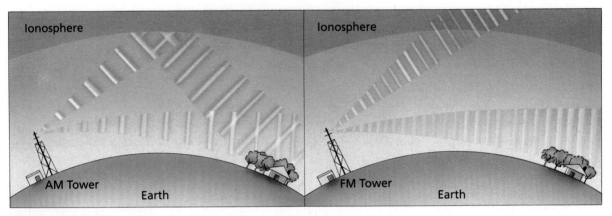

Figure 11–6. AM radio waves have a longer wavelength and are reflected by the ionosphere back to Earth. FM waves have a shorter wavelength, which allows them to pass through the ionosphere.

Television   The communications engineer then tells you that the first public TV broadcast was made in England in 1927. The United States followed with its own broadcast in 1930. However, daily broadcasting did not occur in the United States until 1939. That broadcast was of ceremonies at the World's Fair in New York City.

The engineer explains that television broadcasts are similar to radio broadcasts. Your TV antenna receives the waves and uses them to make electric currents, which are sent to your TV set. There the information is taken from the modulated waves, amplified, and sent to the picture tube.

In a color TV set, the colors are produced on the screen by little glowing dots. These dots are called *phosphor dots*. If you look at a color TV screen through a magnifying glass, you can see these dots. These dots occur in three colors—red, blue, and green—and are arranged in clusters, with one of each color per cluster.

The wave that carries the picture information is coded separately for each of the three colors. The separate information for each color becomes part of the electric signal that the antenna sends to the TV receiver. In the receiver, the information for

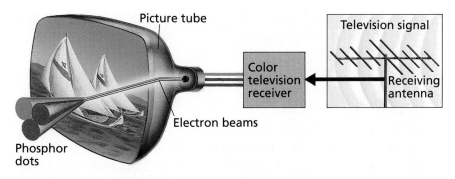

Picture tube

Television signal

Color television receiver

Receiving antenna

Electron beams

Phosphor dots

Figure 11–7. The dots on a color TV picture tube light up at different times to produce the moving television picture.

each color is separated and sent to the picture tube. There, the proper dots on the screen are stimulated, resulting in a color TV picture.

Radar The communications engineer tells you that radar waves occur between radio wave frequencies and IR frequencies. They are often used to track planes in flight, to detect ships at sea, to locate storms, to track space vehicles and satellites in orbit around the earth, and even to detect speeding cars. He explains that the word *radar* is an acronym for **ra**dio **d**irection **a**nd **r**anging. A radar device sends out short pulses of radio waves. Any plane within a certain distance will reflect these

Figure 11–8. Video games would not be possible without the technology of phosphor dots.

SHIELD

Figure 11–9. Among its many other useful applications, radar is used to track ships and airplanes.

waves. The radar receives the reflected waves and calculates the distance between the plane and the receiver. The radar does this by measuring the time it took for the echo of the wave to return to the receiver. The positions of detected objects can be seen on a screen similar to a TV screen.

Recently, engineers have developed "stealth" technology for military applications. This technology combines coatings that absorb commonly used radar frequencies with shapes that scatter the waves. Objects that employ this technology do not reflect commonly used radar frequencies and are therefore not detected by radar. In order to "see" the new stealth aircraft, radar installations would have to have new radar equipment that would detect different frequencies.

 ASK YOURSELF

Compare radio and television broadcasts.

Figure 11–10. This stealth aircraft can "sneak" by normal radar undetected.

Microwaves

The communications engineer asks what you know about the use of *microwaves*—radio waves with wavelengths of about a centimeter. Most students know that these electromagnetic waves are used in microwave ovens; one student describes a microwave transmission tower she once saw while traveling in the Rocky Mountains.

The engineer tells you that in addition to being used to cook food, microwaves are used for **telecommunications**—the science and technology of sending messages over long distances, such as by telephone. Where telephone lines are difficult or impossible to install, microwave dishes are set up to send up to 100 telephone calls at one time. Then he adds that microwaves are used to transmit live broadcasts from foreign countries. International sporting events are often transmitted this way.

Next he explains how the familiar microwave oven works. Microwaves pass through some substances, such as glass, but are absorbed by other substances, such as water. Microwave ovens send microwave radiation through the food placed inside. The water in the food is heated by the microwaves, and the food is cooked. Although the container in which the food is cooked is not usually heated by the microwaves, it may still get hot as it absorbs heat from the food inside.

Figure 11–11.
Microwave transmission towers allow telephone messages to be sent to areas where the installation of telephone lines is difficult or impossible.

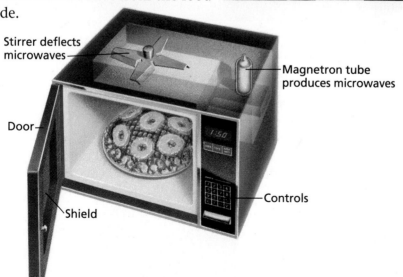

Stirrer deflects microwaves

Magnetron tube produces microwaves

Door

Controls

Shield

Figure 11–12.
Microwave ovens can quickly cook most types of food.

The engineer cautions that overexposure to microwave radiation may be harmful to humans since microwaves can kill healthy cells. They can even cause people who use pacemakers to have erratic heartbeats. Because of the danger, microwave ovens are constructed with shields and tight seals to keep radiation from escaping during use.

 ASK YOURSELF

What are two important uses of microwaves?

Figure 11–13. People who wear pacemakers must be careful about microwave radiation.

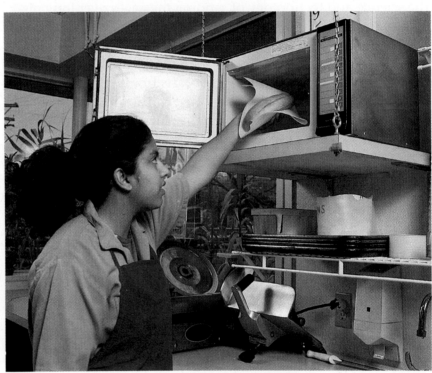

SECTION 2 *REVIEW AND APPLICATION*

Reading Critically

1. In what ways do the different electromagnetic waves resemble each other? What makes them different?

2. What problems can humans have if overexposed to X-rays, UV waves, and microwaves?

3. How can the problems in question 2 be prevented?

Thinking Critically

4. Why do you think radio waves can pass through walls, while radar waves bounce off?

5. What important information could you get from an infrared photograph of a house?

Color

A s you pass by the Art booth at the festival, you remember Ms. Oyama's demonstration about the visual spectrum. You wonder how an artist uses this knowledge to create colorful paintings. You have used pastels to create images of a sunset, blazing in oranges and pinks. At a museum once, you marveled at early paintings by the artist Pablo Picasso (1881–1973), executed almost exclusively in various shades of blue. Near the Picasso blue-period paintings, you were startled by a Vincent van Gogh (1853–1890) self-portrait in striking greens. You imagine a self-portrait you would like to create, with deep purples and dazzling yellows in the background. As much as you savor drawings and compositions in black and white, art would be far less exciting for you without the use of luscious colors, thousands of them!

Objectives

Describe the relationship between color and wavelength.

Relate rods and cones to the way people see color, including primary and secondary colors.

Identify the way in which certain filters polarize light.

Picasso, "The Tragedy" (1903)

Courtesy of the Fogg Art Museum, Harvard University Art Museums, Bequest of collection of Maurice Wertheim, class of 1906.

Vincent van Gogh, "Self Portrait Dedicated to Paul Gauguin, 1888," oil on canvas, 60.5 × 49.4 cm.

Color and Wavelength

You learned in science class that the colors of the visible spectrum differ from one another in terms of wavelength. The commercial artist begins her presentation by saying, "Color is the way your eyes interpret the wavelengths of light. Ordinarily, each combination of wavelengths gives you its own special sense of color. When white light strikes most objects, certain colors are reflected while others are absorbed. The reflected light is what you see as color. So a blue object appears blue because it reflects mostly wavelengths of blue light. Most of the other colors are absorbed by the object. The human eye is capable of distinguishing among 17,000 colors!" Then the artist poses a problem for the students to solve.

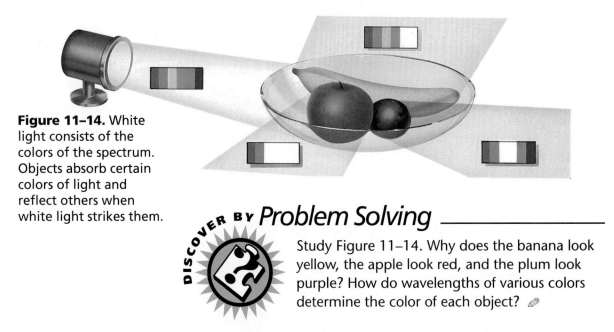

Figure 11–14. White light consists of the colors of the spectrum. Objects absorb certain colors of light and reflect others when white light strikes them.

DISCOVER BY *Problem Solving*

Study Figure 11–14. Why does the banana look yellow, the apple look red, and the plum look purple? How do wavelengths of various colors determine the color of each object? 🖉

After each student responds, the commercial artist explains, "Your ability to distinguish between wavelengths of light is not perfect. The banana is yellow because it reflects almost the full spectrum of light to your eye. Only the shortest wavelengths are missing. The light from a sodium vapor lamp, like those that are now widely used on some streets and highways, is also yellow. That light contains only a single wavelength, yet its color is very similar to the color of the banana, which reflects many different wavelengths of light."

▼ **ASK YOURSELF**

Explain why a cherry looks red and a lemon looks yellow.

How You See Color

You ask the commercial artist how the eye works in order to see color. She explains, "In the retina of the eye, there are two types of cells—rods and cones—that respond to light. The *rods* are light sensitive and allow you to see in dim light. The *cones* detect color and control the sharpness of the image you see. The cones do not work in dim light. This is why you cannot see sharp images or colors when the light is very low."

She displays a graph and continues, "The cone cells of a human eye can be compared to the phosphor dots on a color TV set. Like the dots on the TV screen, there are three kinds of cone cells in the retina. Each kind of cell is sensitive to a different set of wavelengths. One kind of cone cell—the red receptors—receives only the longer wavelengths. The blue receptors respond only to the short wavelengths. The green receptors react to light in the middle of the visible spectrum."

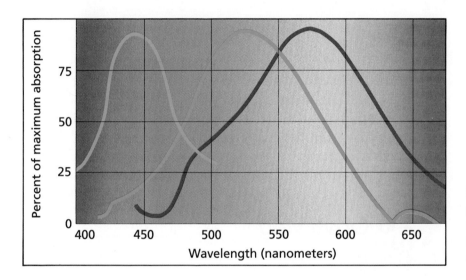

Figure 11–15. This graph shows how the cones in the retina of the human eye absorb colored light.

"If you study the graph (Figure 11–15), you can see that the wavelengths the cones receive overlap. This overlap means that the ability of the cones to receive colors overlaps. For instance, the green receptors are sensitive to some short wavelengths and some long wavelengths. However, they are most sensitive to wavelengths in the middle of the spectrum."

Then she explains that some people are unable to distinguish certain colors; these people are *colorblind*. About seven percent of men and less than one percent of women cannot distinguish between red and green. There is a rare kind of colorblindness in which a person cannot distinguish any colors at all. John Dalton (1766–1844), a British chemist, was the first person to detect colorblindness. In fact he was colorblind himself! Colorblindness, for a time, was also called *Daltonism*. The commercial artist then presents a test for red-green colorblindness.

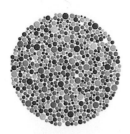

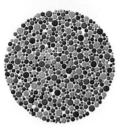

Follow these instructions to find out whether or not your eyes can see green and red. Your color vision is considered normal if you can see a *5* in the top circle and an *8* in the bottom circle. If you do not see a number in the top circle and you see a *3* in the bottom circle, you have red-green colorblindness. What is the result of your test? ✐

► ASK YOURSELF

What function do the rods and cones of the human eye perform?

Figure 11–16. These patterns are used to test for colorblindness.

A Model for Color

Next, the commercial artist shows a diagram of the colors people see. She explains that since the eye has three kinds of cone cells, the colors people see are made from these three colors—red, green, and blue—known as **primary colors.** By studying the three-primary model diagram shown to the students, you can understand primary colors and the colors formed by a combination of two primary colors—known as **secondary colors.**

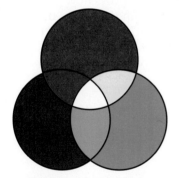

 **Observing** _____

Look at the diagram of primary and secondary colors (Figure 11–17). What two primary colors produce each of the secondary colors? How is white produced? ✐

Figure 11–17. The primary colors of light are red, green, and blue. When two of these colors overlap, secondary colors are produced.

Filters   Then the commercial artist hands out different cellophane filters. When you look at white light through a filter of a primary color, the other two primary colors are absorbed and everything takes on the primary color of the filter. The world becomes totally red with a red filter, blue with a blue filter, and green with a green filter.

One student asks, "What happens if you look through a filter of a secondary color?" She answers, "If a filter of a secondary color is used, one primary color is absorbed—the one that is not necessary to make that color. For instance, a yellow filter, which is composed of green and red, absorbs blue light."

The artist then has everyone look again at the illustration of the banana, apple, and plum from earlier in her presentation.

Figure 11–18. The three primary pigments are yellow, magenta, and cyan. When these colors are combined with black for contrast, as they are when a book is printed, they produce a full-color photograph.

DISCOVER BY *Problem Solving* _____

Look again at Figure 11–17. Imagine that this image is in the form of a slide. When you send white light from the projector through the slide, each filter absorbs some colors and allows the other wavelengths through. Why is the image of the apple red? ✐

Pigments   A student asks the artist how the color of paint is produced. She explains that pigments—colors used to tint other materials—give paints their colors. She tells you that in color film, color printing, and color TV screens, the three-primary model works well because the processes are carefully controlled. The simple three-primary model is not especially useful in mixing paints because the paints reflect many wavelengths of light to produce the color you see. Every artist and house painter has to learn by experience what kinds of pigmented paints to mix in order to get the desired color.

▼ **ASK YOURSELF**

What does the three-primary model tell you about the colors people see?

Polarized Light

Outside the school, after the festival, students talk about future career opportunities. Some of Ms. Oyama's science students tell her what they found out about the ways in which electromagnetic waves are related to many careers and technologies. Ms. Oyama smiles, hiding the pleased look in her eye behind dark sunglasses. A student asks, "Ms. Oyama, do sunglasses really keep the glare of the sun out of a person's eyes?" "Yes," she answers.

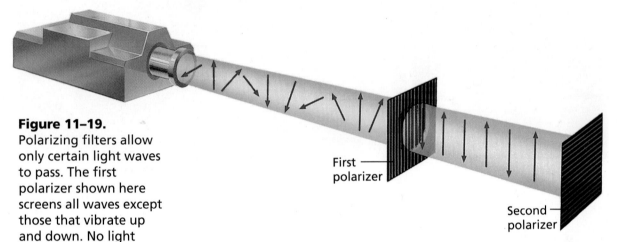

Figure 11–19.
Polarizing filters allow only certain light waves to pass. The first polarizer shown here screens all waves except those that vibrate up and down. No light passes through the second polarizer. Why?

First polarizer

Second polarizer

In class the next day, Ms. Oyama uses a diagram to explain how sunglasses work. She shows her students that *polarized light* is light passed through a polarizing filter. The filter reflects all the waves except those that are vibrating in a particular direction. The light waves that come into a polarizing filter are transverse, like all electromagnetic waves. The vibrations of some of the waves are up-and-down, some are side to side, and others are at different angles. The first filter, however, blocks all waves but those that vibrate in a vertical, or up-and-down, direction. The light passing through the filter is polarized light because its waves only vibrate in one direction.

After being polarized by the first filter, the light then passes to the second filter. The second filter is horizontal, so it blocks all waves but those with horizontal vibrations. Since only light waves with vertical vibrations could pass through the first filter, the second filter allows no light to pass through.

Ms. Oyama shows her sunglasses and explains that they have a vertical polarizer, like the first filter in the diagram. The vertical polarizer blocks the glare from the sun. To help you understand more about polarized light, she has you do the following experiment.

Figure 11–20.
Shown here are two photographs of the same store window. The photograph on the left, however, does not show the glare because it was taken through a polarizing filter.

DISCOVER BY *Doing*

Use a polarizing filter, such as a lens from polarizing sunglasses, to identify polarized light around you. Look through the filter as you rotate it. If rotating the filter dims the light, the light is polarized. If the light does not dim, it is not polarized. ✏

It has been an exciting day. You had no idea that so many careers involved a working knowledge of electromagnetic waves. In fact, until today, you didn't know what an electromagnetic wave was! It seems strange that these invisible things have such an effect on your everyday life. Cooking food with microwaves, seeing, checking for broken bones or cavities in your teeth, rays from the sun—electromagnetic waves are pretty important.

▶ ASK YOURSELF

How do polarizing sunglasses reduce glare?

SECTION 3   *REVIEW AND APPLICATION*

Reading Critically

1. What is the relationship between wavelengths and the way you see different colors?

2. What is colorblindness?

Thinking Critically

3. What color would pass through a combination cyan and magenta filter?

4. Why are polarizing sunglasses especially useful for drivers and pilots?

5. Make a list of 10 ways electromagnetic waves affect your daily life.

INVESTIGATION

*S*tudying the Three-Primary Model

▶ MATERIALS
- lamp with a 15-watt bulb ● color filters: red, green, blue, cyan, yellow, and magenta

▼ PROCEDURE

1. Make a table like the one shown.
2. Look at the lamp through the red filter. What color do you see?
3. Look at the lamp through each of the other filters. Record the colors you see.
4. Take the red filter, and put the green filter behind it so that the light goes first through the red filter and then through the green filter. Look at the lamp through the two filters. What color do you see?
5. Reverse the two filters so that the light goes through the green filter first. What color do you see?
6. Repeat steps 4 and 5, using all the possible combinations of filters. If little or no light comes through the filters, write *black* in the table.

TABLE 1: EFFECTS OF FILTERS ON LIGHT

		Top Filter					
		Red Filter	Green Filter	Blue Filter	Cyan Filter	Yellow Filter	Magenta Filter
Bottom Filter	Red Filter						
	Green Filter						
	Blue Filter						
	Cyan Filter						
	Yellow Filter						
	Magenta Filter						

▶ ANALYSES AND CONCLUSIONS
1. How do the filters affect the colors that you see?
2. How does combining the filters affect the colors that you see?
3. Draw a conclusion about how these filters fit the three-primary color model.

▶ APPLICATION
Sunglasses come in many colors. Using your knowledge of filters, explain what color sunglasses would be best for (a) very sunny days,(b) overcast days, and (c) water activities such as fishing. How would you produce each of these tints?

✳ *Discover More*
Create watercolor paint mixtures with the same combination of colors. Compare the results to the results when colored filters are combined. Explain any differences.

Vincent van Gogh, *The Night Cafe*, oil on canvas, 1888. 28-1/2" × 36-1/4". Yale University Art Gallery, Bequest of Stephen Carlton Clark, B.A., 1903.

The Big Idea

The electromagnetic spectrum consists of electromagnetic waves—transverse waves that carry both electric and magnetic energy. The specific characteristics of these different waves determine their effects on humans and their usefulness in technology.

Electromagnetic waves have many characteristics in common, but they have some unique characteristics as well. Their specific qualities determine how they interact with other parts of the universe, from atoms to humans to planets. Humans can manipulate certain electromagnetic waves to perform useful functions.

For Your Journal

Review the questions you answered in your journal before you read the chapter. How would you answer the questions now? Add an entry to your journal that shows what you have learned from the chapter. Include new ideas about how electromagnetic waves affect your everyday life.

Connecting Ideas

By looking at the diagram of the electromagnetic spectrum and creating illustrations for each type of wave, you can show how the big ideas of this chapter are related. In your journal, illustrate each type of electromagnetic wave, except gamma rays. Write a paragraph that briefly describes each kind of wave you have illustrated.

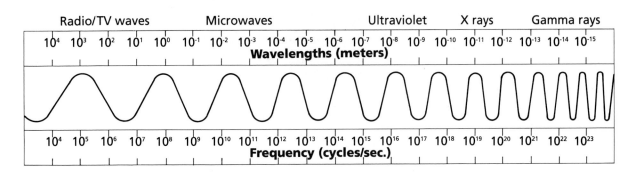

Radio/TV waves   Microwaves   Ultraviolet   X rays   Gamma rays

| 10^4 | 10^3 | 10^2 | 10^1 | 10^0 | 10^{-1} | 10^{-2} | 10^{-3} | 10^{-4} | 10^{-5} | 10^{-6} | 10^{-7} | 10^{-8} | 10^{-9} | 10^{-10} | 10^{-11} | 10^{-12} | 10^{-13} | 10^{-14} | 10^{-15} |

Wavelengths (meters)

| 10^4 | 10^5 | 10^6 | 10^7 | 10^8 | 10^9 | 10^{10} | 10^{11} | 10^{12} | 10^{13} | 10^{14} | 10^{15} | 10^{16} | 10^{17} | 10^{18} | 10^{19} | 10^{20} | 10^{21} | 10^{22} | 10^{23} |

Frequency (cycles/sec.)

REVIEW

Understanding Vocabulary

1. For each set of terms, explain the similarities and differences in their meanings.
   a) electromagnetic waves (284), electromagnetic spectrum (285)
   b) primary colors (302), secondary colors (302)
   c) telecommunications (297), radio waves (292), microwaves (297)

Understanding Concepts

MULTIPLE CHOICE

2. X-rays pass most easily through
   a) bones.   b) metal.
   c) skin.   d) teeth.

3. The electromagnetic waves given off by all warm objects are
   a) microwaves.   b) infrared waves.
   c) ultraviolet waves. d) radio waves.

4. People with defective cone cells might be unable to
   a) protect themselves from uv rays.
   b) distinguish certain colors.
   c) adapt to seeing in bright sunlight.
   d) see in dim light.

5. Radio waves are used in all of the following except
   a) AM broadcasts.
   b) radar.
   c) FM broadcasts.
   d) thermograms.

6. The electromagnetic waves that cause people to tan and burn are
   a) ultraviolet waves. b) visible light.
   c) infrared waves.   d) microwaves.

7. The different kinds of waves in the electromagnetic spectrum all travel
   a) at the speed of sound.
   b) at the speed of satellites.
   c) at different speeds.
   d) at the speed of light.

SHORT ANSWER

8. Name three objects found in the home that produce or use electromagnetic waves. Which type of waves does each object produce or use?

9. What is radar and how is it used? Describe how it has made air and sea navigation safer.

Interpreting Graphics

Photo A is a visible-light view of the Andromeda galaxy. Photo B shows the same galaxy on infrared film similar to a thermogram.

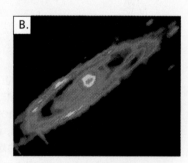

10. Can you see already existing stars in Photo A? Why or why not?

11. Dust and gases that eventually become stars must first get very hot. In which areas in Photo B are stars forming? Explain how you know.

12. Explain why the stars forming in Photo B cannot be seen in Photo A.

Reviewing Themes

13. *Energy*
Why is the interaction between the electromagnetic spectrum and life forms on Earth such a delicate balance?

14. *Technology*
Cancer cells are warmer than healthy cells. Explain how a doctor could use a type of electromagnetic wave to detect cancer cells.

Thinking Critically

15. Suppose you were totally colorblind—you could see only black, white, and shades of gray. How might your colorblindness affect your life?

16. Recall how ultraviolet and X-ray waves affect human beings. What effect do you think gamma rays would have on the human body? Reexamine the electromagnetic spectrum on page 285, if necessary.

17. The sun emits all forms of electromagnetic radiation. Which two forms are the most important for life on Earth? Why?

18. The shiny color, or iridescence, of this soap bubble is caused by reflected light. Light is reflected from the inside and outside surfaces of the bubble, causing two sets of reflections. These reflections can combine to form stronger colors or they might cancel each other out. Using your knowledge of reflection, refraction, interference, and color, explain how iridescence occurs.

Discovery Through Reading

Kesten, Lou. "Future Visions: Tomorrow's Video Technology Looks Out of Sight!" *3-2-1 Contact*. 109 (September 1990) 24–27. Three-dimensional movies? A television that is smaller than a matchbook? Read about these technological advancements that could be closer to reality than you think!

USES FOR LASERS

*L*asers have long appealed to the makers of science fiction. In the old television series Star Trek, Captain Kirk cried "Set phasers on stun!" and zapped his enemies with laser-like beams. The Jedi knights of the Star Wars movies dueled with sabers of laser light.

COURTESY OF LUCASFILM LTD.™ & © Lucasfilm Ltd. (LFL) 1980. All Rights Reserved.

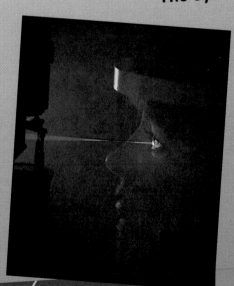

Imagine a movie starring a young doctor whose sweetheart is blinded in an accident. The doctor daringly aims a laser beam at her eye! The eye is repaired, and the sweethearts live happily ever after. Medical science fiction? No, today's physicians use lasers to repair damage to the human retina.

Instead of "Ghostbusters," a future movie might star a team of "Scumbusters." These heroes could travel the world restoring ancient buildings and statues nearly destroyed by air pollution. A laser gun could zap the coating of scum caused by decades of air pollution. The stone beneath the scum would reappear in all its beauty. In fact, such lasers are already being used to help

restore damaged art works. In one restoration, a tiny laser beam cleaned silver threads in a precious drapery. The beam did not touch the fragile cotton threads woven into the same cloth.

Imaginative scientists are only beginning to tap the potential of laser tools. Already, their real progress surpasses some of the wildest inventions of science fiction!

For Your Journal

- *What makes laser light different from ordinary light?*
- *How are lasers used in medicine?*
- *How can lasers be used to carry sound?*

The Laser

Richard loves science fiction stories, especially those with wild, futuristic inventions. One rainy day he went to choose one of the many science fiction books from the family library. When he reached up to pull a book off the shelf, down fell a dusty old folder that had been buried behind the row of books. He blew off the dust, opened it up, and saw the name *Marsha Carroll* on the top sheet of paper. "My grandmother," he thought. "I had no idea she wrote science fiction."

Richard began to read with delight. He realized that his grandmother had written this story many years before the invention of technologies that are common today.

Figure 12–1. Lasers are used to produce light shows like this one at Stone Mountain, Georgia.

Coherent Light

One passage in the story read, "This was not an ordinary flashlight. The beam from this light pierced the air like an arrow, thin and straight. It shone for miles and miles."

"Wow!" thought Richard. "Grandma Marsha's futuristic light sounds like a laser." This was a technology he had been learning about in science class.

The beam of a flashlight may be about 3 cm across when it leaves the flashlight, but it enlarges to about 40 cm by the time it reaches the other end of a room. In contrast, the beam produced by a laser is only 2 mm across, and it stays 2 mm wide even when it reaches the other end of a room.

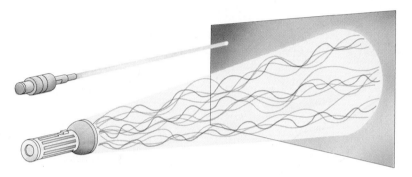

Figure 12–2. Laser light consists of light waves that are usually close in wavelength with a uniform pattern. Ordinary light consists of many different wavelengths that have a random pattern.

Unlike ordinary light from a flashlight, a laser beam's light contains only a few wavelengths. These wavelengths are usually very close together. For example, a helium-neon laser contains wavelengths in the long-wave part of the visible spectrum, that is, in the red part of the spectrum. Therefore, the light looks red. Other materials in lasers produce light of different colors.

Laser light is unique because all the light waves are in step. That is, crests travel with crests and troughs travel with troughs. In a beam of ordinary light many waves overlap, and there is a jumble of crests and troughs. When all the light waves in a beam are in step, the light is called **coherent light.** Coherent light may contain more than one wavelength, but for each wavelength the crests and troughs are in step.

Coherent light produces a very straight, concentrated beam. During research for his science class, Richard found out that when the subway tunnel was built under San Francisco Bay, its course was lined up by a laser beam that went from one side of the bay to the other. He also found out that the most accurate measurement of the distance to the moon was made by shining a laser at a reflector that was placed on the moon by an astronaut. Grandma's light beam was amazing, but not as amazing as the real thing!

Figure 12–3. This low-power helium-neon laser uses neon gas to produce coherent light.

What is the difference between ordinary light and coherent light?

How Lasers Work

In Grandma Marsha's story, a weapon shot out beams of straight light. The weapon looked like an old-fashioned dart gun. Richard learned in science class that lasers are often created in tubes filled with neon. He also learned that the way coherent light is produced from neon in tubes is similar to the way in which a dart gun is loaded for firing.

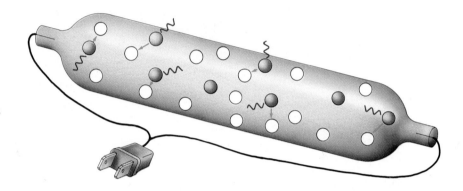

Figure 12–4. In a neon light, electric current raises many atoms to an excited state. The atoms quickly return to their ground state by emitting photons. This makes the bulb glow.

To load a dart gun, you compress the spring. This adds the potential energy of the compressed spring to the system. When you shoot the dart, the spring is released. The energy that was stored in the spring is transferred to the dart, which shoots toward the target.

An atom of a gas acts in a way similar to the spring of the dart gun. The atom can absorb and store energy. When an atom contains stored energy, it is in an **excited state**. In a neon light, the electric current going through the neon gas provides the energy to raise the atoms to an excited state. The atoms in the gas are pumped by the current. **Pumping** is the process of raising atoms to an excited state.

Now the atoms in the neon tube are like a loaded dart gun. Each of the atoms possesses stored energy. But there is no mechanism to hold the neon atoms in their excited state, so they quickly lose their energy. When atoms give off the energy they gained in an excited state, they return to their **ground state**.

When atoms return to their ground state, they release the extra energy by emitting a photon of light. A photon is the small packet of energy in light waves. The photon shoots out of the neon atom just as a dart shoots out of a dart gun. ("Grandma was pretty smart," thought Richard.) In the neon tube, the photons are released in all directions, at all times. This means that the emitted photons do not form a coherent beam of light.

Stimulated Emissions of Photons   A neon tube can be converted into a laser because of a unique property of neon atoms. When an excited neon atom is struck by a photon, the atom immediately releases two photons.

The two photons are identical. They leave the atom together, in the same direction and in step. The process of forcing identical photons into step is called **stimulated emission**. While Grandma probably didn't know about excited or ground states of atoms or stimulated emissions of photons, she was on the right track.

Stimulated emissions from one photon produce two photons. If each photon strikes an excited neon atom, four photons are produced. This process continues and takes place quickly. An excited atom must be stimulated to give off a photon before it has a chance to drop back to its ground state.

Richard decided that, with his parents' permission, he would try to modernize his grandmother's manuscript. How could Richard make a diagram that would help his readers understand the coherent light of lasers? Try the next activity to find out.

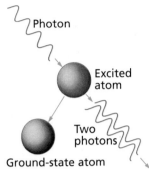

Figure 12–5. When a photon strikes an excited atom, the excited atom will emit two photons in step with one another. This process is called stimulated emission.

DISCOVER BY *Writing* _____

Illustrate the difference between ordinary light and coherent light. Include labels and captions to describe each kind of light. ✎

Making a Laser   Years and years ago, when Grandma Marsha wrote her science fiction story, she had never heard of the word laser, which is an acronym for **l**ight **a**mplification by **s**timulated **e**mission of **r**adiation. If you know what the word means, you can understand how a laser works. To create a laser, you need more than photons that are in step. The ends of the tube of neon gas must be coated with a reflecting surface to cause any photons traveling up and down the tube to bounce back and forth between the ends. The bouncing photons stimulate emission from other atoms as they pass by. Soon, there is an enormous flow of photons back and forth in the tube,

Figure 12–6. Use the key to follow the bouncing photons created in this neon tube from A to D. The photons bounce off the reflective surface at the ends of the tube. One end is only partially mirrored. This is the end from which the laser beam passes out of the tube (D).

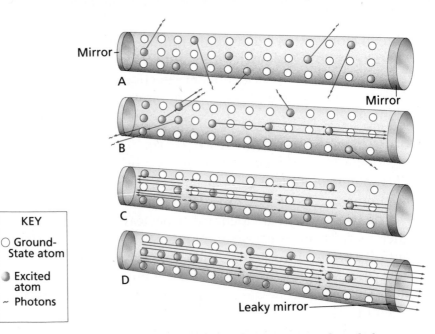

KEY

○ Ground-State atom

◑ Excited atom

~ Photons

Figure 12–7. The laser is being used to cut engine bores.

all in step with one another. If one of the mirrored ends has a thinner coating, some of the light can pass through. From this end, five to ten percent of the coherent light escapes to form a laser beam.

Richard learned that any material that can produce photons in step can produce a laser beam. He also found out that energy other than electric energy, like chemical reactions or flashes of light, can stimulate the production of photons in step. Lasers, Richard discovered, are produced when reflective ends cause a standing wave of light inside a tube. His science teacher explained in class that coherent light has been observed coming from Mars. Scientists hypothesize that it is produced naturally by the action of sunlight on the carbon dioxide in the Martian atmosphere.

Richard writes down this new information. It is giving him great ideas that may help him to modernize Grandma's story.

 ASK YOURSELF

How are bouncing photons produced?

SECTION 1 REVIEW AND APPLICATION

Thinking Critically

1. Explain what coherent light is.

2. Why is the mirror at one end of a laser tube only partly mirrored?

Reading Critically

3. Explain how lasers could create an image of the American flag on a screen.

4. How could a biologist use a laser to destroy the nucleus of a single cell?

Using Lasers

In addition to fantastic futuristic inventions, Richard's grandmother had a knack for creating colorful characters. Her main character, Jack Ratty, had enjoyed adventures around the galaxy before he settled in at Gulch Creek, Earth. One day as he leaned over a simple household gadget, he seriously injured his eye. A visiting venusian friend shone a thin blue beam of light from her forehead into the injured eye. Immediately, Jack regained perfect vision. Later in the manuscript, a small yellow beam of light was struck against a mirror, and galactic music was amplified throughout space.

At the end of the manuscript are the words *The End* and the date: 1940. Grandma's imagination was well ahead of her time.

Objectives

Construct *a chart explaining medical procedures that use lasers.*

Explain *how music is stored and played on a compact disc.*

Lasers and Medicine

If the character Jack Ratty had suffered a detached retina in his eye in the 1990s, an earthling rather than a venusian could have helped him. To treat the condition, laser pulses are fired onto many points around the edge of the detached region. Each pulse produces a tiny bit of scar tissue that holds the retina in place. If he had suffered broken blood vessels in his eyes due to diabetes, a laser could have been focused on a micrometer-sized spot on the retina. A short, intense pulse of light would close the bleeding vessels. No bleeding occurs because the laser heats the tiny spots of tissue so much that it fuses the tissues together.

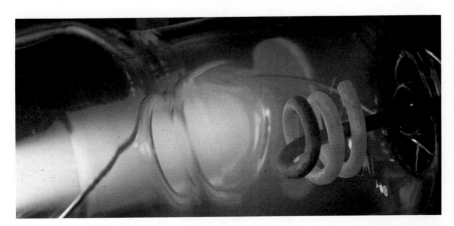

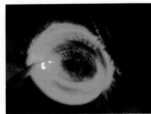

Figure 12–8. An argon gas laser (left) is often used in performing delicate eye surgery (right).

In some cases, lasers can be used instead of scalpels during surgery. An infrared laser made from carbon dioxide can focus continuous light to kill a tumor while sealing blood vessels around the tumor at the same time. A YAG laser, made from a crystal of **y**ttrium **a**luminum **g**arnet, makes an intense infrared beam, focused to a fine point, that can reach areas within the middle ear where a scalpel cannot be used. Since Richard wants to find out about other medical procedures in which lasers are used, he decides to take a trip to the library.

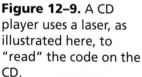

DISCOVER BY Researching _____

Go to the library or talk to a laser surgeon, and collect as much information as possible on the use of lasers in medicine. Make a chart about the procedures in which lasers are used and the specific type of laser that is needed for each procedure. ✎

▶ ASK YOURSELF

Why are lasers used in surgery?

The Compact Disc

Richard's grandmother may have thought that a music machine played by a light beam was a wild idea, considering it was 1940. But CDs, or compact discs, played with laser light, are common today.

The surface of a CD is touched only by a microscopic spot of infrared light. The light produced by a tiny laser is focused

Figure 12–9. A CD player uses a laser, as illustrated here, to "read" the code on the CD.

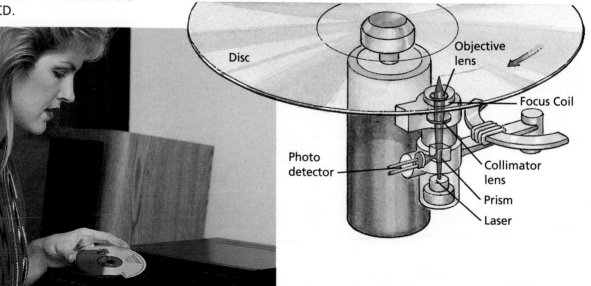

by a lens system to a pinpoint spot. This spot of light reflects from the tiny pits on the mirrored surface of the disc. The reflected light is converted into an electric signal, which is then amplified and sent to the speakers.

Music is recorded on a CD in a digital code. The music comes to the recording microphone as a sound wave. The microphone converts the sound wave into a wavelike electric signal. This signal is coded into a series of tiny pulses. Each pulse stands for the number one. A missing pulse stands for the number zero. Each number is called a *bit*. About 16 billion bits will put one hour of music on the disc.

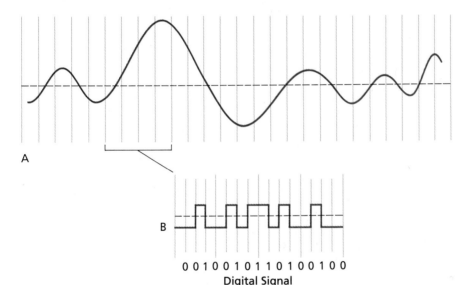

A

B

0 0 1 0 0 1 0 1 1 0 1 0 0 1 0 0
Digital Signal

Figure 12–10. In recording sound on a CD, the sound wave is converted by a microphone into an electric signal (A). This signal is then coded into a series of pulses (B). The digital signal turns the laser beam on and off to record sound on a CD.

 **ASK YOURSELF**
How is music coded onto a CD?

SECTION 2   *REVIEW AND APPLICATION*

Reading Critically
1. How does a laser beam allow us to hear music on a CD?
2. What is a bit?

Thinking Critically
3. A laser can be used to remove a red birthmark called a portwine stain. This type of birthmark contains many blood vessels. Why do you think a laser is used for this procedure?
4. Since a surgical laser uses infrared light, the surgeon cannot see the cutting beam. Suggest a way that another laser might be added to the system to solve the problem. Justify your answer.

Using Light Waves

Describe *the image produced by a hologram.*

Evaluate *the necessary conditions for practical communication by light waves.*

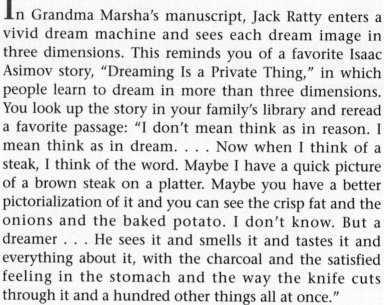

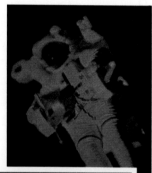

Figure 12–11. The two photos above are the same hologram, viewed from different angles.

In Grandma Marsha's manuscript, Jack Ratty enters a vivid dream machine and sees each dream image in three dimensions. This reminds you of a favorite Isaac Asimov story, "Dreaming Is a Private Thing," in which people learn to dream in more than three dimensions. You look up the story in your family's library and reread a favorite passage: "I don't mean think as in reason. I mean think as in dream. . . . Now when I think of a steak, I think of the word. Maybe I have a quick picture of a brown steak on a platter. Maybe you have a better pictorialization of it and you can see the crisp fat and the onions and the baked potato. I don't know. But a dreamer . . . He sees it and smells it and tastes it and everything about it, with the charcoal and the satisfied feeling in the stomach and the way the knife cuts through it and a hundred other things all at once."

Richard laughs as he thinks about what his grandmother's reaction would be to a holographic image, or even virtual reality machines. Grandma surely would think she was dreaming!

Holograms

Richard saw his first hologram in science class a few weeks ago. The picture showed an astronaut floating in space. By turning the hologram, he could see the figure in three dimensions.

The hologram looked like a sheet of dirty transparent plastic, but it was actually covered with tiny, closely spaced squiggly lines that can only be seen under a microscope. If a light is placed behind the hologram and a person looks through the plastic sheet, a tiny astronaut floating in space appears. The image has a three-dimensional quality, a feeling of reality, that does not exist in other optical images.

Richard learned that a holographic image changes when you look at it from a different angle. From one position, the astronaut's left arm hid the instruments on his abdomen (top photograph). Then, you moved to the left to look behind that arm and gradually saw something different as you moved (bottom photograph). When you moved even farther to the left, the

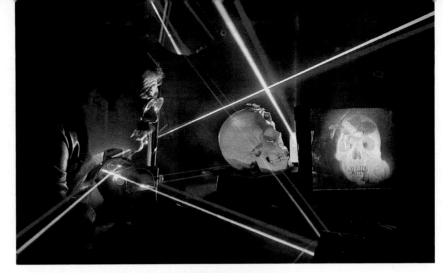

picture changed even more. A hologram can do this because it produces a set of light waves exactly like those coming from a real object.

A **hologram** is an interference pattern created by a laser on a piece of film. The laser beam is split into two parts. One part goes directly to the film. The other part goes to the object being photographed and is reflected onto the film. When the film is developed, a series of dark and light lines appears. The dark lines show where interference of the two beams is constructive. At the light lines, the film was not exposed. There was no light at those points because crests of one beam arrived at the same time as troughs of the other. That is, there was destructive interference.

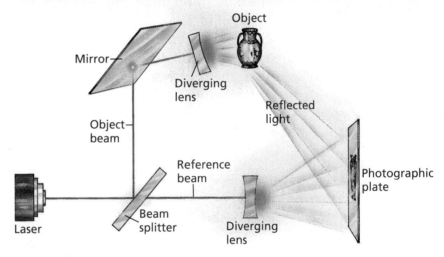

When light comes through the hologram, the interference pattern blocks some of it. What comes through is exactly like the beam that made the hologram. This transmitted light has all the properties of the light reflected from the object.

ASK YOURSELF
How is a hologram made?

Figure 12–12. A hologram is made by splitting a laser beam into two beams and then reuniting the beams on the film. The interference pattern formed when the two beams meet produces the holographic image.

Fiber Optics

There is another futuristic invention Grandma Marsha never even dreamed of—glass fibers that carry telephone conversations. You can find out more about these glass fibers by doing the next activity.

CAUTION: Put on safety goggles to complete this activity. Get a small piece of fiber-optic cable. Carefully take it apart. Draw a diagram of what you find. Be sure to label your drawing correctly. ✏

Light flows through these optical cables. Today, long distance lines transmit conversations in the form of tiny pulses of light. A new kind of laser makes this system work. When light travels through a medium in which it travels more slowly than it would through air, the light may not be able to get out. If the angle of incidence at the surface is too great, the light is reflected inward. This reflection from the inside surface is called **total internal reflection.** When total internal reflection occurs, the light waves act like a rock does when you skip it across a quiet pond. Instead of going through the surface and into the water, the stone is reflected into the air. The light waves travel long distances through the tiny glass fibers because they are continually reflected into the fiber.

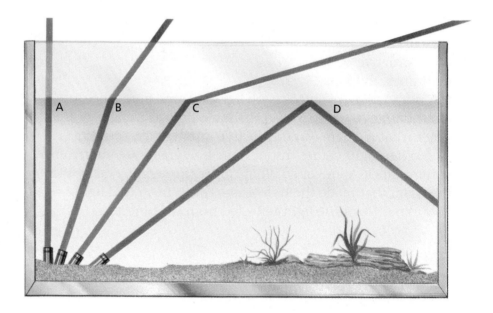

Figure 12–13. This illustration of lights in an aquarium shows how a beam of light reflects from the internal surface of water. When the angle of incidence is great enough there is total internal reflection (D).

Figure 12–14 shows a surgeon viewing the throat of a patient with a light pipe. A **light pipe** is a clear plastic rod that can be bent into a curved shape to carry light into otherwise unreachable places. Total internal reflection is used in the light pipe. As long as the curvature is not too great, the light strikes the surface at a high angle of incidence and is reflected inward.

Richard learned from his science teacher that a long, thin light pipe is called an **optical fiber.** Then he looked up a picture in his science book that shows how many optical fibers make up an image conduit. Each fiber, made of glass and encased in plastic, is considerably thinner than a human hair. Light travels faster in the plastic than it does in the glass, making total internal reflection possible. Each fiber picks up one small portion of the light from the object and carries it to the other end of the conduit. The conduit can be bent and twisted into all sorts of shapes without losing the image. You can see this for yourself in the next activity.

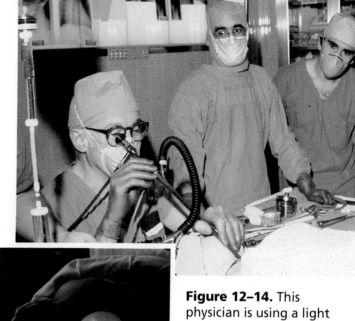

Figure 12–14. This physician is using a light pipe to check a patient's throat. The light pipe provides the view shown on the bottom.

ACTIVITY

How can you use an image conduit?

MATERIALS

safety goggles, laboratory aprons, Bunsen burner, image conduit, triangular file

PROCEDURE

1. **CAUTION: Put on safety goggles and a laboratory apron, and leave them on throughout this activity.** Turn on and adjust the Bunsen burner. Hold the image conduit by the ends, and heat the center of it in the burner flame.

2. When the image conduit softens, bend it, twist it, or stretch it to change its shape. Remove it from the flame, and let it cool.

3. With an edge of the file, cut a nick in the conduit at some point beyond where you have bent it, as shown. Then break the glass at that point.

4. Put one end of the conduit directly on top of the letter *e* somewhere on this page. Look into the other end of the conduit. What do you see?

5. Reverse the conduit so that the other end is on the letter *e*. Now what do you see?

APPLICATION

1. Why is the image of the letter *e* able to come through the conduit undistorted?

2. Suggest some uses for an image conduit.

Figure 12–15. The laser used to send light in an optical fiber is very small. One such laser is shown here—lying on George Washington's eye on this quarter.

Telephone cables made of optical fibers also consist of fine glass fibers encased in plastic. When you talk on the telephone, your voice is converted into an electric signal. At a local telephone station, this signal is changed into a digital code, like the code used on a CD. The electric bits of the code are used to trigger a tiny laser, no bigger than a grain of salt. The laser codes your voice as a series of flashes of infrared light, which travels through the glass fiber.

When you speak, you leave spaces between words and syllables. In the fiber, these intervals are filled in with someone else's conversation! A cable of 12 fibers can carry 50 000 conversations at once.

Grandma Marsha would marvel at the new technologies available to earthlings in the twentieth century. Richard is looking forward to updating her wonderful story.

Figure 12–16. Each of the optical cables (right) consists of 144 individual fibers shown on the left.

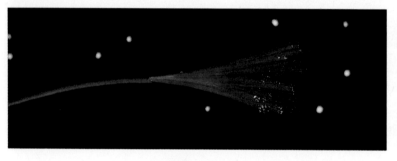

ASK YOURSELF
What is total internal reflection?

SECTION 3 *REVIEW AND APPLICATION*

Reading Critically

1. What are the advantages of optical fibers for transmitting telephone conversations?

2. How is total internal reflection used to transmit telephone conversations?

Thinking Critically

3. Biologists have developed a technique for making holographic pictures through a microscope. How would such pictures be useful in the study of microscopic organisms?

4. Why do CDs, optical fibers, and digital audio tapes all represent sound in the form of bits rather than waves?

SKILL Measuring Angles

► MATERIALS
- protractor ● paper ● ruler

▼ PROCEDURE

1. Look at your protractor carefully. Note that it is shaped like a half circle, with a scale along its rounded edge. The scale indicates that the number of degrees in a half circle is 180. The center of the flat side of your protractor is marked with a short line. (In some protractors there is a hole at the center point.)

2. To measure an angle, place the protractor flat on the paper, with its center at the vertex, or point, of the angle. Move the protractor until the 0° line is on one ray of the angle. Be sure to keep the protractor center on the vertex. The measure of the angle is the point where the other ray intersects the protractor's scale. The number of degrees in the angle can be read directly from the scale.

3. To construct an angle of a specific measurement, use a ruler to draw a line on your paper. On the line, mark the point where you want the vertex of the angle to be. Place the protractor on the paper with its center point at the vertex of the angle. Adjust the protractor until the 0° line is on the line you have drawn. Then mark a dot on the paper at the desired degree position on the scale. Remove the protractor, and draw a straight line through the vertex and the mark you just made.

4. Use a ruler to draw an angle on a sheet of paper. Label the vertex and rays. Measure the angle.

5. Construct five other angles, and measure them. Mark each of your angles with its correct measure in degrees.

6. Construct angles of 25°, 30°, and 60°.

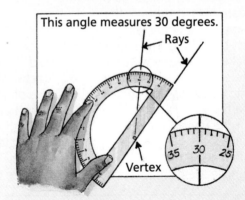

This angle measures 30 degrees.
Rays
Vertex

► APPLICATION

1. Why is it important to locate the vertex of the angle?

2. What is the result if you add an angle of 25° to an angle of 15°? Use a protractor to see whether your answer is correct.

3. How many degrees are there in the angle between the vertical and a road that slopes at 12° from the horizontal? Diagram the road.

✳ Using What You Have Learned

You have decided to enter a contest to design a new science lab. How can knowing how to measure angles help you with your design? Design a new lab to test your ideas.

INVESTIGATION

Determining Total Internal Reflection

▶ MATERIALS

- metric ruler ● white paper ● hemicylindrical plastic dish ● flashlight
- tape ● protractor

▼ PROCEDURE

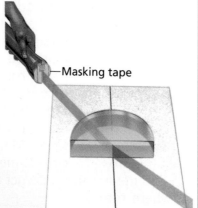

—Masking tape

1. Using a ruler, draw a line down the center of a sheet of white paper. Place the plastic dish on the paper in such a way that the dish is centered on the line. The line on the paper is now the normal to the flat side of the dish.
2. Add water to the dish until it is about three-fourths full.
3. With your ray box, send a beam of light through the curved side of the dish. Adjust the path of the beam to make it leave the water at the exact center of the flat side of the dish, as shown in the figure.
4. On the paper, mark the path of the beam entering and leaving the water. Then remove the dish, and con-

nect the lines. Use the protractor to measure the angles of incidence and refraction at the flat side of the dish. Remember that the angles are measured from the normal.
5. Prepare a chart like the one

shown. Enter the size of the angle of incidence and the size of the angle of refraction in the chart.

TABLE 1: TOTAL INTERNAL REFLECTION	
Angle of Incidence	Angle of Refraction

6. Increase the angle of incidence, and measure the two angles again. Continue taking readings until you find the largest angle of incidence that allows the beam to come out of the water. Record the angle at which total internal reflection begins to occur.

▶ ANALYSES AND CONCLUSIONS

1. How does the angle of refraction compare with the angle of incidence?
2. Why does the beam bend away from the normal as it leaves the water?
3. The *critical angle of incidence* is the largest angle of incidence that will allow the beam to leave the water. What happens to the beam when the angle of incidence is larger than this?
4. From your data, what is the critical angle of incidence of water?

▶ APPLICATION

Relate what you learned in this investigation to the transmission of telephone conversations through optical fibers.

Discover More

Continue the procedure with angles larger than the critical angle of incidence. What are your results? How do the angles of incidence and reflection when total internal reflection occurs compare to the angles of incidence and reflection when total internal reflection does not occur? Why is this so?

*H*IGHLIGHTS

The Big Idea

Light energy has many uses, some of which are made possible by modern technology. The development of new technologies has produced the laser and the optical fiber. Both of these technologies are based on the behavior of light under certain circumstances: coherent light must be produced to create a laser beam; total internal reflection must be achieved to allow optical fibers to carry light. These systems and their interactions have made possible technological advances in medicine, manufacturing, telecommunications, the arts, and many other fields.

For Your Journal

Lasers probably have more effect on your everyday life than you thought. What have you learned about lasers in this chapter? Look back at the ideas you wrote in your journal at the beginning of the chapter. How have your ideas changed? Revise your journal entries to show your new ideas.

Connecting Ideas

The following situations are made possible by either lasers or fiber optics. Decide which. Then write a description in your journal that tells how lasers or fiber optics made the situation possible.

REVIEW

Understanding Vocabulary

1. For each set of terms, tell how they are related.
   a) coherent light (313), optical fiber (323)
   b) pumping (314), excited state (314)
   c) ground state (314), stimulated emission (315)
   d) laser (315), hologram (321)
   e) total internal reflection (322), light pipe (322)

Understanding Concepts

MULTIPLE CHOICE

2. Light does not escape from an optical cable because
   a) it travels more slowly in that medium than it does through air.
   b) it travels faster in that medium than it does through air.
   c) the cable is curved.
   d) the cable is too thin.

3. Coherent light can be produced only by
   a) electric lights.
   b) lasers.
   c) neon tubes.
   d) stars.

4. One characteristic of a laser beam is that
   a) it spreads out the further it travels.
   b) it has more wavelengths than an ordinary beam of light.
   c) it maintains its width no matter how far it travels.
   d) its waves are continuously overlapping.

5. Which of the following is not a type of laser?
   a) quartz crystal
   b) YAG
   c) helium-neon
   d) argon gas

6. Constructive interference and destructive interference is used to a unique advantage in
   a) the taking of X-rays.
   b) the making of a hologram.
   c) laser surgery.
   d) the making of a compact disc (CD).

SHORT ANSWER

7. List and briefly describe as many uses as you can think of for lasers.

8. Explain why the beam of a laser must be split in order to make a hologram.

9. Using what you know about atoms, photons, ground state, and excited state, describe how neon signs work.

Interpreting Graphics

10. Which of these two figures represents coherent light?

11. How many frequencies are represented in Figure A? Figure B?

A.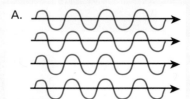

B.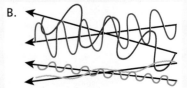

Reviewing Themes

12. *Energy*
Light is a form of energy. Discuss some aspects of light that demonstrate its ability to do work.

13. *Technology*
Most people use a telephone on a daily basis. Without realizing it, we rely on immense communications systems to provide telephone service. Describe the ways in which this system has changed our lives and the ways the telephone system can be used.

Thinking Critically

14. Explain why many credit card companies have included holograms on their credit cards.

15. In addition to the example given in this chapter, what other surgical techniques could be aided by the use of the light pipe? Describe at least three examples.

16. The Northern Lights are rainbow-colored lights appearing at night. They are caused by huge explosions on the sun. These explosions emit millions of particles which travel through space and reach Earth's atmosphere, where they collide with other particles and produce light. Are the Northern Lights an example of coherent light or ordinary light? Explain your answer.

17. Shown here are two examples of laser-cut greeting cards. Many new types of cards and stationery are made using industrial lasers. Using your knowledge of lasers, explain why this type of cut-out card can be made using lasers. How else could cards of this type be made? What is the advantage of a laser over the other methods you can think of?

Discovery Through Reading

Pennisi, E. "Etching Technique Lights Up Porous Silicon Three Ways." *Science News* 141 (June 27, 1992): 423. Read how new techniques are being developed to improve holographic data storage.

ENERGY FROM THE SUN

Waves of energy flow from the sun. All life on Earth depends on the energy of these waves—they allow plants to make their own food through the process of photosynthesis. Without plants, animals would have no food and would die.

Visible and Invisible

The largest part of the energy from the sun is in the form of visible light. However, the energy comes in other forms also. Waves of invisible infrared rays and ultraviolet rays bombard you every minute the sun is shining. While your eyes cannot detect these waves, they affect your body in many ways.

Have you ever seen a cat seek out a sunny spot for sleeping? The warmth the cat feels when it lies in the sunshine comes from infrared

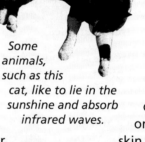

Some animals, such as this cat, like to lie in the sunshine and absorb infrared waves.

waves. These waves are also known as radiant heat. When was the last time you felt the warmth from infrared waves?

Ouch! The ultraviolet waves coming from the sun can cause sunburn. On the other hand, ultraviolet waves do have an important beneficial effect. They act on a chemical in the skin to produce vitamin D. Every child needs vitamin D to form strong bones and teeth. Adults need it to help keep their bones healthy. About 15 minutes a day of bright summer sunshine provides the human body with all the vitamin D it needs.

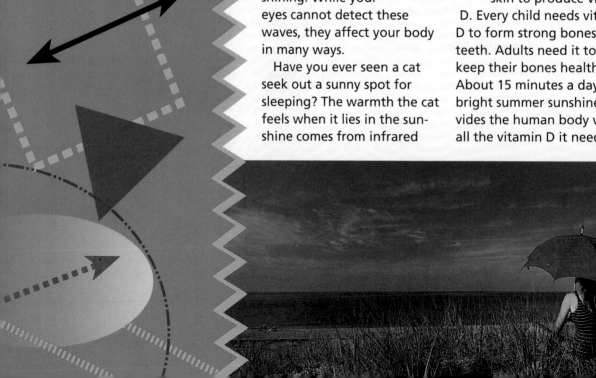

Good (and Bad) Things Come in Small Packages

All the sun's wave energy comes in tiny packages called *photons.* Ultraviolet rays come in two forms: ultraviolet A (UVA) and ultraviolet B (UVB). In the short wavelength ultraviolet, UVB, each photon has a lot of energy. In the longer wavelength ultraviolet, UVA, the energy is divided into many smaller packets. UVB can be much more damaging than UVA. UVB is like a shower of bricks; UVA is more like a shower of sand.

Solar panels convert infrared rays from the sun into heat.

To UVB or Not to UVB

Sunburn is caused by UVB waves. These waves are most intense in the middle of the day. As you probably know, people differ greatly in their susceptibility to sunburn. The dark pigment in the outer layer of the skin offers some protection against sunburn. Fair-skinned, freckled redheads have practically no dark pigment; therefore, any exposure to UVB can be dangerous to them. On the other hand, very dark-skinned people are much less likely to get sunburned. Between these two skin variations are people with some pigment in their skin. Whether the skin tans or burns depends on the amount of pigment in the skin and on the length of exposure to the sun.

Exposure to the sun in the morning or late afternoon, when there are few UVB waves in the light, produces a UVA tan. Sunscreens and tanning lotions promote a UVA tan. Most of these lotions contain a chemical that blocks the UVB and lets the UVA get through to the lower living layers of the skin. Since UVA waves are less damaging than UVB waves, some people believe that a UVA tan is healthy.

Is a "healthy tan" really healthy? NO! Even mild UVA waves can damage the connective tissue fibers in the lower layers of the skin. UVB waves can produce even worse damage: wrinkling, coarseness, mottling, and roughness. Every suntan you see is a step toward skin growing old too soon. Worse yet, repeated exposure to sunshine may lead to cancer of the skin.

The next time you see a sunburned person, what will you think about? Think about using common sense and sunscreen. Both of these things will help you enjoy the sunshine without endangering your life. ◆

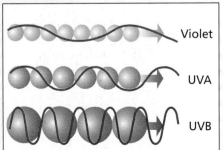

Violet

UVA

UVB

UVA consists of more photons than UVB, but each photon carries less energy.

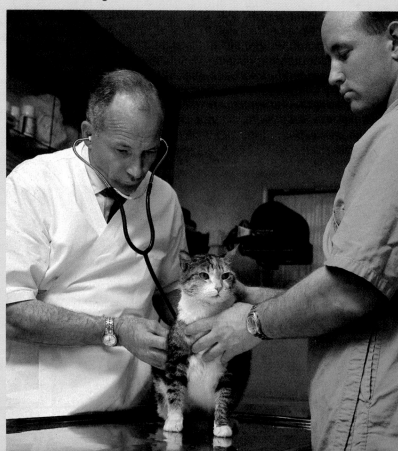
CAT Scans

High-Tech Medicine Can Save Animals, Too

by Lisa Feder-Feitel

from *3-2-1 CONTACT*

Marjorie Shaw was upset. Her normally active, bright-eyed 13-year-old was wandering around the house, bumping into furniture and walls. When it was clear that her teenager couldn't see, Marjorie rushed her to the hospital.

There, a team of medical experts used a CAT scan—a machine that makes computerized images of the inside of the body—to check out the patient's brain. As the teen was wheeled into the CAT scan room, a doctor placed a hand on her forehead and comforted her. Within moments—and without pain—the scan was complete. It showed doctors that the teenager had been suddenly blinded by a brain tumor.

Several times over the next few weeks, the patient lay on a table as invisible cobalt rays destroyed her tumor. By the end of the treatments, the patient had gotten her sight back.

This kind of modern medical "miracle" is not so unusual. What is unusual is where it took place: The patient's sight was saved at the Animal Medical Center in New York City. Marjorie Shaw's 13-year-old was her dog, Echo.

First-Class Care for Fido

Echo's story is not unusual. The same treatments, high-tech machines and medicines that were developed to help save people's lives are now being used to help save the lives of animals. So, animals, as well as humans, are living longer, healthier lives.

For many people—young and old alike—a pet is a member of the family. Because of this, many owners spend a great deal of money making sure that their pets stay healthy.

332

"Veterinary medicine can do anything on cats and dogs that we can do on human patients. I like to say we work out the 'bugs' on humans. Then when it's safe enough, we find ways to use the same procedures on animals," jokes Dr. Michael Garvey. He is a veterinarian at the Animal Medical Center in New York City.

Many of the latest medical techniques used on pets go beyond the emergency care that Echo received. Your dog may have her cataracts (a cloudiness that fills the lens of the eye, causing blindness) removed with laser beams, just as humans do. Or your cat (like people) may have an artificial hip placed in his body if arthritis (a painful disease of the joints) sets in.

For pups whose hearts don't beat to a regular rhythm, there are human pacemakers. Pacemakers are tiny, battery-operated machines placed in a person's body to keep the heart beating in a strong and regular rhythm. The exact same machine—and procedure—is now used to keep a dog's heart in tip-top shape. In St. Petersburg, FL, the Pinellas Foundation gives donated human pacemakers to vets, who use them on needy dogs.

Not all pets are physically ill, however. Some have emotional problems or fears. If a pet's problem is psychological, there are therapists to handle that, too. Does your dog hide and shake under the bed during thunderstorms? Suzanne Johnson of Beaverdam, VA, might be able to calm the pooch down.

Ms. Johnson is one of about 40 animal psychologists in the U.S. who treat fearful, aggressive or unhappy pets. To treat a fear of thunder, for instance, she may put the dog in a room and play a tape recording of a storm. She'd start at low volume, and increase it until the dog hardly notices the booming noise. It may frighten the owner, though!

Good Health at Great Cost

Now that veterinarians are using high-tech equipment, advanced techniques and new medicines, pets are living longer and healthier lives. That's great news for animals and their owners. But there's also a down side to this story. A pet's good health can carry a very high price tag.

Marjorie Shaw's bill for Echo came to $3,000. Cataract surgery can cost between $600 and $1,200 an eye. A pacemaker can be fitted for $1,500. Americans spend nearly $6 billion a year on veterinary services.

Laser surgery on cats (and dogs) with eye problems is common. An infrared light is used as part of the operation to remove cataracts.

Troubling Questions

These higher costs and new treatments may also force veterinarians and pet owners to make difficult choices: When should care for the pet continue, and when is it more kind to the pet to stop it?

Dr. Garvey of New York City's Animal Medical Center uses the example of a 15-year-old cat whose kidneys stopped working. "The animal may respond to treatment and do very well for months to years," he says. "On the other hand, the cat may live only for a few days or weeks. It will cost the owner several hundred dollars to find out. What should be done?"

Dr. Mike Shires of the University of Tennessee at Knoxville has dealt with another troubling problem: an owner's inability to see what is best for his or her pet.

"Many people seek all sorts of extreme medical treatments for their pets that weren't available a few years ago," he says. In cases where continued treatment is cruel, he recommends that the pet be put to sleep. If an owner still insists, he says, "we tell them to take their pet some place else."

Some doctors believe in treating pets and owners with equal care. Dr. Jane Mason, from Chantilly, VA, says, "I don't push the medical advances on owners with older pets or ones whose chances of a pain-free life are not good."

Deciding how far a person should go in saving his or her pet is becoming a common question. Humans who are seriously ill can usually say how much they want done to help them. But animals can't talk. The choice is ultimately up to the animal's owner.

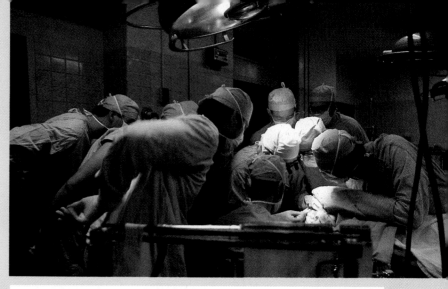

An operating room in an animal hospital looks very much like one in a hospital for humans. High-tech medical equipment that was developed for people is used to treat animals.

A veterinarian checks instruments as she prepares her animal patient for an operation.

Health Food Heals Pets

Kiss the canned dog food good-bye. Instead, treat your pup to a dish of yogurt or steamed broccoli. Or if your pet has fleas, try soaking its collar in oil made from eucalyptus (*say: you-kah-LIP-tus*) leaves or sprinkling garlic on its supper. Yes, garlic!

Healthful foods make healthy pets, says Dr. Monique Maniet, a veterinarian in Takoma Park, MD. She practices a kind of care for pets that some doctors use on humans. It's about as far from high technology as Maryland is from Madagascar. Called holistic medicine, it encourages the use of natural foods and soothing herbs.

Vaccines and antibiotics do have a place in Dr. Maniet's medicine cabinet, however. But she prefers to use more natural cures. If your pet suffers from swollen joints, for example, she might inject it with some honeybee venom. You might say that the poor pet gets stung twice, but Dr. Maniet says it helps a lot more than it hurts!

Should people use holistic or high-tech treatments with their sick pets? Always discuss it with your veterinarian. And never try any treatments, or give your pet any medicines, yourself. Always put your pet under a doctor's care. ◆

Thomas Young (1773-1829)

Today's scientist is expected to be a specialist, but scientists in the 1800s were mostly amateurs. That means that they did science for fun! Thomas Young was one example—he was a practicing physician, but he was interested in many other academic fields.

Young's skill as a linguist appeared early. By the age of 14 he was fluent in Latin, Greek, French, Italian, Hebrew, Persian, and Arabic. His language skills were later put to use when the famous Rosetta stone was found in Egypt. This stone, which contained an inscription in Egyptian hieroglyphics and a Greek translation, was the key to deciphering the Egyptian written language. Young was one of the first to make progress understanding it.

Young is best known, however, for reviving the wave theory of light. He conducted an experiment in which he showed that separate bands of light appeared where there should have been nothing but shadow. These bands of light diffracted around corners and could not be explained by particle theory. This experiment allowed Young to calculate the wavelength of visible light and determine what kinds of waves they were.

Thomas Young was also interested in other forms of energy. In fact, he initiated the use of the word *energy* in the modern sense we use today. During his lifetime, Young established his reputation as a distinguished linguist, physician, physicist, and writer. ◆

Jean M. Bennett (1930-)

Jean M. Bennett is a member of the Research Department of the Naval Air Warfare Center at China Lake, California. In 1979 she was elected Woman Scientist of the Year at the Center. She was chosen to be a Senior Fellow for her scientific contributions to the Center in 1989.

Bennett was the first woman to receive a doctorate in physics from Pennsylvania State University in 1955, and has done important research in several fields, including the physics of the atmosphere and infrared spectroscopy. She has written two books about her current specialty, studies of the surface roughness and scattering of optical light.

Her main work in recent years has been in the field of optical surfaces and coatings. Most lenses in optical equipment are polished to a very smooth finish and are coated with thin films that prevent ghost reflections from interfering with the sharpness of the image. The effectiveness of the coatings is limited because they are not perfectly smooth. Irregularities in the surfaces tend to scatter some of the light.

Bennett has made many contributions to the understanding of light scattering and has worked with manufacturers to make smoother surfaces, and coatings that have lower scatter levels. ◆

SHRINIVAS KULKARNI
Astronomer

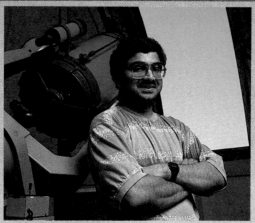

"I don't remember even once looking at the sky and saying, 'Wow, look at the stars…,'" Shrinivas Kulkarni says. While growing up in southern India, Kulkarni never thought that one day he would be an astronomer. In college he decided to study physics and engineering.

Kulkarni discovered astronomy by accident. He had to choose a place for summer study. India can be very hot in the summer, so he chose a school in a cool place. Astronomy was taught at that school. "I went there, and I instantly liked it," explains Kulkarni. After college in India, he came to the United States to study astronomy and has lived here ever since.

Kulkarni says that astronomy may have begun with people looking at the sky. Perhaps you think that this is all modern-day astronomers do, but it isn't. "People think that I go out on a cold night and look through a telescope," Kulkarni says. In fact, astronomers do not always look through the large telescopes that they use. They can program a computer to tell the telescope what to look for. Then astronomers look at the pictures and information that the telescope collected.

Shrinivas Kulkarni's special interest is how stars die. A star dies with a big explosion and becomes a neutron star. Neutron stars give off powerful radio waves as they rotate. These stars are called *pulsars* because the telescopes on Earth record something like pulses of light as the neutron stars rotate.

"A pulsar is like a lighthouse," Kulkarni explains. A lighthouse sends out a beam of light as the lamp rotates. You see the light each time the lamp turns and points toward you. Each time a pulsar turns, telescopes can record the radio waves from the pulsar. The study of pulsars is called *radio astronomy*.

Most known pulsars rotate 30 times a second or less. At least, that was the speed of those that had been discovered. Kulkarni decided to look for a pulsar that rotated 1000 times a second. He thought that he could find more slowly rotating pulsars by slowing down his computer data. Kulkarni meant to slow down the data when he looked at it the next day—but he forgot. Because he forgot, Kulkarni found the pulsar he was looking for; it was rotating at a rate of 642 times a second!

Today, Kulkarni is a professor of astronomy at the California Institute of Technology in Pasadena, California, where he continues his research on pulsars. ◆

Discover More

For more information about careers in astronomy, write to the

Astronomical Society of the Pacific
390 Ashton Avenue
San Francisco, CA 94112

Laser Fusion

Someday, scientists hypothesize, we will have a power source as powerful as the sun and as inexhaustible as the sea. In the next century, scientists expect that nuclear fusion will produce a large part of the world's electric energy. Today's most promising experiments involve the fusion of molecules by laser beams.

Explosive Energy!

Nuclear fusion happens when small atoms, such as hydrogen, suddenly combine to form larger atoms, such as helium. The fusion releases enormous amounts of energy, as we see in the explosion of a hydrogen bomb. Of course, that kind of explosive energy is not useful for generating electricity!

The major problem with fusion energy is the very high temperatures required to fuse atoms. Fusion of certain small nuclei takes place at almost 100 million degrees Celsius. In a hydrogen bomb, this temperature is reached by exploding a fission bomb.

The Promise of Fusion

The first usable fusion energy will probably come from a combination of laser technology and fusion science. In laser fusion, scientists begin with small pellets of deuterium and tritium. Deuterium and tritium are "heavy" hydrogen atoms found in ordinary sea water. To heat the deuterium fuel, scientists use a powerful laser beam. The laser beam is split into several separate beams, and each beam is amplified. When the amplified beams converge on the fusion pellet, they heat its nuclei enough to cause fusion and release heat energy.

This computer simulation shows the core of a fusion fuel pellet just before fusion. Fusion occurs at the core, where the fuel is hottest and most dense.

Today's laser fusion devices all consume more energy than they create. However, using computer simulations, scientists hypothesize that a laser fusion reactor could produce 100 times more energy than it uses. A reactor would ignite ten deuterium fuel pellets every second and generate a billion watts of electricity. Rocket scientists also hope that brief bursts of fusion energy can be harnessed to power rockets. If the promise of fusion energy is someday realized, the world may never again suffer an energy shortage. ◆

In this painting you can see the laser beams being focused on a fuel pellet. The fuel pellet is the small, bright sphere at the center of the painting.

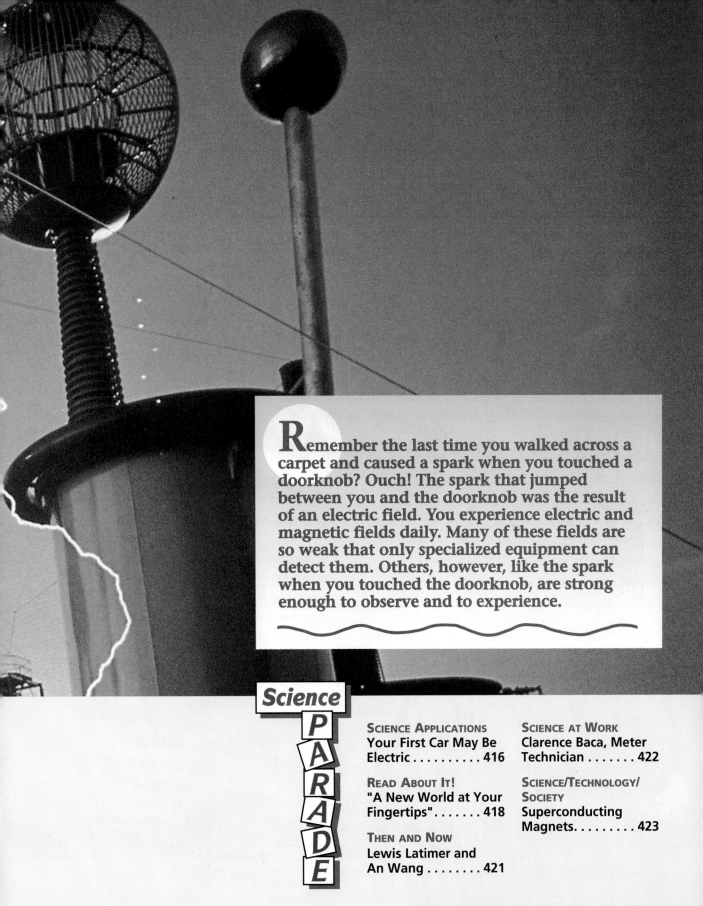

R emember the last time you walked across a carpet and caused a spark when you touched a doorknob? Ouch! The spark that jumped between you and the doorknob was the result of an electric field. You experience electric and magnetic fields daily. Many of these fields are so weak that only specialized equipment can detect them. Others, however, like the spark when you touched the doorknob, are strong enough to observe and to experience.

Science PARADE

CHAPTER 13

ELECTRICITY

Neon griffins, by Len Davidson, at the Philadelphia Museum of Art

How would you make an electrical announcement that glowed from the rooftops? How about an electrical sales pitch that could draw customers from sidewalks and roadways? You might say it in neon lights!

Neon light is the brightly colored, high-voltage lighting that decorates buildings from Tokyo to Las Vegas. In 1910 the inventor of neon lights pitched them as "the latest and most artistic form of electrical advertising." He had filled thin tubes with inert gases and then excited the gas with high-voltage electricity. Ionized by the electricity, neon gas glowed orange-red and mercury gas glowed blue.

Neon quickly became a brilliant commercial art form. The glass tubes could be bent into writing or intricate designs. At the height of its neon glory, New York's Times Square buildings glowed with flapping neon eagles and steaming neon coffee cups. Around the world custom-made neon signs became a trademark of restaurants, shoe stores, and theaters.

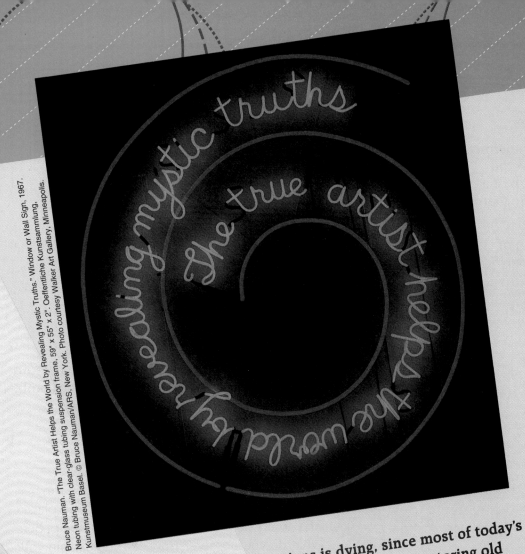

Today the art of custom-made neon signs is dying, since most of today's signs are mass-produced. However, some artisans are busy restoring old neon signs. Neon enthusiasts comb the city and highway landscape looking for rusting neon classics.

At the same time, artists are making new artistic statements in electric light. Artists like neon light because it combines the bright colors of painting and the shapeability of sculpture. One neon artist includes the lights' high-voltage electric transformers in his neon creation. He treats the transformer's hum as part of the art work. Neon lights enable artists to make vivid statements about their work. One neon work reads, "The true artist helps the world by revealing mystic truths." It's not the same as "Eat at Joe's," but it's written in the same glowing electric script.

For Your Journal

- ✎ List as many characteristics of electricity as you can think of.
- ✎ What can electricity do?
- ✎ How do you use electricity every day?

Static Electricity

Describe the effects produced by the accumulation of a static electric charge.

Apply the law of conservation of charge to specific cases.

Y ou've probably heard someone say that clothes dryers eat socks. You may have even thought so yourself as you sorted clean clothes. However, the socks usually turn up, stubbornly clinging to a shirt or towel. The crackling sound that you hear when peeling the clothes apart is entertaining though. Okay, so the dryer hasn't eaten the socks, but something got into them!

Electric Charge

You may have heard that crackling sound before. But where? You've probably heard that noise while combing your hair. Your hair wants to stand up instead of lie down or cling to the comb. You can investigate a comb's mysterious attractive force for yourself.

DISCOVER BY Observing

Tear some paper into small pieces. Then run a comb through your hair a few times, and place the comb close to, but not touching, the paper. What do you observe? How can you explain your observations? ✎

Figure 13–1. The cat's fur is attracted to the charged balloon.

In the activity, you discovered that the comb picks up the bits of paper. It can do this because it has an electric charge. Many objects can become electrically charged. For example, have you noticed how a nylon shirt wants to cling to your body? Notice the effect of the charged balloon on the cat's fur in the photograph. All of these effects are due to the accumulation of an electric charge by one object, which causes another object to be attracted to it.

Charge It!   An electric charge can be transferred from one object to another. For example, a vinyl strip stroked with a piece of wool will gain an electric charge. The charge has been rubbed off the wool. When you touch a graphite-covered plastic-foam ball with the vinyl strip, the ball jumps away. If you repeat the exercise using an acetate strip and a piece of plastic food wrap, the same reaction occurs.

Now forget about the ball, and try touching the vinyl and the acetate together. What happens? Instead of jumping away, the two strips are attracted to each other. All charges are *not* the same; some charges repel objects while others attract them. There are two kinds of electric charge. The charge on the vinyl is a negative charge (-), and the charge on the acetate is a positive charge (+). The general rule, as shown in the illustration, is this: Like charges repel one another, and unlike charges attract one another.

A Mutual Attraction   Most things don't have any charge in their normal state. But objects can become charged by static electricity. *Static* means "not moving." **Static electricity** is a buildup of an electric charge on an object; it is one form of electric energy. You can discover one way that objects acquire static electricity.

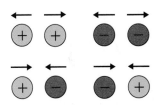

Figure 13–2. Only opposites attract.

DISCOVER BY Doing _____

Put five wheat puffs in a balloon, and blow it up. Rub the balloon on your shirt; then hold it close to your hair. What do you observe? Try rubbing the balloon with different materials, and test its "attraction" each time. In your journal, explain what is happening in terms of positive and negative charges. 🖉

You just found that a balloon can do the same thing to hair that a comb sometimes does! So when you have trouble getting your hair just right, don't blame the comb. Blame static electricity. To investigate further, do the following activity.

ACTIVITY

MATERIALS

buret clamp, ring stand, wooden rod, thread (2 pieces), vinyl strip, wool, graphite-coated plastic-foam balls (2), acetate strip, plastic food wrap

PROCEDURE

1. Set up the apparatus as shown.

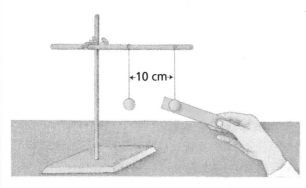

 ←10 cm→

2. Stroke the vinyl strip with the wool a few times. Then bring the strip near one of the foam balls. What happens?

3. Stroke a foam ball with the charged vinyl strip. Is the ball now charged? Explain.

4. Touch the same ball with your hand. What happens?

5. Repeat steps 2 through 4 using the acetate strip after it has been stroked with plastic food wrap to redistribute the electric charges.

6. Charge one ball from the vinyl and the other from the acetate. Slide the threads closer together without touching the balls. What happens? Explain the results.

7. After the balls touch, test them with the vinyl and acetate strips to see whether they are charged.

APPLICATION

1. How did you charge the foam balls?

2. How could you tell when two of the objects had opposite charges?

3. Explain why each ball has the same charge as the strip from which it was charged.

▶ **ASK YOURSELF**

How would two negatively charged objects act if brought near each other?

Conservation of Charge

Now you know that static electricity can make your hair do strange things. Does the static electricity stay in your hair forever? What is there about rubbing one kind of material against another that produces an electric charge? When you stroked the vinyl strip with the wool, the strip became negatively charged. Something else also happened—the wool became positively charged.

It would appear that stroking a vinyl strip with wool creates two kinds of charges. Actually, rubbing does not produce any charges—it just moves them around. The object that loses negative particles becomes positively charged because its charges are no longer completely balanced. The object that gains negative particles becomes negatively charged because it now has more negative particles than positive particles.

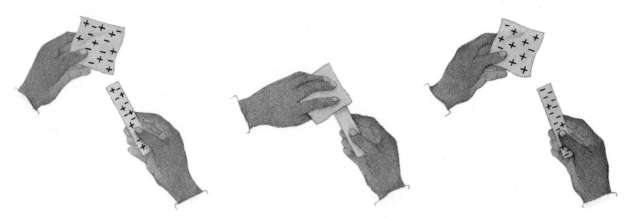

Figure 13–3. The placement of particles, or electrons, has been redistributed.

As shown in the illustration, wool and vinyl become charged when they are rubbed together because particles move from the wool to the vinyl. The particles that move are called *electrons*. The vinyl becomes negative when it gains electrons because the electrons have a negative charge. The wool becomes positive when it loses some electrons.

Keeping It Equal   The *law of conservation of charge* states that an electric charge cannot be created or destroyed. The atoms that make up all materials have positive and negative charges. When something is uncharged, its positive charge is equal to its negative charge. The positive and negative charges cancel each other, so that the total charge is zero. An object with zero charge is said to be neutral. This is when your hair behaves and your socks don't stray.

When an object *is* charged, it may have either a negative or a positive charge. An object has a positive charge if it has given some of its negative electrons to another object. The object that receives the extra electrons has a negative charge. Remember, only the negatively charged electrons move.

You'll Get a Charge Out of This   Just when you've pulled yourself together and put your socks on and combed your hair down, the car attacks you! You reached to open the

Figure 13–4. An electric discharge neutralizes both objects.

car door and suddenly received a shock. If it had been dark enough, you might have seen the spark that caused the "Ouch."

When a negatively charged object and a positively charged object are brought together, the effects of both charges are cancelled out. Electrons move from one object to another until both objects are neutral. If the charge is strong enough, the objects do not even have to be touching for this exchange to take place. Electrons can jump a gap between two oppositely charged objects. When this happens, the charge heats the air enough to make a spark. After the spark, both objects are electrically neutral. This process of transferring electric charges is called *electric discharge.*

Static electricity may be discharged if a charged object comes in contact with an object that will accept the charge. When you shuffle across a carpet and touch a doorknob, the static electricity is discharged and in the process you receive a small electric shock. You may even see a small spark as the static electricity is discharged.

The Biggest Shock of All   While lightning is certainly fascinating and dangerous, it's just a spark. Okay, so it's a really

Figure 13–5. This student is experiencing static electricity firsthand!

big spark. How is this enormous spark, or electric discharge, produced? Obviously no one is rubbing wool or plastic wrap over the clouds.

Electric charges in a thunderhead become redistributed as rapidly rising air currents carry water drops and hailstones upward. The upper part of the cloud in the illustration has a positive charge, and the lower part has a negative charge. The negative charge of the lower portion of the cloud becomes strong enough to make charges in the earth redistribute themselves. What type of charge must the earth's surface have?

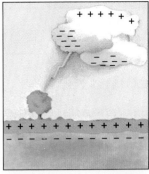

A lightning bolt occurs when the charge is neutralized. The negatively charged bottom of the cloud discharges either to the positively charged upper part of the cloud or to the positively charged earth. How do these types of discharges affect where the lightning appears?

While the illustration shows how lightning strikes the ground, it is important to remember that lightning may also go from cloud to cloud or even from the ground to the cloud!

Figure 13–6. Lightning is a spark that forms when opposite charges are neutralized during a storm.

 ASK YOURSELF

How does a lightning bolt demonstrate the law of conservation of charge?

SECTION 1 *REVIEW AND APPLICATION*

Reading Critically

1. When you stroke an inflated balloon on your hair, the balloon acquires a negative charge. How is your hair affected?
2. When wool is stroked with vinyl, the wool acquires a positive charge. How does the law of conservation of charge apply to the vinyl and the wool?

Thinking Critically

3. How can you determine whether or not a balloon has an electric charge?
4. You discover your pen can pick up scraps of paper. Tell how you can determine whether the charge on the pen is positive or negative, using only a rubber rod, a woolen glove, and a suspended plastic-foam ball.
5. Explain how electric charges act with respect to one another.

Electric Current

Explain what electric current is and what the function of a battery is.

Measure the units in which electric current and potential difference are measured.

Utilize equations to determine the values of electric power and energy.

Although lightning may seem like a giant Terminator, you now know that a more accurate nickname would be the Equalizer. Lightning can knock down power lines, leaving you without electricity. Are you really without electricity?

What about a flashlight? It doesn't have a cord and a plug, but still uses electricity to produce light. How? Inside the flashlight are batteries that generate electric charges. It is the flow of these electric charges that causes the flashlight to light.

Producing Electric Current

As you found out when you touched the car, too many electrons in one place produce an electric charge. The clouds in a thunderstorm develop electric charges. When the charges neutralize, electrons move, or flow, from one place to another. The flow of electrons is called **electric current.** Electric current is a form of electric energy. A bolt of lightning is a large electric current flowing through the air.

Figure 13–7. Lightning is one type of electric discharge.

The energy from electric currents can be used for many purposes. As you switch on your flashlight, you are using the electric current produced by an *electric cell*. An electric cell converts chemical energy into electric energy and contains two terminals (positive and negative) and an electrolyte. An *electrolyte* is a substance that allows an electric current to be conducted.

The cells in a battery may be either dry cells or wet cells. A dry cell, like those used in flashlights, contains a pastelike electrolyte. A wet cell, like those used in some car batteries, contains a liquid electrolyte. Examine a dry cell for yourself.

DISCOVER BY *Doing*

Caution: Do not break open the dry cell, since the electrolyte can injure skin tissue. Obtain a dry cell. Compare the dry cell to the drawing, and identify the positive and negative terminals. Get a flashlight, and place two dry cells in it. Examine the way the dry cells contact the parts of the flashlight. Make a sketch in your journal to show how electricity flows from the cells through the bulb and back to the cells.

This diagram shows the inside of a dry cell. The positive and negative terminals of the cell are made of carbon and zinc. An electrolyte separates the two terminals. When the battery is attached to a device such as a light bulb, the electrolyte reacts with the negative terminal of the battery to release electrons. The electrons flow from the negative terminal through the bulb and back to the positive terminal. Remember only the electrons move, so the flow has to be from where there are more electrons to where there are fewer electrons. The current will flow until the chemicals in the cell are used up. When this happens, you will be in the dark.

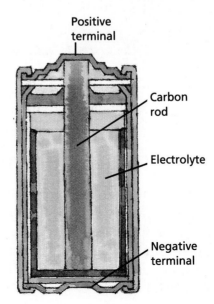

Positive terminal

Carbon rod

Electrolyte

Negative terminal

Figure 13–8. A cross section of a dry cell

How much more energy is in a lightning bolt than in a flashlight battery? Lightning is hard to measure, but electric current can be measured with an instrument called an *ammeter*. The ammeter determines how much charge passes through it every second. This movement, or current, is measured in units called *amperes* (A). A device using 2 amperes is using electricity twice as fast as a device using 1 ampere. Twice as many electrons flow through the 2-ampere device every second.

The current produced by a battery flows steadily in one direction from the negative terminal to the positive terminal. This kind of current is called **direct current (DC).** Does this mean that a power station is nothing more than a gigantic battery? No. There are other sources of current besides batteries. Your electric company produces electricity with generators. This kind of current is not DC. Instead of flowing in one direction, the current repeatedly changes direction. The current is alternating its direction, so it is called **alternating current (AC).**

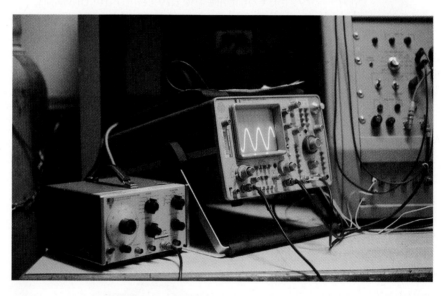

Figure 13–9. This oscilloscope is being used to monitor AC current. The crests of the wave indicate current traveling in one direction. The troughs indicate current traveling in the opposite direction.

 **ASK YOURSELF**

What is the major difference between DC and AC?

Potential Difference

Sitting in the dark on a stormy night may seem fun for a while, but it can quickly become a bore. If you never got the electricity back, how would this affect your life? Electricity is used everywhere. What is there about electricity that enables it to carry energy? Perhaps an experiment with magnets can help to answer the question.

Obtain two bar magnets. Bring the south pole of one magnet close to the north pole of the second magnet. As with opposite electric charges, the opposite poles attract. Slowly separate the magnets. Do you feel the force of attraction between the poles? How do you know that energy is required to separate the opposite poles? ✐

You supplied the energy to separate the magnets. Since negative and positive electric charges attract each other, it takes energy to keep them apart. This energy is stored any time negative and positive charges are separated from each other. In ordinary matter, electrons carry the negative charge and atomic nuclei carry the positive charge. (You will learn more about this when you study Chapter 16.) Negatively charged electrons can move easily, but atomic nuclei cannot. In a battery, a generator, or a thunderhead, electrons are moved away from positively charged nuclei. When they are separated from the nuclei, the electrons have stored energy. This situation is analogous to a rock being raised above the earth. The rock has stored energy because of earth's gravitational attraction for the rock. The electrons have stored energy because of the electrical attraction between them and the positive nuclei. The rock's stored energy is called gravitational potential energy; the electrons' stored energy is called *electrical potential energy.*

In a thunderhead, the lightning flash reunites the separated charges and produces a spectacular display of light, heat, and noise—enough to shake you up. When negative electrons move toward and reunite with positive nuclei, electrical potential energy is converted into other forms of energy. In other words, moving electrons can be used to do work, just like falling water can.

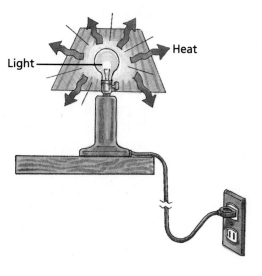

Light

Heat

Figure 13–10. When energy passes through a lamp or an appliance in the form of moving electrons, some energy is converted into heat and light.

The strength of batteries, generators, or thunderclouds is measured by a value related to the electrical potential energy of the electrons called **potential difference.** Although you can calculate potential difference if you know what the electrical potential energy is, it is a lot easier to measure it directly. Potential difference is expressed in *volts* (V) and can be measured by a voltmeter.

Potential difference is created by having extra electrons in one place and missing electrons in another. The negative terminal of a battery has extra electrons, for example, and the positive terminal is missing electrons. This can be compared to having a higher level of water on one side of a dam than the other. A battery is like an "electron dam" that keeps more electrons at the negative terminal than at the positive terminal. Just as the flow of water over a dam can be used to drive a water wheel, a flow of electrons can be used to light a flashlight bulb.

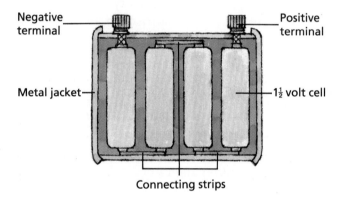

Figure 13–11. This is a cutaway view of a dry cell battery containing four 1.5-volt cells.

A dry cell has a potential difference of about 1.5 volts. A voltmeter will read 1.5 volts whether it is connected to the terminals of a flashlight battery, a large laboratory cell, or a tiny cell that operates a watch. All these cells give about the same energy to the electrons.

To get a larger potential difference, several cells may be combined to form a battery. An electron that is boosted to 1.5 volts can be sent into another cell that will raise its energy an additional 1.5 volts. This is similar to raising a rock higher and higher above the ground. The higher you raise the rock, the more energy it will have when it falls. Look at the dry cell battery shown on this page. Each cell supplies a potential difference of 1.5 volts. What total voltage is supplied by the battery?

▼ ASK YOURSELF

How many cells are contained in a 9-volt battery? Explain your answer.

Keeping Up with Current Events

You know that light is a form of energy. The rate at which energy is delivered is called *power*. The rate at which electrical energy is delivered to an appliance is the power the appliance uses. Electric power depends on two things: the potential difference (in volts) and the amount of charge that goes through the appliance each second, or current (in amperes). Power can be calculated by using the following formula.

power = current × potential difference   OR $P = I \times V$

If current is measured in amperes and potential difference is measured in volts, the power is expressed in *watts* (W). You may recall that mechanical power is also measured in watts. Try the next activity to find out more about power.

DISCOVER BY Calculating

Different appliances and light bulbs in your home use different amounts of power. Check several bulbs and appliances, including a toaster, TV, radio, and a room air conditioner, if you use one, for information on the power (watts) that they use. All of these appliances are likely to be connected to 120-volt lines. Calculate the current that each appliance and bulb uses. In your journal, make a table showing the power and current usage for each.

◤ ASK YOURSELF

How can you calculate the amount of electric energy you use in your home?

SECTION 2   REVIEW AND APPLICATION

Reading Critically

1. Compare the amount of charge per second passing a point when the current is 2 amperes and when it is 10 amperes.

2. What is the function of a battery?

Thinking Critically

3. The electric system of a small boat operates from a 12-volt battery. What do you think would happen if you used a household light bulb on this boat?

4. A toaster on a 120-volt line uses 7.5 amperes of current. How much power is the toaster using?

Electric Circuits

Objectives

Describe the parts that make up a simple electric circuit.

Compare and **contrast** series and parallel circuits in terms of the way currents and potentials are distributed in them.

Perhaps you have seen an electric train set. The engine races around a track, pulling the cars behind it. If one section of track becomes disconnected, even if the train is on a section where the track is complete, the train comes to a stop! Why does the train stop just because one piece of track is broken? The train set is still plugged in. Try the next activity to find some answers.

Figure 13–12. Plugging in the train set provides current. So why isn't this train moving?

DISCOVER BY Problem Solving

Obtain a 1.5-volt battery and holder, a switch, three wires with alligator clips, and a 1.5-volt light bulb. Connect the items in such a way as to make the bulb turn on when you throw the switch. Make a sketch of your setup. Why doesn't the bulb light if the switch is open? ✎

Making an Electric Circuit

If you were able to light the bulb in the last activity, then you made a complete path along which electricity could flow. To make current flow through the bulb, two things are needed: a potential difference and a complete circuit. The potential difference can be supplied by a battery. An **electric circuit** is the path electricity follows. A complete circuit from one terminal of the battery to the other is necessary for electricity to flow.

In the circuit shown in illustration 13–13, the electrons flow out of the negative terminal of the battery and through a wire to the switch. They pass through the closed switch into another wire, through the bulb, and back through another wire into the battery. Compare this illustration to the sketch you made when you did the last activity. Explain whether or not you made a complete circuit.

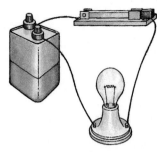

Figure 13–13. A battery, a switch, a light bulb, and some wire are all you need for an electric circuit.

Electricity has an easier time moving through some materials than others. A material through which electric current can flow is called a **conductor.** Metals, especially silver, copper, aluminum, and gold, are good conductors of electricity.

Just as some materials allow the energy flow of electricity, others prevent it. An **insulator** is a material that will not carry current. Rubber and plastic are good insulators. An electric cord contains metal wire, usually aluminum or copper, surrounded by an insulator, usually rubber. What could happen if part of the metal wire were exposed because the insulator wore away?

A switch contains a conductor and an insulator. A switch is an electric device that is used to start and stop the flow of current through a conductor. When the switch is open, no current can flow. What part of the switch is an insulator?

A good way to describe what is happening in a circuit is to use a circuit diagram. A circuit diagram is an illustration that shows how the parts of a circuit are connected. The circuit diagram shown here uses standard symbols for a battery, a switch, and a light bulb. Wire connections are indicated by straight lines. The diagram shows clearly that there is a complete conducting path around the circuit as long as the switch is closed. The next activity gives you some practice using circuit diagrams.

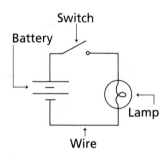

Figure 13–14. A circuit diagram consists of standard symbols that represent the actual parts of a circuit.

Discover by Doing

In your journal, make a circuit diagram of the circuit you made in the Discover by Problem Solving on page 354. Refer to the original sketch you made of your circuit. Be sure to use the standard symbols shown in Figure 13–14 as you complete your circuit diagram. ✎

If you compare the circuit diagram you've drawn with the nonworking train track, you can spot the train's problem. The same idea applies to the electric circuits in your home. The electricity supplied to your home must flow in a complete circuit, just as the electricity flows in the circuit you made. An electric cord leading to a lamp or an appliance contains two wires that carry electrons to and from the appliance.

 ASK YOURSELF

How are insulators used to make an electric circuit safe?

Current and Resistance

Several things affect how much electricity can flow through a system. One way the flow of electricity can be lessened is through resistance. **Resistance** is a measure of how much a substance opposes the flow of electricity. Resistance in conductors causes electric energy to change to heat energy. Resistance can be thought of as electric friction. Just as friction opposes the movement of any rubbing surfaces, it also opposes the movement of flowing electrons. And just as friction produces heat when you rub your hands together, resistance in circuits causes some of the energy of moving electrons to be changed to heat energy. Resistance is measured in *ohms* (Ω), named after Georg Ohm, a German scientist who studied electricity.

Different materials provide different levels of resistance. For example, a copper wire, a good conductor of electricity, has little resistance to the flow of current. It does not oppose the current, which is why copper is commonly used in wiring.

Since resistance converts electric energy to other forms of energy, such as heat or light, you may want a material that offers high resistance. For example, the purpose of a light bulb is to provide light. The filament of the bulb is made from tungsten, a resistant material. Its resistance produces so much heat that the filament glows white hot. It radiates light as well as heat as current flows through it. As resistance in a circuit increases, the flow of current through the circuit decreases. By working with the right materials, you can make electric energy do many things.

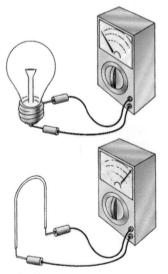

Figure 13–15. These ohmmeters check the resistance of the tungsten filament in a light bulb (top) and of an ordinary copper wire (bottom).

 ASK YOURSELF

What causes resistance?

Series Circuits

Ornamental twinkle lights are often used to decorate public places year round. At night, they twinkle, adding mystery and glamor. In the daylight, they just look like longs strings of wire and bulbs strewn about the tree branches. Years ago, these strings weren't used as much because they had a problem—one bad bulb would cause the entire string of lights to go out. It was difficult to find the bad bulb.

The reason all the bulbs went dark was that they were connected in *series*, as shown in the illustration. In a **series circuit**, there is only one path for the current to follow. The current passes through the first bulb and then goes through each bulb in turn. A burned-out bulb is like an open switch: there is no other path for the electricity to follow. Like the train tracks, if one part is unconnected, nothing happens. The current cannot pass through it, so the circuit is incomplete. In a circuit, a switch must be able to stop the flow of electricity to the device it turns on and off. For that reason, any switch must be wired in a series circuit with the device it controls.

On the Straight and Narrow The current is the same everywhere in a series circuit. All electrons must follow the same pathway. Look at the circuit diagram of a series circuit in Figure 13–17. The squiggly lines represent any sort of electric device. Ammeters are shown at several places, and the current is the same at each one.

Figure 13–16. If one bulb in a series circuit burns out, all the bulbs will go out.

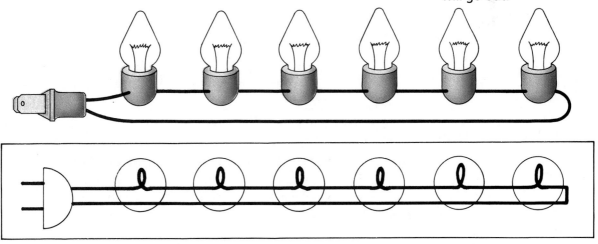

Sharing the Load As the electrons pass through each of the light bulbs in Figure 13–16, the electrons give up some of their energy. If all the bulbs are identical, they use equal amounts of energy. With six bulbs, each bulb gets one-sixth of

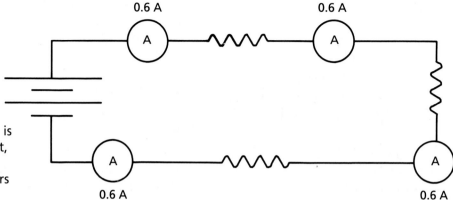

Figure 13–17. In a series circuit, current is the same throughout, as you can see by reading the ammeters shown here.

0.6 A 0.6 A

0.6 A 0.6 A

the energy of the electrons. Remember, the energy supplied by an electron is equal to the potential difference across the bulb. This is measured by connecting a voltmeter from one side of the bulb to the other. If the outlet supplies a potential difference of 120 volts, the voltmeters across each of the six bulbs will read 20 volts. You just divide the total 120 volts by the 6 bulbs that must share it to find each bulb getting 20 volts.

The potential differences in a series circuit have to add up to the potential difference of the whole circuit. Figure 13–18 shows the same circuit as Figure 13–17, but with voltmeters added to measure potential differences. Add up the potential differences across the resistances. Their sum must equal the potential difference supplied by the battery.

Figure 13–18. The potential difference within a series circuit varies with the amount of resistance. The sum of the potential differences equals the total potential supplied to the circuit.

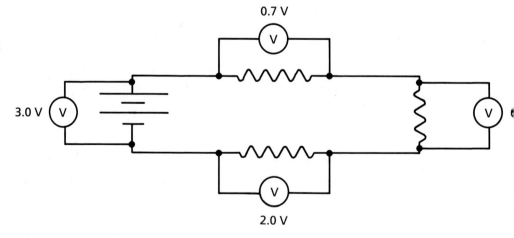

0.7 V

3.0 V

2.0 V

ASK YOURSELF

If you inserted two ammeters in a 120-volt, 1.5-amp series circuit, what would the reading on each ammeter be? Explain.

Parallel Circuits

As you've learned, a string of 100 twinkle lights in a series circuit with one bad bulb is a problem. The circuit diagram in Figure 13–19 shows how a string of ornamental lights is usually arranged today—in a parallel circuit. A **parallel circuit** has two or more separate paths through which electricity can flow. The current enters the circuit and travels to the first branching point. Some of the current goes through the first bulb. The rest continues past the branching point to the rest of the bulbs. At the second bulb, the circuit branches again. Eventually, all the current comes back together again before it goes back into the wall outlet. As you can see, if one bulb burns out, there is a path for the electricity to take around the open part of the circuit and the remaining bulbs stay lit. The electric outlets around your house and school are in a parallel circuit with each other. This is why you can use any outlet you want and it will provide the flow of electricity to power your appliance. In the next activity, you'll have a chance to make an electric circuit and to determine whether it is a series or parallel circuit.

Figure 13–19. If one bulb in this parallel circuit burns out, the others will continue operating.

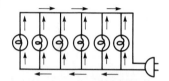

DISCOVER BY Doing

Gather two flashlight bulbs, a battery, a switch, and some wire. Put them together in such a way that both bulbs light up. Make a diagram of your circuit. Is it a series or parallel circuit? Now try to make the other type of circuit.

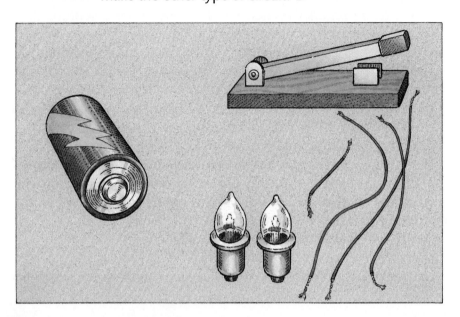

▼ **ASK YOURSELF**

As you read, the outlets around your house are wired in parallel circuits. Why is that important? How would they work if they were all wired together in a series circuit?

SECTION 3 *REVIEW AND APPLICATION*

Reading Critically

1. How could you determine whether a string of lights is in a series circuit or a parallel circuit?

2. Three devices are connected in series to a 12-volt battery. The potential differences across the first two devices are 3 volts and 5 volts, respectively. What is the potential difference across the third device?

Thinking Critically

3. Under what conditions would you want to use a wire that has a very thin copper core and an extremely thick insulating cover?

4. Three electric devices are connected in series to a 120-volt circuit. In the first device, the current is 2 amps and the potential difference is 40 volts. The potential difference across the second device is 20 volts. How much power is used by the third device?

SKILL Interpreting a Circuit Diagram

▶ **MATERIALS**
- pencil • paper

▼ **PROCEDURE**

This circuit diagram shows a battery, three electric devices, four ammeters, and two switches in a circuit. The circuit is a combination of series and parallel arrangements. Study and analyze the circuit carefully.

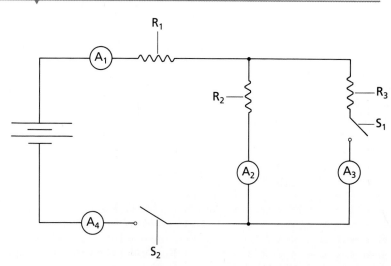

▶ **APPLICATION**

Answer the following questions about the diagram.

1. If both switches are closed, which ammeter must have the same reading as ammeter A_1?
2. If switch S_1 is opened and switch S_2 is closed, which two devices will be in a series?
3. What will the ammeters read if switch S_2 is opened?
4. Which device is rated for the same potential difference as R_2?

5. Which other ammeter reading must be added to the reading on A_3 to give the reading on A_1?
6. Which device is in a parallel circuit with R_3?

✳ ***Using What You Have Learned***

Two bulbs in a lamp are to be operated at their rated value. Both must be controlled by the same switch so that both of them are always either on or off. Draw a circuit diagram showing how the bulbs and the switch are connected to the wall outlet.

INVESTIGATION

Making Series and Parallel Circuits

▶ MATERIALS
- 6-volt lantern battery ● 6-volt bulbs (2) ● sockets for flashlight bulbs (2)
- key switches or knife switches (2) ● wire leads with alligator clips (8)

▼ PROCEDURE

1. Follow circuit diagram A to connect the bulbs to the battery in a parallel circuit. Leave the switches open.

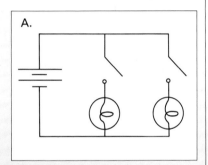

A.

2. Close one switch. What happens? How many complete

circuits do you have? List the elements in the circuit.

3. Predict the effect of closing the second switch on the brightness of the first bulb. Now close the second switch. What happens? How many complete circuits do you have?

4. Now disconnect all the wires. Reconnect all the elements in a series circuit, following circuit diagram B. Leave the switches open.

5. Predict the brightness of the bulbs in this circuit compared with the brightness of

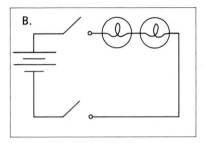

B.

the bulbs in the parallel circuit. Close one switch. What happens? How many complete circuits do you have?

6. Close the other switch. What happens? How many complete circuits do you have?

▶ ANALYSES AND CONCLUSIONS

1. Both the battery and the bulbs are rated at 6 volts. In which of the two circuits do the bulbs receive the potential difference for which they are designed?

2. When the bulbs are connected in series, what is the potential difference across each bulb?

3. What is the evidence that the voltage across each bulb in the series circuit is less than the voltage across each bulb in the parallel circuit?

4. In the series circuit, why do you have to close both switches to make the bulbs light?

▶ APPLICATION

In designing an experiment, an engineer finds that she must have exactly the same current in four different devices. How should she connect them? Explain.

✳ Discover More

Obtain a third bulb and socket. Predict the effect on the brightness of the bulbs if you include this bulb in a series circuit and then in a parallel circuit. Test your prediction.

*H*IGHLIGHTS

The Big Idea

Energy takes many forms and has many roles. Both static electricity and electric current are forms of electric energy, one more usable than the other. Electric energy involves the buildup (static) or flow (current) of electric charges. Both types are the same in that the total charge is always conserved. The electrons themselves can be neither created nor destroyed. Electricity is a transformable energy that can be put to almost limitless use.

For Your Journal

Look back at the ideas you wrote in your journal at the beginning of the chapter. How have your ideas changed? Revise your journal entry to include more characteristics of electricity. What can electricity do, and how is it used?

Connecting Ideas

This concept map shows how the ideas in this chapter are related. Copy the concept map into your journal and complete it.

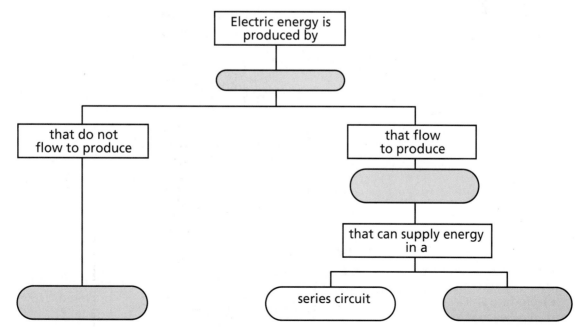

REVIEW

Understanding Vocabulary

1. Compare the meanings of the terms in each set.
   a) static electricity (343), electric current (348)
   b) direct current (350), alternating current (350), potential difference (352)
   c) resistance (356), conductor (355), insulator (355)
   d) electric circuit (355), series circuit (357), parallel circuit (359)

Understanding Concepts

MULTIPLE CHOICE

2. An electric iron is rated at 800 watts. This tells you the
   a) potential difference to which it should be connected.
   b) amount of current it takes.
   c) rate at which it uses electric energy.
   d) total amount of energy it uses.

3. When a vinyl strip is stroked with wool
   a) new negative charges are produced on the wool.
   b) new negative charges are produced on the vinyl.
   c) new positive charges are produced on the wool.
   d) the total amount of charge does not change.

4. Which unit measures the number of electrons passing through an electrical appliance per second?
   a) watt
   b) volt
   c) ampere
   d) ohm

5. If two positively charged objects are put close together, they will
   a) attract each other.
   b) have no charge.
   c) produce an electric current.
   d) repel each other.

6. The reason a switch must be placed in series with the device it controls is that
   a) when a series circuit is opened, the current stops.
   b) the switch must use some of the energy of the circuit.
   c) there must be a high potential difference across a closed switch.
   d) there must be no current passing through the switch.

SHORT ANSWER

7. Explain why a radio that works on 12 volts has to use several dry-cell batteries.

8. A microwave uses 500 watts of power on a 120-volt circuit. What is the microwave's amperage?

Interpreting Graphics

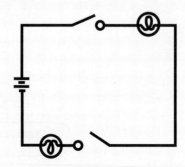

9. Look at the circuit diagram. What kind of circuit is shown?

10. How many lamps are in the circuit?

11. Are the switches open or closed? How can you tell?

Reviewing Themes

12. *Energy*
Three identical light bulbs are connected in a series circuit. What fraction of the total energy of the electrons in the circuit does each light bulb use?

13. *Systems and Structures*
Explain how resistance in conductors produces heat energy and light energy.

Thinking Critically

14. Explain how it is possible to make someone's hair stand on end by giving the person an electric charge.

15. Two lamps in a circuit carry currents of 1 ampere and 2 amperes, respectively. In what kind of circuit must the lamps be connected?

16. Explain how a toaster toasts bread.

17. How can you determine whether the charge on a pen is positive or negative, using only a vinyl strip, a piece of wool, and a suspended plastic-foam ball?

18. This transmission line carries current at a potential difference of 240 000 volts. The line is made of bare wire, kept isolated from the steel tower by ceramic insulators. Why isn't this wire covered with insulation? Birds can sit on this line and nothing will happen to them. Why is this so?

Discovery Through Reading

Elmer-Dewitt, Philip. "Mystery—and Maybe Danger—in the Air." *Time* 136 (December 24, 1990): 67–69. This article discusses evidence which suggests that exposure to electric devices can cause cancer in humans.

MAGNETISM

THE WONDER of the AGE !!
INSTANTANEOUS COMMUNICATION.

Under the special Patronage of Her Majesty & H.R.H. Prince Albert.

THE GALVANIC AND ELECTRO-MAGNETIC

TELEGRAPHS,
ON THE
GT. WESTERN RAILWAY.

May be seen in constant operation, daily, (Sundays excepted) from 9 till 8, at the
TELEGRAPH OFFICE, LONDON TERMINUS, PADDINGTON
AND TELEGRAPH COTTAGE, SLOUGH STATION.

Under the Special Patronage of Her Majesty

And H. R. H. Prince Albert
GALVANIC AND MAGNETO

ELECTRIC TELEGRAPH,
GT. WESTERN RAILWAY.

The Public are respectfully informed that this
interesting & most extraordinary Apparatus, by
which upwards of 50 SIGNALS can be
transmitted to a Distance of 280,000 MILES
in ONE MINUTE,

May be seen in operation, daily, (Sundays excepted,) from 9 till 8, at the
Telegraph Office, Paddington,
AND TELEGRAPH COTTAGE, SLOUGH.

ADMISSION 1s.

The Electric Fluid travels at the
rate of 280,000 Miles per Second.

ADMISSION ONE SHILLING.

The telegraph was such a novel invention that people
paid admission to watch the operator tap out messages.

At first people did not
know what to do with this
"wonder of the age"
except watch the opera-
tors with their needles and
wires. Soon, however, the
power of "instantaneous
communication" was to
be demonstrated in a
most impressive way.
On 1 January 1845 the
operator at Paddington
received this telegram:

*"TELEGRAPH WIRES
CATCH MURDERER!"
Those might have
been the London
headlines in 1845
when a telegraph and
its electromagnetic
needles helped catch
a dangerous criminal.
The following is a
historian's account of
how the telegraph
helped nab a murderer.*

a murder has just been commit-
ted at salthill and the suspected
murderer was seen to take a
first class ticket for london by
the trains which left slough at
7:42 a.m. he is in the garb of a
kwaker with a brown greatcoat
on which reaches nearly down
to his feet. he is in the last compartment of the second first
class carriage.

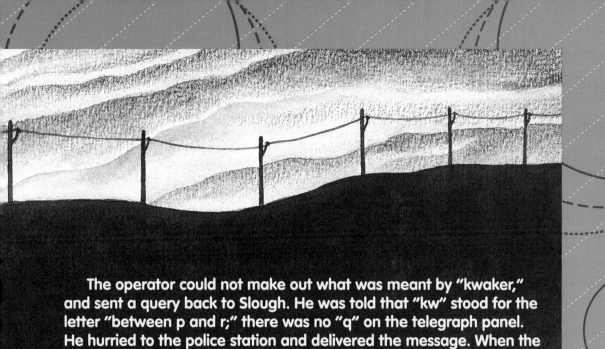

The operator could not make out what was meant by "kwaker," and sent a query back to Slough. He was told that "kw" stood for the letter "between p and r;" there was no "q" on the telegraph panel. He hurried to the police station and delivered the message. When the train arrived at Paddington, two plain-clothes policemen were on the look-out for the "kwaker." They shadowed him across London on a horse-bus, and finally arrested him.

The murder trial of John Tawell was the sensation of 1845. The policemen gave evidence how the telegraph delivered the suspect into their hands. Tawell confessed and was executed, and Londoners said: "Them cords [the telegraph wires] have hung John Tawell!"

from *A History of Invention*
by Egon Larsen

The first transatlantic cable successfully carried a message from Queen Victoria of England to U.S. President James Buchanan.

Permanent Magnets

Objectives

Describe the properties of a magnet.

List some uses of permanent magnets.

Use a model to explain the properties of magnets.

When Christopher Columbus first set out on his voyages, he was looking for a shortcut to Asia. He didn't even know he was lost when he found the Americas. How did any of these early explorers find anything?

The explorers did have a few things going for them. One of these was the lodestone, a natural magnet that was used on sailing ships as part of a compass. By looking at the compass, Columbus knew immediately which direction was north, even if he didn't know precisely where he was.

Figure 14–1. Columbus used a compass as he sailed west in search of Asia.

A Magnetic Attraction

A compass is just a small magnet. A magnet is a material that attracts certain materials, such as iron, nickel, and cobalt. Any bar magnet that is free to rotate will act like a compass. The magnet will swing back and forth until it comes to rest lined up approximately north and south. For example, a bar magnet demonstrates magnetic force, or **magnetism,** a force of repulsion or attraction between like or unlike poles. The end that points north is called the north-seeking pole, or north pole, and is marked "N." The south-seeking pole, or south pole, is marked "S." You can investigate the properties of magnetism by trying the next activity.

Obtain two bar magnets and some iron nails. Observe how the two north poles of the bar magnets behave when brought near each other. Do the same with the south poles. Now see what happens when a north pole and a south pole are brought close together. Pick up a single nail with one pole of a magnet so that only one end of the nail is attached to the magnet. Then try picking up a second nail by touching it with the hanging end of the first nail. What can you infer about a nail that is touching a magnet? Record your observations in your journal. ✏

In the activity, you should have discovered that an N pole and an S pole attract each other. On the other hand, the N pole of one magnet repels the N pole of another. Likewise, an S pole repels another S pole. Magnetic force both attracts and repels.

▶ ASK YOURSELF

Why are the poles of a magnet marked "N" and "S"?

Making Magnets

Where would an explorer be without a compass? He or she would probably be lost! If Columbus had lost his compass, could he have made a new one? Yes, if he knew what materials he needed to make a magnet. Try making a magnet yourself.

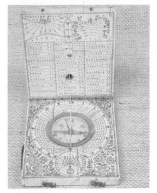

Figure 14–2. A variety of historical compasses

Obtain a bar magnet, an iron nail, and some paper clips. Stroke the iron nail with one end of the bar magnet. Stroke the nail only in one direction. Do this at least 50 times. Then see if the nail will attract the paper clips. How many paper clips can you lift with the nail?

As you discovered, some metals, such as soft iron, can easily be changed into magnets. However, because they also lose their magnetism easily, they are known as *temporary magnets.* Harder metals, such as steel, are more difficult to magnetize but tend to keep their magnetism better. A magnet made of materials that keep their magnetism is called a *permanent magnet.* Permanent magnets are made of a mixture of iron, aluminum, nickel, and cobalt.

Ceramic magnets are made of mixed oxides of iron and other metals. They are light in weight, and they break easily. Soft, flexible magnets, such as the ones you may have on your refrigerator, are made of powdered magnetic ceramics embedded in rubber.

How might Columbus have made a magnet that would stay magnetized long enough to get him home? In the following activity you can find out.

Figure 14–3. This map shows the location of Magnesia, where lodestone, also known as magnetite, was discovered. Lodestone is a naturally occurring magnetic material made of mixed oxides of iron.

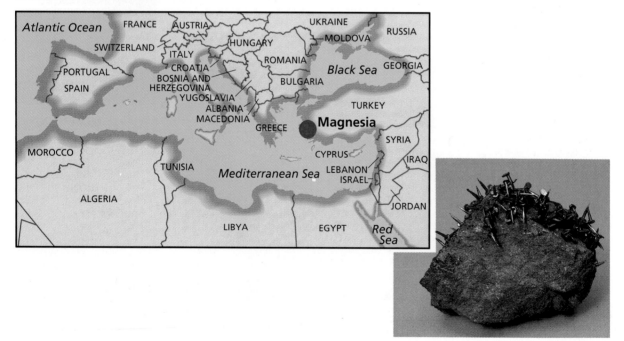

ACTIVITY

How can you make a permanent magnet?

MATERIALS

hacksaw blade, mounted compass needle, permanent magnet, paint

PROCEDURE

1. Test the hacksaw blade to see whether it is magnetized. You can do this by bringing one end of it near the compass needle. If it attracts, try the other end. Magnetization is indicated when you find repulsion.
2. Stroke the hacksaw blade in one direction only against one of the poles of the permanent magnet. Test it again to see whether the blade is now magnetized.
3. Using the compass needle as a reference, determine which end of the blade is the N pole. Mark the N pole with a spot of paint. (Remember, the N of the compass is the north-seeking end and will be attracted to the south pole of your hacksaw-blade magnet.)
4. Break the hacksaw blade in half. Mark the end that is closest to the blade's N pole.
5. Test the piece of blade with the marked pole against the compass needle. What kinds of poles does it now have?
6. Test the other half of the hacksaw blade.
7. Repeat steps 4, 5, and 6.

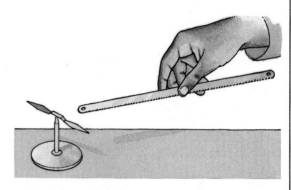

APPLICATION

1. In your journal, describe the poles that are always present after you have broken the magnetized blade. Label all the ends of the broken pieces "N" or "S."
2. Why is it necessary to use something made of steel for this demonstration rather than iron or copper?

 **ASK YOURSELF**

What is magnetism?

Inside Magnets

If you dip a bar magnet into a box of small iron nails, it will come out with nails clinging to it. Notice in the photograph that many nails are clinging to one end of the bar magnet.

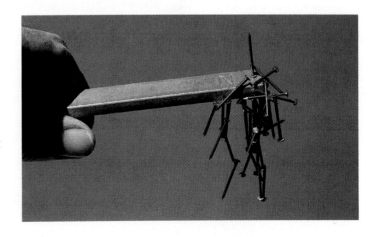

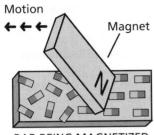

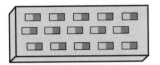

BAR BEING MAGNETIZED

BAR MAGNETIZED

Figure 14–4. Iron bars contain atoms that have magnetic poles. When a magnet is brought close to an iron bar, the poles of the atoms all point in the same way.

The model shown here explains why a magnet's force is strongest at the poles. Within a magnetized rod, there are many N and S poles facing each other. They attract each other but have little effect on any outside object. At one end of the magnetized rod, there are many N poles that are not facing any S poles; this produces a strong N pole at the end of the rod. At the other end, lots of S poles face nothing but air. This end becomes the S pole of the rod. If the rod is cut, its atoms, still lined up in the same way, will continue to produce N and S poles at the ends of the shorter pieces.

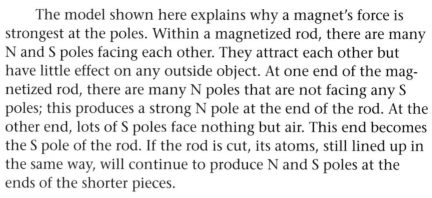

ASK YOURSELF

Why does every magnet have two poles?

Magnets in Use

Columbus needed a magnet for his compass, and of course you need magnets to hold notes and coupons on the refrigerator. Are magnets really that important or useful today?

You may be surprised to find out that a magnet might have provided you with a quick, hot meal recently. Every microwave oven contains a magnet. The poles of the magnet face each other, and a vacuum tube that produces electrons is between the poles. The magnetic field causes the electrons to follow

a circular path. As they travel, they lose kinetic energy. This energy is used to produce the microwaves that cook food.

Your home is probably full of magnets, and you don't even know it. Electric clocks, motors, stereos, loudspeakers, and televisions all contain magnets. One magnet that is easy to find is the magnetic catch that holds your refrigerator door closed. In

Figure 14–5. Microwave ovens contain strong magnets that guide the flowing electrons that produce the microwaves.

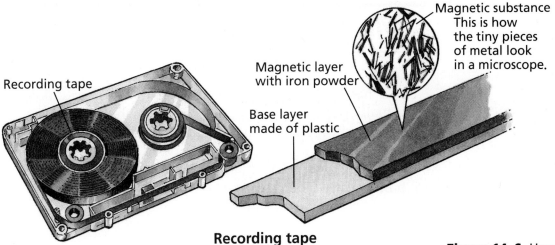

Recording tape

Magnetic substance
This is how the tiny pieces of metal look in a microscope.

Magnetic layer with iron powder

Base layer made of plastic

Recording tape

Figure 14–6. How magnets are used in recording tape

fact, magnets are much more important today than they were in Columbus's time. Try the following activity to get an idea of how you would be affected if all magnets suddenly disappeared.

DISCOVER BY Writing

Think about the uses that you just read about for magnets around the home. Find other ways that you use magnets in your home. Then, in your journal, write about an imaginary day when none of the magnets in your home worked. How were you affected? What discoveries about magnetism did you make?

▶ ASK YOURSELF

What are some uses for magnets in the home?

SECTION 1   REVIEW AND APPLICATION

Reading Critically

1. If you take a bar magnet and cut it into three pieces, what is the total number of magnetic poles you will have? Explain.

2. What kinds of magnetic poles repel each other? What kinds of poles are attracted to each other? Why?

Thinking Critically

3. Suppose you want to test an iron rod with a compass to find out whether it is magnetized. Why would you have to show repulsion before you can be sure the rod is a magnet?

4. Suppose you had a bar magnet and you drilled a hole in it. Into the hole you put a compass. Toward which pole of the magnet would the N pole of the compass point? Explain your answer.

5. Why should you never expose a magnet to high temperatures?

Magnetic Fields

Explain *what is meant by a magnetic field and how it is detected.*

Describe *the magnetic field of the earth.*

Diagram *the magnetic fields around different kinds of permanent magnets.*

What is so important about north? Columbus didn't actually want to go north, but that is where the compass always pointed. Why do compasses point north? What is attracting the magnet in the compass? It must be something big that could make every compass everywhere point north. There really is only one answer. Only the earth itself could affect all magnets everywhere.

The Magnetic Earth

When Columbus returned from his voyages, he noted that in the lands he had visited, his compasses did not point directly north. The "north" that Columbus meant was north according to the direction a compass in Europe pointed. It seemed that north was a somewhat different direction in the lands Columbus visited.

Today it is known that compasses do not point to the *geographic* North Pole. Earth's axis, the imaginary line it spins around, goes through the geographic North Pole. Compasses point to Earth's north *magnetic* pole. This point is in northern Canada more than 2000 km away from the geographic pole. And the direction to the north magnetic pole depends on your location on Earth. You can see what difference, if any, exists between true north and magnetic north for your location by doing the following activity.

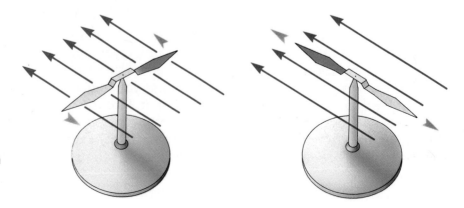

Figure 14–7. The needle of a compass will stop turning only when it is aligned with the earth's magnetic field.

Obtain a compass. Use it outdoors at night to find magnetic north. Use a stick to mark the direction of magnetic north. Look at the diagram shown here, and then find Polaris, the North Star. Polaris is directly over the geographic North Pole, or true north. Over the stick that indicates magnetic north, place a second stick to mark the direction of geographic north. Compare the two sticks. Use a protractor to measure the number of degrees difference between magnetic north and geographic north. Write your measurements in your journal. What do you know now about the difference between magnetic north and geographic north? ✐

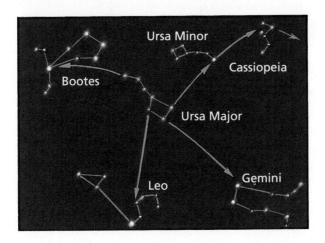

The Two Norths

In many places the difference between the two norths is so slight that it can be ignored when using a compass. But in some places the difference is great enough to cause problems in navigation. Imagine how annoyed explorers Admiral Peary and Matthew Henson would have been if they had not reached the geographic North Pole but the magnetic north pole—a difference of 2000 km! Fortunately, maps used for navigation and in hiking usually indicate this difference so that adjustments can be made when choosing a direction.

Feeling the Pull

No matter where you are on Earth, something causes the N pole of a compass to point one way and the S pole to point the other way. Any region in which magnetic forces are present is a **magnetic field.** The earth itself is surrounded by a magnetic field. When a compass needle comes to rest, its poles are lined up with Earth's magnetic field.

Magnetic fields are represented in diagrams by magnetic lines of force. These lines identify the position and strength of

Figure 14–8. A compass rose shows magnetic north.

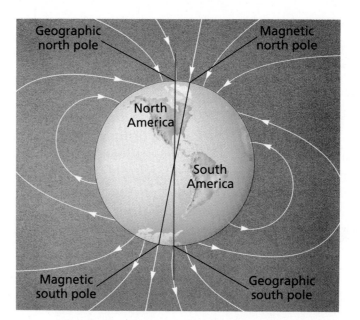

Figure 14–9.
Geographic north and magnetic north are in different locations. Geographic north is at the earth's point of rotation.

the magnetic field around an object. Figure 14–9 shows magnetic lines of force drawn to show the magnetic field around Earth. They tell which way the N pole of a compass would point from any given location. They also tell how strong the field is. The field is strongest where the lines are closest together. Where is Earth's magnetic field strongest?

 ASK YOURSELF

How would a compass behave south of the equator? Explain your answer.

The Fields Around Magnets

If all magnets are affected by the earth's magnetic field, then what happens to each magnet's own magnetic field? If you move a compass around a magnet, the compass needle turns. This happens because the compass is affected by the magnetic field produced by the magnet. At any location in the field, the compass needle comes to rest in line with the magnetic lines of force. Look at the photograph that shows several compasses around a bar magnet. Notice that each needle points toward a pole of the bar magnet.

Figure 14–10. The needle of each compass shows the direction of the magnetic field around the magnet.

Like the earth's field, the magnetic field around a bar magnet can be represented by magnetic lines of force. These lines are not visible, yet they can be used to predict which way a compass needle will point. You can investigate the magnetic lines of force around a bar magnet in the following activity.

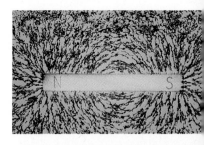

Discover by Doing

Obtain a bar magnet, a sheet of paper, and some iron filings. Place the bar magnet under the sheet of paper, and sprinkle iron filings on the paper. Gently tap the paper so that the iron filings line up with the magnetic lines of force of the magnet. In your journal, make a sketch showing the pattern made by the iron filings. ✏

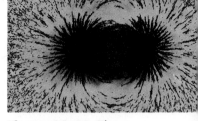

Figure 14–11. These iron filings show the magnetic field that surrounds the magnet.

Is the Force Field Really There?

The magnetic force of the bar magnet made little magnets of the iron filings. These magnetic filings lined up, forming long, thin chains in the magnetic field of the bar magnet.

The illustration below shows a diagram of the field around a bar magnet and a horseshoe magnet. The arrowheads show the direction of the magnetic lines of force, which come out of the N poles of the magnets and enter the S poles. The concentration of magnetic lines of force at the poles shows that the field is strongest there.

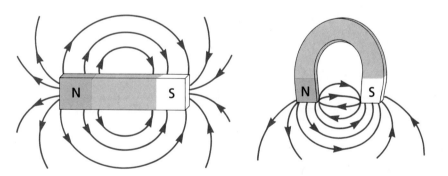

Figure 14–12. These diagrams show the magnetic lines of force around a bar magnet and around a horseshoe magnet.

Do Not Cross!

Magnetic lines of force never cross each other. When two or more magnets produce fields that overlap, the result is a combined field. Figure 14–13 shows the fields produced when the poles of two bar magnets are brought near each other. If opposite poles are aligned, most of the lines run from the N pole of one to the S pole of the other. Investigate the overlap of magnetic fields for yourself in the next activity.

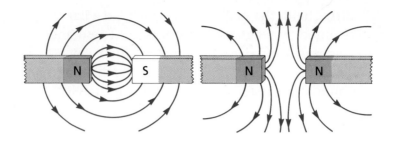

Figure 14–13.
Magnetic lines of force never overlap.

Obtain two bar magnets, a sheet of paper, and some iron filings. Place the bar magnets on a table with their north poles facing each other about 2 cm apart. Place the paper over the magnets. Sprinkle iron filings on the paper. Tap the paper so that the filings align along the magnetic lines of force. In your journal, make a sketch of the magnetic fields produced by the two magnets. Compare this sketch with the one that you did earlier. How do they differ? ✎

What were you able to tell about the magnetic lines of force of the two bar magnets? Try placing the bar magnets in different locations and determining the magnetic fields. Be sure to report your findings in your journal.

▸ **ASK YOURSELF**

From what parts of a magnet do most of the magnetic lines of force extend? Why?

SECTION 2   *REVIEW AND APPLICATION*

Reading Critically

1. If a bar magnet were bent into the shape of a horseshoe, how would its magnetic field be changed?

2. Why doesn't a compass point to geographic north?

3. How can you tell by looking at magnetic lines of force where the field is strongest?

Thinking Critically

4. How do scientists know that the earth's north magnetic pole is an S pole?

5. The earth's magnetic field is much like the field of a bar magnet about 2200 km long. Where would such a magnet have to be located to produce the earth's magnetic field? Explain your answer.

SKILL Constructing Models

▶ **MATERIALS**
● paper ● pencil

If you understand what a magnetic field is and how magnetic lines of force are used to represent the field, you should be able to construct on paper the magnetic lines of force that exist around a magnet.

▼ **PROCEDURE**

1. Choose a title for your diagram.
2. Then draw the shape of your magnet. Label one end "N" and the other "S."
3. Start by drawing short lines, about 1 cm long, from the surface of the magnet. Follow these guidelines as you work:
   - The lines should be crowded together at the poles, with about half of the total number of lines coming straight out from the poles. The rest of the lines touch the sides of the magnet.

- Draw arrowheads on the lines, showing that the field comes out of the N pole and goes into the S pole.
- Complete the lines, making them run from one pole to the other. Remember that the lines are smooth curves. Lines must never cross each other. All lines spread out from each other as they go away from the magnet.

▶ **APPLICATION**
Using the information in the procedure, draw and label a diagram of a bar magnet and a horseshoe magnet. There is one important difference between the two. In a horseshoe magnet, some of the lines from the poles travel through the arch of the magnet.

✳ *Using What You Have Learned*
Use the skills you learned in this activity to construct a diagram of your school, showing the routes you travel during the course of your day.

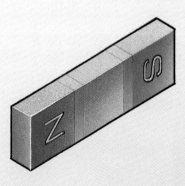

Magnetism and Electricity

Thanks to a compass and Earth's magnetic field, Columbus was able to make several voyages to the Americas. Not only was he able to get back home, but he also was able to find the Americas more than once. Compasses are still useful today, but there are more powerful magnets that are even more useful. It was 300 years after Columbus sailed the Atlantic that these powerful magnets were discovered.

Electricity Makes Magnetism

In 1820 Professor Hans Christian Oersted of Denmark was giving a lecture on electricity to his students. He happened to have a compass nearby. When he closed a switch to demonstrate the flow of current, he noticed that the compass needle turned. That seemed odd to Oersted. After all, weren't magnets affected only by other magnetic fields? Oersted realized that the electric current must be generating a magnetic field.

Look at the setup in the illustration to see how this magnetic field can be detected. Before the wires are connected to the battery, all of the compasses point in the same direction. To what magnetic field are the compasses responding? When the wires are connected to the battery, current flows in the wires, moving upward through the tube in the platform. The compasses on the horizontal platform show the direction of the magnetic field at various points. The magnetic lines of force are shown in red. They form circles around the wire. Notice what happens when the connections to the terminals of the battery are reversed. How would you explain this in terms of magnetic poles?

Figure 14–14. Compasses are used here to detect the magnetic field around an electric current. Notice how the compasses behave when the battery is disconnected or connected as shown.

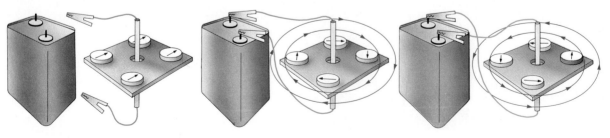

Cranking It Up   To produce an even stronger magnetic field, the wire can be wound into a spiral. A wire spiral through which a current can flow is called a **solenoid.** When a current flows through a solenoid, the magnetic fields produced by the coils of wire combine. The result is a strong magnetic field in the center of the solenoid. Increasing the number of coils in the solenoid increases the strength of the magnetic field. Increasing the current flowing through the coils also strengthens the magnetic field.

Look at the photograph of a solenoid and the diagram of its magnetic field. If you compare the solenoid's magnetic field with the field around a bar magnet, you will see that the fields are exactly alike. The lines of force come out of the N pole of a solenoid and go in at the S pole.

Electromagnets   If an iron core is placed inside a solenoid and a current is passed through the coil, the field of the solenoid acts on the atoms in the core. They line up with the solenoid's field, making the field hundreds of times stronger. A solenoid with a core is called an **electromagnet.** If the core is made of a material that can be made into a permanent magnet, the solenoid's field will turn it into one.

Walking on the ceiling, like the person in this photograph, might be fun, although it is not particularly useful. Turning a magnetic field on and off at will is useful, though. That's really one of the things you're doing when you turn on an appliance. Electromagnets operate motors that run appliances. Electromagnets are also used in television sets, circuit breakers, and meters that measure the use of electric energy.

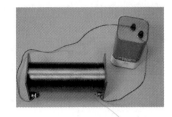

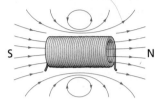

Figure 14–15. When current passes through a solenoid (top), the current produces a strong magnetic field (bottom).

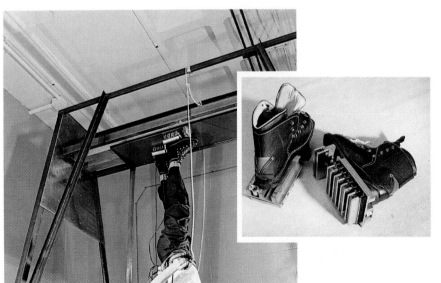

Figure 14–16. When electric current is passed through the electromagnets in these boots, they produce a magnetic field strong enough to allow a person to walk on the underside of a steel beam.

Figure 14–17.
The ability to turn magnetism off and on has many uses.

To see a powerful electromagnet in action, visit a junkyard. The crane that moves junked automobiles from one place to another has a large, strong electromagnet. You can make a small electromagnet in the following activity.

DISCOVER BY *Doing*

Obtain a 6-volt dry cell, an iron nail at least 15 cm long, and about 75 cm of insulated bell wire. Remove some of the insulation from each end of the wire. Leaving about 20 cm of wire free at one end, begin wrapping the wire around the nail, starting just under the head. When the wire is about 2 cm from the nail's tip, begin wrapping upward, toward the head. Wrap wire about halfway to the top of the nail. Connect the ends of the wire to the dry cell. Get some paper clips and other metallic objects. See how strong your electromagnetic crane is. ✐

So, how successful were you in building an electromagnet? Ready to go into the junk business? Or are you ready to junk your electromagnet?

◀ ASK YOURSELF
What is a solenoid?

Magnetism Makes Electricity

If electricity can create a magnetic field, can a magnetic field create an electric current? Your local energy plant can demonstrate that relationship.

In electric generating plants, huge generators are operated by engines. These generators make the electric energy network function and can be used to produce an electric current.

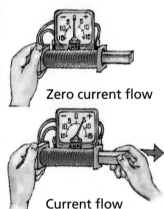

Zero current flow

Current flow

Figure 14–18. Moving a bar magnet within a coil of wire produces an electric current by electromagnetic induction. When the bar is stationary, no current is induced.

The process by which a magnetic field produces an electric current is called **electromagnetic induction.** You can see this process at work by moving a bar magnet through the coils of a solenoid. As shown in the photograph and the illustration, if the ends of the solenoid are connected by wires to a sensitive ammeter, the ammeter will indicate the current in the solenoid.

Current will be induced in the solenoid whenever there is a *changing* magnetic field in the solenoid. If you move a bar magnet into the solenoid, the ammeter will show a current as long as the magnet is moving. As soon as you stop moving the bar magnet, the current will stop. Thus it is not just a magnetic field but a changing magnetic field that induces the flow of electricity. The greater the number of coils in the solenoid, the stronger the current that will be induced. The more powerful the magnet, the greater the current that will be induced.

 ASK YOURSELF

How can a magnetic field be used to produce electric current?

SECTION 3   REVIEW AND APPLICATION

Reading Critically

1. How might electromagnets be useful in industry?
2. How does an electromagnet demonstrate that an electric current produces a magnetic field?
3. Why does an iron core make the field of a solenoid so much stronger?

Thinking Critically

4. You find that an electromagnet you are using is not strong enough. Suggest two ways in which you might make a stronger electromagnet.
5. When an electromagnet is at rest inside a solenoid, no current flows in the solenoid. If the core is pulled out of the electromagnet, current will flow in the solenoid while the core is being removed. Explain why this happens.

INVESTIGATION

Studying Electricity and Magnetism

▶ **MATERIALS**
- nail ● wire, 1.5 m ● batteries (2) ● wire connectors (2)
- flashlight bulb and socket ● paper clip ● compass ● tape

▼ **PROCEDURE**

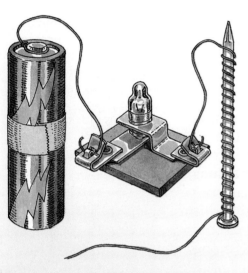

1. Wrap 30 turns of wire around a nail and set up an electric circuit as shown in the diagram. Before making the final connection, bring the nail head near a paper clip. Does anything happen to the clip?

2. Now make the final connection to complete the circuit. How can you tell that a current is flowing?

3. While the current is still flowing, bring the nail near the paper clip again. What happens to the clip this time?

4. Next, with the circuit still complete, bring the head of the nail near the compass. Does it attract or repel the south pole of the compass needle?

5. Repeat step 4 using the point of the nail. What happens this time?

6. Remove the wrapped nail from the circuit and set it aside.

7. Now wrap some fine insulated wire around the compass about 30 times along its north-south axis. What would you call the device you just made?

8. Line up the compass needle under the wire and connect the ends of the wire to the battery. What happens to the needle when the circuit is completed?

9. Reverse the connections on the battery. What happens to the needle this time?

▶ **ANALYSES AND CONCLUSIONS**

1. What does the wire-wrapped nail become once it is connected to an electric circuit?

2. What was demonstrated by wrapping a coil of wire around the compass and connecting it to the battery? What does this confirm about electricity?

▶ **APPLICATION**

Using the materials in this Investigation and a piece of thread, explain how you could make a compass.

 Discover More

A galvanometer contains a coil of wire and a permanent magnet. When it is connected to a source of current, its needle is deflected to one or the other side of zero. Explain how this works.

HIGHLIGHTS

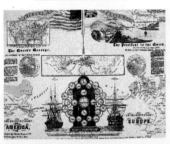

The Big Idea

The discovery of magnetic forces led to many useful applications. Magnets are used in many technologies including robotics, manufacturing, engineering, and the television and recording industries. Electric currents produce magnetic fields, and changing magnetic fields produce electric currents. This interaction between electricity and magnetism plays a key role in many systems. The magnetic field produced by an electromagnet operates doorbells, motors, and many other electrical devices. The electric current produced by changing magnetic fields is the primary source of electricity for homes and industry.

Look back at the list of uses for magnets you wrote in your journal. What new uses can you add to your list as you complete this chapter? You may wish to do research on electromagnets and continue to add to your list.

Connecting Ideas

This concept map shows how ideas about magnetism are related. Copy the concept map into your journal, and extend the concept map to include ideas that link magnetism and electricity.

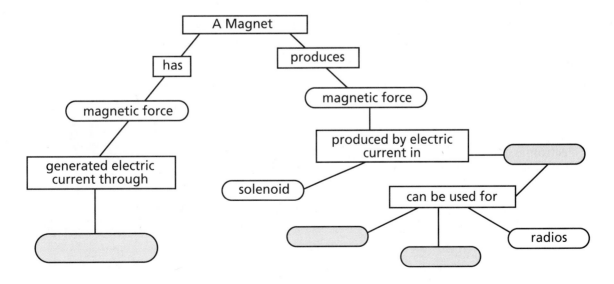

A Magnet

has

produces

magnetic force

magnetic force

generated electric current through

produced by electric current in

solenoid

can be used for

radios

REVIEW

Understanding Vocabulary

1. Compare and contrast the meanings of the terms in each set.
   a) magnetism (368), magnetic field (375)
   b) solenoid (381), electromagnet (381), electromagnetic induction (383)

Understanding Concepts

MULTIPLE CHOICE

2. The N pole of a compass needle points to the north magnetic pole of Earth because Earth's magnetic pole is
   a) a positive pole.
   b) an S pole.
   c) an N pole.
   d) a negative pole.

3. If the poles of two magnets repel each other,
   a) both poles must be S poles.
   b) both poles must be N poles.
   c) one is an S pole and the other is an N pole.
   d) both poles are either S or N poles.

4. Magnetizing a piece of iron is a process by which
   a) the poles of existing atoms are brought into line.
   b) magnetic atoms are added to the iron.
   c) each atom in the iron is converted into a magnet.
   d) magnetic lines of force are brought into line.

5. An electric motor uses an electromagnet to change
   a) mechanical energy into electrical energy.
   b) magnetic fields in the motor.
   c) magnetic poles in the motor.
   d) electrical energy into mechanical energy.

6. The ability of any object to generate a magnetic field depends upon its
   a) size.
   b) location.
   c) composition.
   d) direction.

SHORT ANSWER

7. You bring a piece of steel near a compass and find that the steel attracts the N pole of the compass. Explain why this does not prove that the steel is magnetized.

8. Why does an iron core make the magnetic field of a solenoid stronger?

Interpreting Graphics

9. Write a statement that might be a conclusion you can draw from Figure A.

10. Write a statement that might be a conclusion you can draw from Figure B.

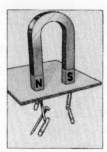

A.

B.

Reviewing Themes

11. Environmental Interactions
In order to produce an electric current, what must be true of the magnetic field in a solenoid?

12. Systems and Structures
Suppose a salvage yard has an electromagnet attached to the end of a crane. An automobile that has been crushed for scrap metal must be placed in a railroad car. Explain how the crane operator can move the crushed automobile from the ground and place it in the railroad car for transport.

Thinking Critically

13. Why do you think a compass does not work at the magnetic poles of Earth?

14. Many libraries use a security system based on magnetism for checking out books. Magnetic strips are put into each book. When books are checked out, they are demagnetized by being run across or through a device that sits on the check-out desk. Then the books are again magnetized when they are returned. (A sensor system in the door buzzes when books are taken from the library without this processing.)

Why do you think library audio cassettes and videotapes are generally not run through this magnetizing/demagnetizing system?

15. On this wind farm, every windmill is attached to a generator. In each generator, a coil spinning in a magnetic field produces electricity. Other methods of producing electricity are also being tried. What do you think are the advantages and disadvantages of using the wind to make electricity?

Discovery Through Reading

Normile, Dennis. "Superconductivity Goes to Sea." *Popular Science* 241 (November 1992): 80–85. The *Yamato 1*, an ocean-going ship with a magnet-driven propulsion system, took its first voyage in the summer of 1992. Read this article to find out how it works.

ELECTRICITY IN USE

*I*f the supply of electricity to your home fails for some reason, all the lights go out. You probably fumble to find a flashlight.

Now imagine such a blackout darkening the entire northeastern United States during evening rush hour. And imagine all of New York City waiting for the lights to come on. In 1965, it really happened!

On November 9, 1965—in Buffalo, in Albany, in Boston, in New York City—the clocks in the Megalopolis sputtered to a standstill. Lights blinked and dimmed and went out. Skyscrapers towered black against a cold November sky, mere artifacts lit only by the moon. Elevators hung immobile in their shafts. Subways ground dead in their tunnels. Streetcars froze in their tracks. Street lights and traffic signals went out—and with them the best-laid plans of the traffic engineers. Airports shut down. Mail stacked up in blacked-out post offices. Computers lost their memories. TV pictures darkened and died. Business stopped. Food started souring in refrigerators.

For the very young and the very old, it remained a fearful night—but numberless others seemed to be enjoying themselves. Strangers helped one another in the dark and sent thank-you notes and flowers the next morning. An off-Broadway show went on, by candle-light, to a gallery of seven people and two dogs.

But the experience was also sobering to the Megalopolitans—a rather unsettling lesson in how totally their lives are wired to electricity. Black Tuesday had been an epiphany of the push-button age, a demonstration of the final vulnerability of the mightiest society on earth. Could it happen again? Maybe so, maybe not—but it was a memory that all the lights in all the cities could not dim.

from *Newsweek* magazine
November 22, 1965

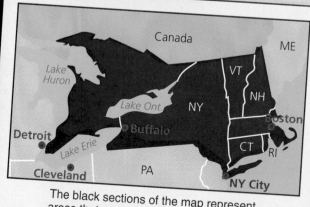

The black sections of the map represent areas that were affected by the blackout.

For Your Journal

🖊 Why is it important to be able to transform electricity for long-distance transmission?

🖊 How does an electric generator work?

🖊 List as many electrical safety features as you can think of.

The Power Grid

Objectives

Apply *the principle of electromagnetic induction to the production of electric energy in generators.*

Compare *and* **contrast** *a generator with a motor.*

Explain *why it is important to be able to transform electric potentials for long-distance transmission.*

Roberto dropped his books on the counter as he got home from school. "Ugh! I hope we move fast through this electricity stuff we are studying. It's so boring."

His father was home early from work. That was unusual. "I hope you're learning as much as you can about electricity," said his dad. "Over the next few days it might be important. By the way, pack a bag with clothes to last you several days."

"Wow!" Roberto jumped, "Are we going on vacation?"

Roberto's dad looked at him very seriously. "No, son," he replied. "We've been asked to evacuate this area because there is a hurricane headed our way."

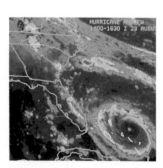

Figure 15–1. Though these electronic signals clearly showed the immensity of the storm, meteorologists could not predict how many lives and how much property would be affected once Hurricane Andrew hit the coast of Florida.

Fleeing from a Hurricane

Roberto and his family live in Florida. In August 1992, south Dade County, the area around Miami, was targeted by a fast-approaching hurricane—Hurricane Andrew.

"Roberto, bring along your science book," his dad continued. "I can help you study about electricity while we drive to your sister's in Orlando." A groan was heard from the hallway as Roberto headed for his room.

Finally, all was packed in the car, and they were ready to go. Roberto knew that hurricanes were dangerous, but he had never seen one, and he wished they could stay. His parents were very worried, but he thought all this was exciting. Just before they left the house, Roberto's dad switched off the main circuit breaker to the house. "Why did you do that, Dad?" Roberto asked.

"Just as a safety precaution," his dad replied. "I'll explain later. Let's hit the road."

Once they were underway, his dad relaxed a little and returned to the subject. "Roberto, you probably don't think about it when you flick a light switch, but do you realize how much our family depends on electricity? Think of how different your life would be if we had to live even one week without it. Why don't you read aloud from your lesson on generating electricity. We can discuss any part you don't understand. It'll be bumper-to-bumper traffic all the way to Orlando, so we'll have lots of time to talk." Roberto opened his science book and began to read.

Thousands of powerful generators stretch kilometer after kilometer across the country. They produce the electricity that is needed for use in homes, schools, and businesses. This energy is fed through a network of transmission lines called a **power grid**. You have probably seen these transmission lines in your neighborhood.

 ASK YOURSELF

How different would your life be if you had to live without electricity for one week?

Figure 15–2. These electric power lines make up a small fraction of the power grid that feeds electric current to all parts of the United States.

Generators

Generators use electromagnetic induction to produce electricity. Recall that this process was explained in the previous chapter. Faraday's law of induction summarizes the principle. This law states that a current may be induced in a loop of wire by changing the magnetic field that is passing through the loop. The illustration shows how an electric generator works. In this case, the loops of the wire coil are moved through the magnetic field between the magnet's north and south poles. Whether the loop or the magnet does the moving, the effect is the same: a current flows through the wire.

Figure 15–3. A simple generator

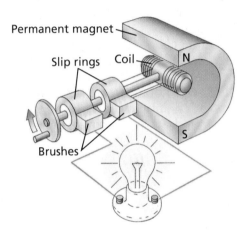

A simple generator consists of a loop of wire or a coil of several loops placed in a magnetic field. The coil shown in Figure 15–3 is between the north and south poles of a magnet. As the coil rotates, it cuts through the magnetic lines of force. This results in a current flowing through the wire. If the loop spins continuously, the orientation of the coils in the magnetic field changes constantly. Current flows in the loop, changing direction every half cycle. The flowing current passes through the slip rings and brushes and is then transmitted to the power user. In the diagram, the power user is the light bulb. A much larger generator would transmit the electric current through the power grid of an entire city or region.

Figure 15–4. Huge generators are used in electric energy plants to produce the electric current used in homes and in industry.

Figure 15–5. Small generators are available for home use in case of power failure or in rural areas without access to a power grid.

Roberto interrupted his reading to ask, "Dad, you bought a generator for the house this week, didn't you?"

"Yes," his father replied. "I thought we might need one if the hurricane came close to Miami and the power went out. I thought we could run some lights with it."

Roberto looked at the diagram of the electric generator in his science book again. "Dad, this generator reminds me of Grandma's hand-cranked ice-cream maker. Would we have to take turns cranking a handle to keep the coil in the generator rotating?"

Roberto's dad chuckled, "No, it has a small gasoline engine that will do that for us. It will run for about eight hours on one tank of fuel."

DISCOVER BY *Doing*

Obtain a small toy motor, two wires, and a galvanometer. A galvanometer is used to measure current. Attach one end of each wire to the terminals of the galvanometer. Then attach one of these wires to one lead of the motor and the other to the other lead. Spin the motor shaft between your fingers, first in one direction and then in the other. What do you observe about the galvanometer's readings? What have you turned the motor into? What energy changes are taking place? Record your answers in your journal. ✎

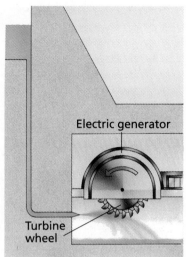

Figure 15–6. The generators of a hydroelectric-energy plant are operated by large turbines. The force of the water draining from the dam causes the turbines to spin.

Electric generator

Turbine wheel

Most home generators produce from 200 to 5000 watts (W) of power. The large generators in power plants may produce 34 000 kilowatts (kW) or more. In these large generators, the coil is stationary and the magnetic field rotates. In most large power plants, a burning fuel, such as coal, or a nuclear reaction is used to turn water into steam. The moving steam causes blades in a turbine to spin. This in turn moves the magnets. In a hydroelectric plant, falling water causes the turbine blades to spin.

 ASK YOURSELF

In a generator, how is current in a wire produced?

Motors

"Dad," Roberto asked, "How can a motor run our generator to make electricity when we need electricity to run the motors in things like the refrigerator, the vacuum cleaner, and the food processor?"

"What you call the motor on the generator is really a small engine," said his dad. "People usually use the term *motor* when the power source is electricity. Actually, an electric motor works a lot like a generator—but in reverse. There's probably something about it in your book. Keep reading."

A diagram of an electric motor would appear very similar to that of a generator. An **electric motor** is a device that converts electric energy into mechanical energy, or movement. A generator does the opposite: it turns mechanical energy into electric energy. That is, a generator uses the circular movement of a wire to produce a current, while a motor uses a current to produce a circular movement.

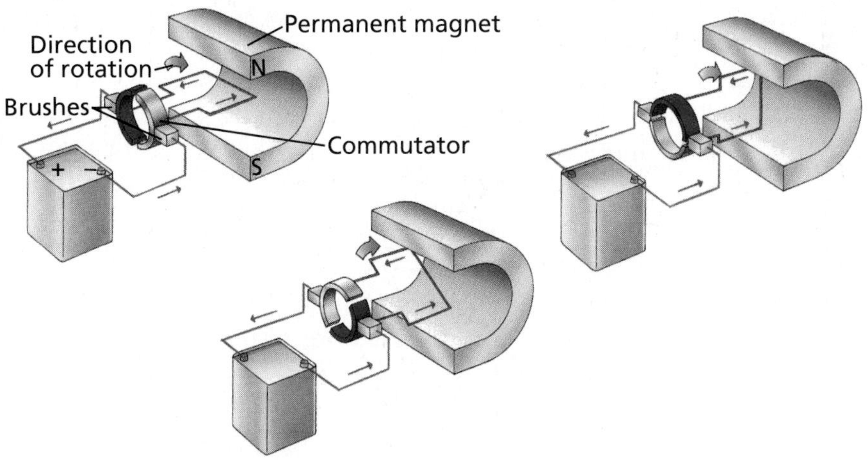

Figure 15–7. A simple electric motor

Direction of rotation
Permanent magnet
N
Brushes
Commutator
S
+

Electric motors operate as a result of the interaction of two magnetic fields. In the simple motor shown in the illustration, there is a permanent magnet with a wire loop between its poles. As a current moves through the wire, it creates an electromagnet that generates its own magnetic field. The magnet exerts an upward, attractive force on one side of the loop and pushes down on the other side of the loop. This makes the loop rotate. The loop acts in the same way as a compass needle. It aligns itself with the magnetic field of the magnet.

In order to be a motor, the loop must turn all the time. The loop in a motor keeps turning because the current flowing into the loop keeps changing direction. Circuits in homes and businesses are supplied with alternating current (AC), which reverses continuously. Many appliances contain AC motors, in which the direction of the current, and therefore the magnetic poles, constantly change. This switching of magnetic poles causes the coil of wire to spin inside the magnet. This produces the motion of the shaft that protrudes from the motor.

DISCOVER BY Doing

Place a compass on top of a wire approximately 60 cm long. The wire should have alligator clips at each end. Align the wire so that it is parallel to the compass needle. Connect the ends of the wire to a 6-V battery, and observe what happens to the compass needle. Without moving the compass, switch the alligator clips so that the one attached to the positive terminal is now on the negative terminal and vice versa. What happens to the compass needle? How is this similar to what happens in a motor? 🖉

It was late at night now, and the traffic was still heavy. Even stopping for gas was a problem. So many people were fleeing north from the hurricane that the family had to wait in a long line of cars at the gas station on the turnpike. Roberto decided to pass the time by figuring out the motor diagram in his science book.

"Dad," he asked, "some of the motors we have at home run on batteries. This book even shows a motor connected to a battery. But batteries make direct current. Direct current doesn't alternate, so how can it run a motor?"

Roberto's dad examined the diagram. "Look right here," he said, pointing to Figure 15–7. "The ends of the loop are connected to a split ring called a *commutator*. Two brushes touch the commutator. The battery's current feeds in through one brush and leaves through the other. From the 'in' brush, current flows through one-half of the ring and on through the loop. The magnetic field this produces causes the loop to turn. This brings the other half of the ring into contact with the 'in' brush, sending current into the loop from the opposite direction and reversing its magnetic field. In a DC motor, the commutator changes the direction of the current. As a result, the loop spins just as it does in the AC motor."

Roberto's father continued, "A single loop of wire like the one in the diagram would carry only a small amount of current. Real motors have many loops of wire that can carry a large amount of current. Also, most motors use electromagnets to generate the magnetic field."

"I get the picture," said Roberto. "Then the spinning loop can be connected to a shaft, causing it to spin. The spinning shaft can be connected to other parts, like the blades in a mixer, causing them to turn to do work. Pretty cool."

Figure 15–8. This cutaway view shows the inside of an electric motor. Notice that many loops of wire are used instead of the single loop shown in Figure 15–7.

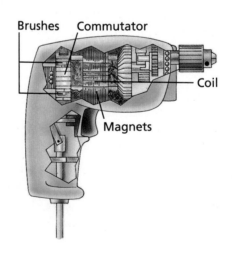

Brushes   Commutator

Coil

Magnets

 **ASK YOURSELF**

In what way does the loop of a motor behave like a compass needle?

Transformers

Roberto's family reached Orlando after a tiring seven-hour drive that would normally take only four hours. Even though it was late, Roberto and his parents could not fall asleep. They listened to the news on television and watched the weather radar that showed Hurricane Andrew moving closer to Miami. The newscaster warned people not to go out after the storm because there might be dangerous high-voltage lines lying on the ground.

Figure 15–9 Some of these lines might have an electric current of 30 000 volts (V) or more.

"Dad," Roberto said, "I don't understand something. If the electricity that gets to our house has 30 000 volts, why don't all our appliances get burned to a crisp? They all say they run on either 220 or 110 volts."

"Good question, Roberto," his Dad replied. "Have you ever heard of transformers?"

"Sure," Roberto grinned. "When I was younger, I had a car that transformed into an awesome robot!"

"No, not those transformers," his father smiled. "Let's look in your science book and read about them."

A **transformer** is a device that uses electromagnetic induction to change the electric potential (voltage) of alternating current. In practical terms this means that wire coils of different sizes can be used to either increase or decrease voltage. You may have seen transformers attached to electric poles. They look like large, gray cylinders. It would be harmful to send thousands of volts into your household circuits. Appliances would burn out, and fires would start. A *step-down* transformer decreases voltage so that electricity reaches your home in a manageable form.

Figure 15–10. This particular transformer lowers the voltage of electric current from about 30 000 V to the 120 V you receive in your home.

There are also *step-up* transformers. A step-up transformer increases voltage so it can be used for long-distance transmission. A power plant generates electricity at a relatively low voltage. A step-up transformer changes it to many thousands of volts before sending it to a city many kilometers away. This is done to reduce the amount of energy that is wasted during transmission through the wires.

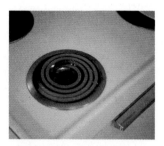

Figure 15–11. The glowing of this heating element is due to the resistance of the burner coil.

Power, Voltage, and Current   In a previous chapter you learned about power, voltage, and current. Recall that $P = V \times I$. That means that in order to increase the power, either voltage or current must be increased For example, if you wanted to generate 60 W of power, here are two possibilities:

$$P = V \times I$$

60 watts = 6 V × 10 A OR 60 watts = 10 V × 6 A

The equation is like a seesaw. When the voltage of a current is increased by a transformer, the current is reduced. Likewise, when the voltage is reduced, the current increases.

There is one problem, and that is resistance. A wire offers resistance to the flow of electrons through it. This resistance makes the wire heat up. You make use of this fact when you cook on an electric stove. The burner glows red due to the resistance of the burner coil. When electricity is sent long distances, some of the electrical energy is converted to heat energy. This heat energy is wasted since it is not put to any practical use. The amount of energy lost in transmission lines depends on the amount of current in the lines. More current in the wires means more energy is wasted due to resistance. Power companies need to limit the amount of energy wasted when electricity is transmitted over long distances. So the power is sent using very high voltage with very small currents. When the energy reaches its destination, the voltage is dropped and the current is raised so it can be used in homes and businesses.

Figure 15–12. These are examples of step-up and step-down transformers.

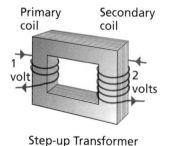

Step-up Transformer

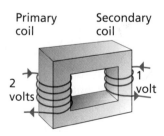

Step-down Transformer

What's Inside a Transformer?

As you can see in 15–12, a transformer is an iron core with coils of wire wrapped around it. Alternating current is sent into the primary coil. Recall that a constantly changing current will produce a changing magnetic field. The iron core magnifies this change because it has magnetic properties. This constantly changing magnetic field induces current in the secondary coil. If the secondary coil has more loops of wire than the primary coil, the induced voltage in the secondary coil will be correspondingly greater than the voltage in the primary coil. For example, if the secondary coil has twice as many loops as the primary coil, the voltage on the secondary coil will be twice the voltage on the primary coil. This is a step-up transformer. As you might infer, reversing this system produces a step-down transformer.

Suppose that a power plant produces electricity at 22 000 V. A step-up transformer changes it to 250 000 V for long-distance transmission. Your local station uses a step-down transformer to reduce the potential to 30 000 V for distribution to homes. Finally, a transformer near each home steps down the voltage again to 110 V. As you can see, electricity can be manipulated to provide the best possible efficiency for the job to be done.

Figure 15–13. Thomas Edison's first generator had no transformers. The current generated was the current delivered to the homes.

 ASK YOURSELF

Why is it not practical for a power plant to send electricity to your house with a high current and a low voltage?

SECTION 1   REVIEW AND APPLICATION

Reading Critically

1. How can turning a loop of wire generate electricity?
2. How are the primary and secondary coils of a step-up transformer different from those of a step-down transformer?
3. Why is the current in a typical generator coil always AC?

Thinking Critically

4. Explain why a transformer will not operate on DC.
5. Why do you think high potentials are used to transmit electric power over long distances?
6. Look again at Figure 15–13. How do you think the scene of Edison's generator would compare with a modern generating plant?
7. How is an electric motor similar to a generator? How is it different?

INVESTIGATION

Testing Electromagnetic Induction

▶ MATERIALS
- galvanometer ● solenoid ● insulated wire leads (2) ● bar magnets (3)

▼ PROCEDURE

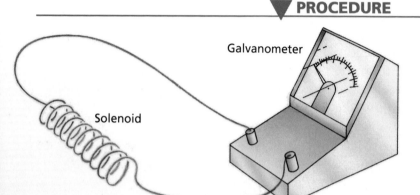

Galvanometer

Solenoid

1. Examine the galvanometer. If the galvanometer has more than one setting, be sure it is set to its most sensitive range.
2. Set up the apparatus as shown. With this arrangement, any current induced in the solenoid will pass through the galvanometer. The galvanometer measures the amount of current.

3. Take one of the bar magnets, and insert its N pole into the solenoid, while observing the galvanometer needle. What happens?
4. Pull the magnet out while observing the galvanometer. What happens?
5. Turn the magnet around and insert its S pole into the solenoid. Compare the result of moving the S pole

in and out with that of moving the N pole in and out. Record both results.
6. Vary the speed with which you move the magnet. Does the speed of the magnet affect the amount of current induced? What happens if you do not move the magnet at all?
7. Create a stronger magnetic field by putting two bar magnets alongside each other, with N poles and S poles together. Test this stronger magnetic field. Record your observations. Then increase the strength of the magnetic field again by using three magnets. How does the amount of current induced depend on the strength of the magnetic field?

▶ ANALYSES AND CONCLUSIONS
1. What evidence did you find that current is induced by a changing magnetic field?
2. What two kinds of evidence did you find to show that more current is induced if the field changes rapidly?

▶ APPLICATION
How is the principle you have just investigated used in making a generator that produces alternating current?

※ *Discover More*

Hold the bar magnet steady within the solenoid, and, while observing the galvanometer, move the solenoid back and forth. What happens? How is this result similar to that of the Investigation you carried out?

Household Circuits

After the hurricane had crossed the state of Florida, Roberto and his family headed for home. Everyone in the car was in a serious mood. They had seen the television reports of the hurricane damage. It showed areas of southern Dade county totally destroyed, with block after block of flattened houses looking as if they had been bombed. Roberto's family wondered whether they still had a home to go to.

The newscaster had shown pictures of power lines down, sparking on the wet ground. The preliminary reports said that some areas would be without electricity for weeks. Roberto realized now why his dad had asked him to think about how he would live without electricity. Imagine having no air conditioning or refrigeration with daytime temperatures in the 90s!

Roberto remembered that his dad had switched off the main circuit breaker before they left. Now he wanted to know more about how that device worked and how electricity moved around his house. Maybe he could help his family make repairs. He decided to look at his science book again.

Objectives

Describe the components of a typical household electric circuit.

Compare and **contrast** the roles of fuses, circuit breakers, ground wires, and grounded appliances as safety features.

Figure 15–14.
Hurricane damage in Homestead, Florida

Circuits

You are familiar with electric outlets, ceiling fixtures, and wall switches. You might also have seen a box with fuses or circuit breakers somewhere in your home. All of these devices are connected to wires inside the walls.

A typical cable used in wiring houses is composed of a group of three wires enclosed in a plastic casing. Two of the wires, one covered with gray insulation and one covered with blue insulation, carry alternating current to the outlets. Touching either of these wires could be dangerous. The third wire is not insulated and acts as a safety feature. If there is any leakage of current, the electricity will flow to the earth, or ground, through the ground wire rather than through you.

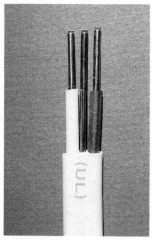

Figure 15–15. Electric cable such as that shown here is used to carry electric current in homes and other buildings.

Roberto wondered whether his house had been damaged by any of the nearby trees falling on it. If one of them had broken some of the electric wiring, maybe his video games had been ruined. He expressed his fears to his dad.

"Roberto, that is the least of your worries," his father said. "Before we left, I turned off the main circuit breaker so that no electricity could enter the house, even if lightning struck a nearby electrical pole. Your mom and I also unplugged all of the appliances. The house is protected against short circuits."

Roberto wondered what a short circuit was, but his dad was now talking to his mom about insurance, so he decided to look in his science book.

Accidental Connections   Worn insulation or a poor connection can create a short circuit. A **short circuit** is any acci-

dental connection that allows current to take an unintended path instead of passing through an appliance. It gets its name because the circuit the electricity takes is shorter than its intended path. For example, a short circuit in a radio would keep the radio from working even when it's turned on. The current would flow through the short circuit instead of the radio's electronic components and speakers.

A 110-V potential difference can produce an enormous current if there is no appliance in the circuit to provide resistance. If a short circuit occurs, the wires carry more current than they were designed to carry. The wires can overheat and cause fires.

Safety Features   In modern homes, the ground wire in the electric cable is designed to protect people against short circuits. The metal shell of an appliance is connected to the ground wire through the round, third terminal of the plug. If the "hot" wire touches the shell, the current goes directly to ground through this low-resistance path. If you touch the shell of a grounded appliance, very little of the current will go through your body because the appliance is already grounded. An appliance with a plastic shell insulates the user from the current. Such an appliance does not need a grounding plug. You can find out whether the appliances in your home have this safety feature by trying the next activity.

ᴰᴵˢᶜᴼᵛᴱᴿ ᴮʏ *Observing* _____

Examine the plugs on the appliances you have at home. Which of them have a third, round prong? Which of your appliances do not need a third prong? Why? Do you own some without a third prong that should have one? What could you do about this? 🖉

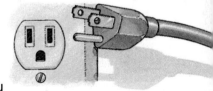

◥ ASK YOURSELF
Why are short circuits considered a safety concern?

Overloads

ꜱ   As Roberto's family approached their house, Roberto couldn't believe the destruction. Most trees were uprooted or simply gone. Those that remained were stripped of leaves. All the road signs had blown away. It was difficult to recognize the once-familiar routes. Along the way Roberto saw transformers that had been destroyed. His dad told him that the damage to the transformers could have been caused by lightning or by flying debris. After their conversation, Roberto began to read the section about overloads.

Have you ever plugged in too many appliances at once? If you have, you may have overloaded the circuit. Your home is probably wired with four, five, or more parallel circuits. Several outlets are connected to each circuit. If too many appliances on one circuit are turned on at the same time, too much electric current moves through the wire at once. This can cause an overload, and that circuit will shut down, cutting off the electricity.

Too much current can cause wires to heat up and melt their insulation. Then current can contact the material of the walls and cause a fire. Building codes require that homes be equipped with safety devices to prevent this from happening. These devices are designed to turn off the current if an overload occurs, before an electric fire can start.

Figure 15–16. These are fuses found in some buildings.

Protecting the Circuits   Protection against overloads may be provided by a fuse. A **fuse** is a safety device containing a short strip of metal with a low melting point. If too much current passes through the metal, it melts, or "blows." This causes a break in the circuit, and the current can no longer flow. This is your signal to find and correct the overload and to replace the fuse. Why should you correct the overload before you replace the fuse? Try the next activity to find out.

ACTIVITY
How does a fuse work?

MATERIALS
safety goggles,
15-A plug fuse, table lamp
CAUTION: Wear safety goggles throughout this activity.

PROCEDURE
1. Examine the fuse. Where are the terminals? Look in the little window to see the fuse wire.
2. Turn off the lamp, and remove its plug from the wall outlet.
3. Screw the fuse into the lamp socket, and plug the lamp into the wall outlet.
4. While wearing safety goggles, look carefully at the fuse wire, and, at the same time, turn on the lamp.

APPLICATION
1. Explain what happened to the fuse wire in step 4.
2. Why is it that nothing happens to the fuse until you turn the lamp on?

Break It Up   Another device that protects circuits from overloads is a **circuit breaker**. One type of circuit breaker is a switch attached to a bimetallic strip. When the metal gets hot it bends, which opens, or "trips," the circuit. This action does not harm the circuit breaker. After the overload has been corrected, the circuit breaker can be reset.

Roberto's dad had switched off the main circuit breaker before they left the house. This may have protected Roberto's video games from an overload. He remembered that the electrical box in the garage contained many circuit breakers. They were labeled according to the appliances or areas of the house they protected. Once when his mom had connected too many appliances to the same outlet, one of the circuit breakers had tripped. When she unplugged two of the appliances and switched the tripped circuit breaker back on, everything was fine.

Figure 15–17. Circuit breakers interrupt the flow of current.

ᴅɪꜱᴄᴏᴠᴇʀ ʙʏ *Doing* _____

Have a parent or other adult help you locate and examine the fuse box or circuit-breaker box in your home. How many separate circuits supply your home? How many of them are 15 A? How many are 20 A? Each time a fuse blows or a circuit breaker trips, note the outlets that are affected. In your journal, make a table showing which outlets in your home are on the same circuit. ✐

▶ ASK YOURSELF
What is the advantage of a circuit breaker over a fuse?

SECTION 2   *REVIEW AND APPLICATION*

Reading Critically
1. Why are houses wired with a ground wire?

2. What steps must you take if a circuit breaker in your home trips?

3. Why is a grounding wire unnecessary in an electric mixer that has a plastic casing?

Thinking Critically
4. Explain why the third prong of a grounded plug should not be removed to make the plug fit a two-pronged outlet.

5. What is the largest number of 100-W bulbs that can be turned on in a 115-V circuit without blowing a 15-A fuse?

6. If you were an insurance agent examining a building someone wanted you to insure, what would you look for in its electric system?

SKILL Calculating Your Electric Bill

An electric utility company charges customers according to the amount of electric energy they use. Electric power, measured in watts, is the amount of energy used per second.

The unit of measure for electric energy is the *kilowatt-hour* (kWh). A kilowatt hour represents 1000 watts of power used for an hour.

The energy used in a household is measured by a kilowatt-hour meter. It is usually located either in the basement or outside the house. A meter reader comes once a month or every other month to read the meter so that the company can calculate the bill.

▼ PROCEDURE

1. To find the amount of electric energy used, multiply the power (in kW) used by the length of time (in hours) it was used.
2. The illustration shows the dials on a typical kWh meter. Each dial stands for a digit of a 4-digit number. Note that all dials are read clockwise. The dials in the top meter read 3964 kWh. Why is the third dial read at six instead of seven?

▶ APPLICATION

1. Read the meter at the bottom, and record the reading.
2. Suppose the rate the electric company charges for energy is $0.08 per kWh. If a meter reads as in Figure A at one time and as in Figure B later, how much will the electric bill be for the time between the readings?
3. How much does it cost to run a 350-W television for an average of four hours every day for the month of March?
4. The three bulbs in a kitchen light fixture burn for an average of three hours a day. How much could be saved every month by reducing the power of the bulbs from 100 W each to 60 W each?

✳ *Using What You Have Learned*

1. Look at the electric bill for your home. What rate does your family pay for electric energy per kWh?
2. Read your home's electric meter every day for five days. Using your electric power company's table of rates, calculate your bill for each day. What do you think your bill will be for the entire month?

Electronics

When Roberto's family reached their own neighborhood, he noticed that most of the electric poles around the neighborhood were down. He knew it would be a long time before electricity was restored. At least they still had a roof over their house. The house next door was missing half of its roof. The yard was filled with debris—pieces of other houses, trees, and scraps of metal. They couldn't get the car into the garage because debris was in the way. Roberto got out of the car and began to clear the driveway.

Later, when Roberto walked into the house, he was amazed to hear the telephone ringing. His sister was calling from Orlando to see how things were. "But Mom," said Roberto, "how can the phone be working if there is no electricity?"

Objectives

Distinguish *between an analog signal and a digital signal.*

Describe *the operation of a telephone and a television.*

Compare *electronic circuits with other types of electric circuits.*

Hello, Is Anyone There?

Roberto's mom told him that the telephone network is separate from the electric power grid. "The telephone lines in this neighborhood are underground," she explained, "so they did not suffer any damage. I'm sure that's not the case in many places. We were lucky." Roberto remembered what his book said about how a telephone works.

Converting Sound Waves   Once electricity reaches your home, it is changed by various appliances into heat, light, and motion. Electricity is also used by circuits in electronic devices such as telephones, VCRs, computers, and televisions.

The telephone converts the sound waves produced by your voice to an audio signal. The audio signal is a series of electric impulses. This signal is an analog of the sound wave. An **analog** is something that is a substitute for something else. Let's say your mom writes a check to pay a bill. The check is not cold cash, but it does represent money. It can be converted into real dollars. In the same way, an analog stands for and can be converted into something else, such as sound.

When you speak into a phone, the sound waves you produce cause a metal disk in the telephone to vibrate. These vibrations produce an audio signal that is the analog of your voice. This audio signal is then amplified and directed through a switching system to your friend's telephone. There it causes another disk in the earpiece to vibrate, producing the same pattern of sound waves that was received by your mouthpiece.

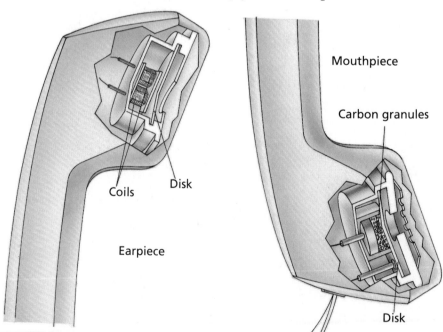

Figure 15–18. These cutaway views of a telephone show the internal parts of the mouthpiece and earpiece.

Mouthpiece

Carbon granules

Coils

Disk

Earpiece

Disk

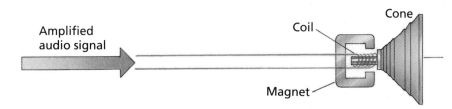

You can see a similar procedure in a large speaker. Changes in electric current cause a coil in back of the speaker cone to be attracted to and repelled from a permanent magnet. You can actually see the speaker cone in a large speaker vibrate back and forth. Go to your nearest electronics store, and check it out.

As telephones have become more complex and their use has increased, more efficient methods of transmission have become necessary. Analog signals carry some unavoidable noise. Every amplifier and electronic circuit adds more noise to the signal. Perhaps you have occasionally noticed the static on the line during a telephone call.

Digital Signals   Technology has made it possible to replace analog signals with digital signals. A **digital signal** is a series of electric pulses that stand for two digits, 0 and 1. These digits make up binary numbers. Each of the digits in a binary number is called a *bit*. An electrical signal can carry bits in the form of a string of pulses. Each pulse stands for the number 1; a missing pulse stands for the number 0. Information is coded by the sequence of 1s and 0s.

Sailors use a flashing light to communicate from ship to ship. They use a lantern that can click on and off at different speeds. One quick flash followed by a longer signal could mean hello. They can carry on long-distance conversations as long as they are able to decode the sequence of light bits. This is similar to a digital signal. A digital telephone is connected directly to an electronic circuit that produces a string of bits.

Digital signals offer many advantages over analog signals. There is no noise in a digital signal, because digital information is a count, not a measurement. Many instruments use digital signals in their electronic circuits. Videodisc players, computers, and fax machines are examples. Many telephone companies have already changed their systems to digital signals. This change has permitted information to be faxed through phone lines. It also allows for computers to transmit data via modem through phone lines.

Figure 15–19. The basic makeup of each of these speakers is shown in the diagram. The reaction between the magnetic field of the coil and that of the magnet causes the cone to vibrate to produce sound.

ASK YOURSELF
What is a digital signal?

The Number Please

To understand the difference between analog and digital signals, consider grapes and grape juice. If you measure the volume of a bottle of grape juice, the answer you get is an analog of the juice. On the other hand, if you count how many grapes there are in a bag, the answer is a whole number, a digital statement. Now suppose you pour the juice from one bottle into another. You will introduce some inaccuracy, some "noise." You might leave the original bottle wet; there might have been some dirt in the second bottle. If you move some grapes from one bag to another, you might introduce some noise. Maybe the grapes were wet, or the bag was dirty. However, in the digital signal the noise does not matter. The number of grapes, the digital information, has not changed.

DISCOVER BY Writing

The letters of the alphabet can be expressed digitally, as a series of 1s and 0s. Use the library to locate a source that shows the alphabet in binary form. Write a brief letter in your journal, and translate it into digital code. See whether a classmate can decode the letter. ✍

ASK YOURSELF

Why are digital signals being used more and more in preference to analog signals?

Figure 15–20. An analog signal can be compared to a pitcher of grape juice. A digital signal can be compared to a bunch of grapes. A quantity of juice must be measured. However, a quantity of grapes can be counted.

Radio and Television

It had been two weeks since Hurricane Andrew struck southern Florida. In that time Roberto and his family had cleared all the debris from around the house and made some minor repairs on the roof. They were able to run a few things from their generator, but the area still had no electricity. The latest estimate was that it would take at least another two weeks to restore power. People tried to avoid going out to buy food or fuel. Without traffic lights, the whole area was in a constant traffic jam. It took two hours to get to the nearest supermarket and back. The heat and the mosquitos made the evenings hard to bear.

Being without electricity showed Roberto how important it is. Anything electronic now interested him. He had begun to enjoy finding out how things work. "Maybe," he thought, "I'd like to become an electrician or an electrical engineer."

Electromagnetic Waves   While Roberto's school was closed, his only source of amusement or information was a battery-powered radio. "Dad," he said one hot afternoon, "I bet the radio station we're listening to uses audio signals the same way a phone does. But how do these signals get to us since the radio isn't connected to any wires?"

"Audio signals are converted to electromagnetic waves," his dad explained. "Your radio antenna picks up these waves and changes them back into electric signals. The audio signal is then amplified and sent to the speakers."

"What about a television, Dad?" Roberto asked. "Besides the air conditioning, I miss television most of all. How does it work?"

"Roberto," his dad replied, "let's look together at your science book to find some answers."

You probably don't realize that the pictures you see on television are painted by electrons. The moving pictures are produced on the inside of picture tubes by beams of electrons. In a black-and-white television, a single electron beam is produced inside the picture tube and aimed at the screen. The screen is coated on the inside with chemicals called **phosphors.** When the electron beam strikes the phosphors, they glow. The electron beam very quickly scans the inside surface of the screen.

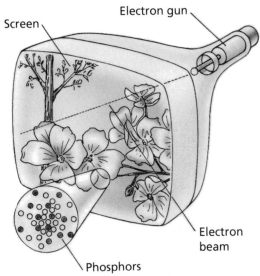

Screen

Electron gun

Electron beam

Phosphors

Figure 15–21. An electron beam causes phosphors on the screen to glow.

Phosphor Dots Television signals travel to houses from the station by electromagnetic waves. But this wave carries both the audio signal and the video signal. The video signal causes fluctuations in the strength of the electron beam. How much the phosphors glow depend on the strength of the electron beam at that particular instant. The changing brightness of the phosphors is what creates an image on the television screen.

In a color television, the screen has phosphors of three different colors. These *phosphor dots* are arranged in groups of three. Each group, called a *pixel*, contains a red dot, blue dot, and green dot. There are three electron guns in a color television. One scans only the red phosphor dots in each pixel, another the blue, and the third the green as their beams move across the screen. The signal coming into the television controls the fluctuations of each of the electron beams. You can actually see the small phosphor dots by looking at a television screen with a magnifying glass. Because the dots are so small, your eyes see blends of colors rather than individual colored dots.

Figure 15–22. Televisions that use digital signals can show more than one program on the screen.

While Roberto's parents went to help their neighbor repair his roof, he sat thinking about how devastating this hurricane had been. "I wouldn't want to go through this again," Roberto said to himself, "but I certainly learned a lot the hard way—about electricity and its uses!"

ASK YOURSELF

In what way are electronic systems, such as that of a telephone, different from other kinds of electric systems?

SECTION 3   REVIEW AND APPLICATION

Reading Critically

1. Explain the difference between an analog and a digital signal.
2. How are black-and-white televisions different from color sets?

Thinking Critically

3. Why can a photograph be considered an analog?
4. When you shop, some items are charged by analog measurements and some are charged digitally. Name three examples of each kind of item.

The Big Idea

In today's world, we depend on many forms of energy: motion, heat, light, and sound. Our many modern conveniences and connections to the world depend on energy conversions. It all begins with coils of wire moving through magnetic fields. After that it's just a matter of harnessing the energy so that we can run motors, heat and light our homes, and enjoy many forms of communication.

For Your Journal

Think about what you have learned from this chapter about electricity in use. Review what you have written in your journal, and revise or add to the entries. Include information about the importance of electricity to your everyday life.

Connecting Ideas

Copy the concept map into your journal. Fill in the missing types of energy to show how electricity is transformed.

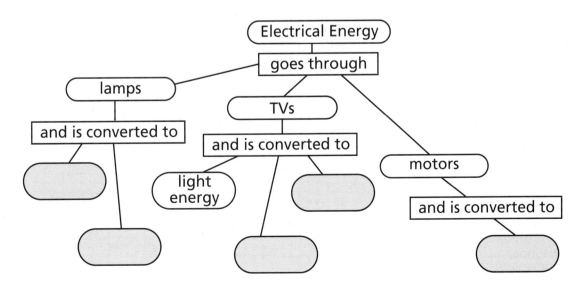

CHAPTER 15

REVIEW

Understanding Vocabulary

1. For each set of terms, explain the similarities and differences in their meanings.
   a) short circuit (403), circuit breaker (405), fuse (404)
   b) power grid (391), transformer (397), electric motor (394)
   c) analog (408), digital signal (409), phosphors (411)

Understanding Concepts

MULTIPLE CHOICE

2. The wire loop of an electric motor spins because
   a) the current flowing through the wire is DC.
   b) its magnetic poles are attracted to the opposite poles of the magnet.
   c) a turbine rotates the loop.
   d) its magnetic poles do not reverse.

3. What device changes the voltage of current entering your home?
   a) step-up transformer
   b) power grid
   c) circuit breaker
   d) step-down transformer

4. A device that opens a circuit in case of overheating is a
   a) conductor.
   b) fuse.
   c) transformer.
   d) short circuit.

5. Which of the following converts electrical energy into mechanical energy?
   a) generator
   b) electric motor
   c) transformer
   d) circuit breaker

SHORT ANSWER

6. Explain how hydroelectric power plants produce electricity.

7. Electrical cables used to wire buildings have three wires. Describe the function of each wire.

Interpreting Graphics

8. Explain the movement of the wire loop in the illustration of the electric motor.

9. How might you make the motor turn in the opposite direction?

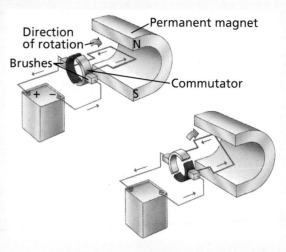

Reviewing Themes

10. *Energy*
Does a generator create energy? Explain your answer.

11. *Technology*
Explain why AC is sometimes more useful than DC.

Thinking Critically

12. Why is it unnecessary for the grounding wire of a plastic-insulated cable to have its own separate insulation?

13. How can wind power be used to generate electricity?

14. Explain why a home's electrical supply system must be connected to the ground at some point.

15. How is an electromagnet used in the earpiece of a telephone to produce sound waves?

16. Would a generator produce electricity if both the coil and the magnet moved simultaneously in the same direction? Explain.

17. Why does a color-TV screen need only three different colors of dots—red, blue, and green?

18. Should a fuse be connected in series or in parallel with the appliances in a circuit? Explain your answer.

19. The nerves in your body send messages to your muscles by using electrical signals. When the nerves are damaged, they cannot send the proper signals. The woman in the photograph is paralyzed from the waist down, yet she is able to ride a bicycle with the aid of computer output devices attached to her legs. Explain how the computer can "talk" to the muscles so that this woman can ride a bike and walk.

Discovery Through Reading

Vizard, Frank. "Previewing Widescreen TV." *Popular Mechanics* 169 (May 1992): 126–131. This article discusses television technology that produces large, detailed images and digital sound.

Science

PARADE

Your First Car May Be Electric

Imagine commuting to your first day of college. You and a companion zip along the expressway at 88 kilometers per hour. The only sound from your car's motor is a high-pitched electric whine. There are no gears to shift, only buttons to push for forward, neutral, and reverse. Arriving on campus, you pull into a parking space and plug your car into a recharger outlet. By the time classes are over, your sporty electric car will be recharged for the drive home.

No More Pollution?

Concerned citizens, auto makers, and government officials have been working to increase the chance that your first car will be electric. Electric cars make sense. They don't consume gas, and they produce no exhaust pollutants. California has already passed a law that goes into effect in 1998. The law will require that 2 percent of the vehicles sold produce "no

This car, called the Impact*, will be ideal for commuting to work or running errands.*

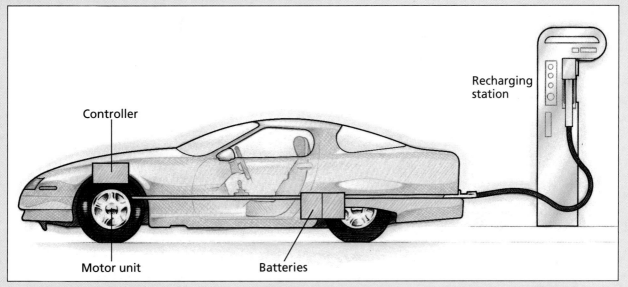

This drawing shows how an electric car's battery might be recharged. Electricity from the battery flows to the controller. The controller regulates the amount of electricity that goes to the electric motor that moves the front wheels.

tailpipe emissions," meaning electric cars. By the year 2003, at least 10 percent of the vehicles sold must meet the new standards. Auto makers have responded to this new legislation with intense efforts to produce a practical electric car.

Actually, an electric car is an old idea. During the auto industry's early years, battery-powered electric cars competed with gasoline engines and steam engines. However, the performance and convenience of the gasoline-powered cars proved too compelling. By 1924 the steam engine car was scrapped, and by 1930 electric cars had virtually disappeared.

The Return of the Electric Car

Scientists and engineers are currently trying to produce an electric car that matches gasoline-powered cars in performance and cost. One of the toughest problems in designing an electric auto is the battery power source. The batteries in electric cars are similar to the batteries in today's cars, only much bigger. They weigh 355 kilograms as compared with a gasoline-powered car's battery that weighs only about 20 kilograms! They are also expensive and must be replaced every

40 000 kilometers. Every battery design involves trade-offs between power, long life, and low cost.

Ladies and Gentlemen— Recharge Your Batteries!

One type of battery being tested in electric cars is made of lead plates, sulfuric acid, and water. It produces electricity by means of a chemical reaction. The sulfuric acid reacts with the lead to produce lead sulfate and extra electrons. The electrons flow from one set of lead plates to another, and an electric current is produced. The battery runs down when all the sulfuric acid is used up but can be recharged by hooking it up to a recharger. Unfortunately, the battery must be recharged every 190 kilometers.

The Car of the Future

Although electric cars have limitations, engineers continue to refine their prototypes. Some of the new technology developed for electric cars may even be useful for gasoline-powered cars. By the year 1998, electric cars will certainly be more than a curiosity. They may be the urban commuter's first choice! ◆

A New World at Your Fingertips

from *National Geographic World*

It has many faces and almost as many names. Scientists and computer hackers call it "artificial reality," "cyberspace," "telepresence," and "virtual reality." "Weird but fun" is what Janaea Commodore, of Novato, California, calls it.

Janaea uses wired goggles and gloves to explore a world that exists only in computers. When she moves her gloved hand, the glove instantly sends information about her movements to a computer. The robot hand, projected here on a giant screen, shadows her movements.

On the page the robot hand looks flat, or two-dimensional. But Janaea sees the action on two small television screens inside goggles that block her view of the real world. Each screen shows a slightly different view of the action. As a result, Janaea sees her computer world in three dimensions. It has height, width, and depth, so it looks like the real world. Well, *somewhat* like the real world. . . .

Janaea describes it like this: "I was in a room and could go through doors and windows. I tried to catch balls and pick up objects. At first it was hard. You don't know where your hand is going. But it got easier."

To grab something, Janaea made a fist. To let go, she opened her hand. "When you point your finger, you fly," she says. "That's really cool!"

Janaea spent only about 20 minutes in the wacky world of computer-based reality. She got the chance to take this fantasy trip because her mother works for a company that makes virtual reality software for computers.

The software can do a lot more than provide spacey and spectacular fun. Doctors can use it to travel inside the human body. Architects can walk through buildings that haven't been built. A weather forecaster can fly above the earth. Space scientists can skim the surface of distant planets. Pilots can safely train for emergency situations.

Virtual reality has a lot in common with computer games. But it goes much further. A person actually seems to enter another real world. "In a virtual reality experience you are totally or at least partially in a 3-D world," explains Howard Rheingold, the author of a book on the subject. "You have the ability to navigate within that world and to interact with the objects in it."

To interact—touch or move objects, change your position, or create sounds—you might put on goggles and a wired glove, just as Janaea did. You might climb into a suit that senses every move you make or walk on a wired treadmill. You might stand in front of specialized video cameras, pull objects out of a framed screen, use a joystick, or simply speak.

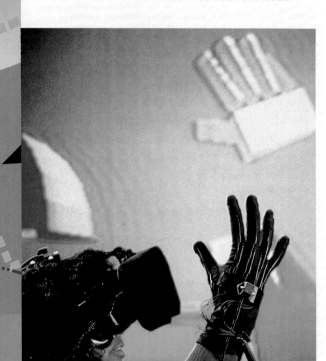

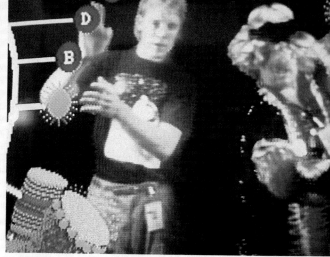

Different Drums. At a computer show, two men drum up sounds in space. You see the drummers as they really are. You see them as they appear on a video screen. A camera feeds their images into a computer. It compares those images with its stored images of drums. A "hit" produces a computer sound.

One of the best things about artificial space is that two people who are really far apart can share it. For example, with their computers linked, they can practice a surgical procedure or play racquetball together.

Artificial reality is still so new that no one can be sure what forms it will take. And no one can predict all the things it may help us do in the future.

Right now the equipment is bulky and expensive. But people who work with computers think that will change within a few years. Someday you may spend part of your time in the real world and part of it in a virtual world that you and computers create together.

Power Play. Using a special glove, Jack Menzel, of Napa, California, plays a ball game on his television screen. "To catch the ball, you make a fist," Jack says. "A hand on the screen makes a fist when you do." When you use such a glove, a flat screen seems more like virtual reality. Although you don't see the action in three dimensions, you can move objects simply by moving your hand. You don't need a computer mouse. Jack has also tried the "goggles-and-gloves" virtual reality. He describes it as "a little confusing, but neat."

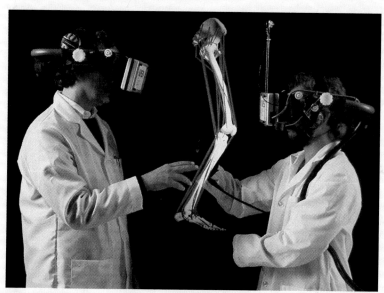

Busy Bodies can be created by computers. This "virtual body" will help medical students learn more about how parts of a real body work. The leg image projected on the screen shows up in three dimensions inside the goggles. Students can look at the leg from all angles and move parts of it around. Surgeons could use a virtual body to practice difficult operations before performing them on real patients.

Making A Hit with A Computer

In the drawing, a girl plays "virtual racquet-ball." Her equipment: a racket, a glove with sensors, 3-D goggles, and a means of linking everything to a computer—in this case, wiring.

The girl uses the racket in her hand to try to hit a ball generated by the computer. Sensors on her wired glove pick up information about her movements and send it to the computer. The racket on the computer screen duplicates each motion of her racket.

Information about the moving ball and the racquetball court have been programmed into the computer. The player can change her position on the court by moving her head. Sensors in the goggles will tell the computer to adjust the scene as she moves forward, backs off, or turns her head.

Two small television screens in her goggles, one for each eye, show slightly different versions of the action. Just as they do in the real world, her eyes combine those two images and see in three dimensions—height, width, and depth.

The player will try to coordinate what her eyes see with what her hand does. If she succeeds, the computer will respond with the sound of a racket striking a ball. And if her aim is good, the ball will zoom off in the proper direction. ◆

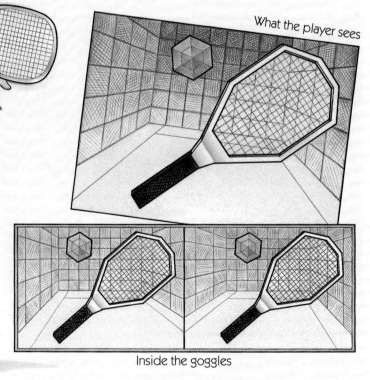

What the player sees

Inside the goggles

420

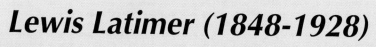
Lewis Latimer (1848-1928)

The son of a fugitive slave, Lewis Latimer was a pioneer in the great age of American invention during the late 1800s. Latimer is best known for his refinement of Thomas Edison's incandescent light bulb. He also assisted in Alexander Graham Bell's invention of the telephone.

Lewis Latimer was born in Massachusetts in 1848. He served in the Union navy during the Civil War. In Boston after the war, Latimer taught himself drawing. Despite discrimination against African Americans, he found work as a draftsperson, often drawing other people's inventions.

Latimer was befriended by a local teacher who was designing a device to help his hearing-impaired wife communicate. The teacher, Alexander Graham Bell, employed Latimer to make drawings and descriptions of his new "telephone." Latimer's sketches helped gain the patent for Bell's first telephone, issued in 1876.

Latimer's work attracted the attention of Thomas Edison, who in 1879 had invented the incandescent light bulb. Latimer went to work for Edison and soon made his most important contribution. Latimer devised a method for manufacturing carbon filaments that lasted much longer than Edison's first efforts. Soon, Latimer-style electric lamps were in wide use. In 1890 he published a textbook on electric lighting and continued to patent his own inventions. ◆

An Wang (1920-1990)

An Wang was an electronics wizard whose company pioneered the computer and communications systems now common in offices. More important to sports fans, Wang's firm designed the first computerized scoreboard. The electronic messages and images now commonplace in arenas and stadiums can be traced to this inventor.

Wang was born in Shanghai, China, and came to the United States in 1945 to study applied physics. At that time, computers were enormous clattering machines that filled entire rooms. Wang revolutionized computer memory by inventing a magnetic device for storing data. His doughnut-shaped magnetic core was the forerunner of today's semiconductor chips.

With $600 in savings, Wang founded his own company, Wang Laboratories, in 1951.

His tiny firm concentrated on special applications of computer technology. One famous example of this technology was the digital scoreboard at New York's Shea Stadium, which opened in 1964. Instead of mechanically placing the numbers and letters to show the score, the electronic scoreboard flashed scores and customized messages.

Wang's company became a pioneer in office automation, bringing electronics to the ordinary workplace. He credited his success to a combination of ancient Chinese wisdom and the American spirit of invention.◆

CLARENCE BACA
Meter Technician

When you turn on the light switch in your classroom, you expect the lights to come on. At home you expect the refrigerator to be cold, the television to work, and the house to be warm on a winter day.

Clarence Baca is one of the people who makes all that possible. Baca is a meter technician. He makes it possible for the lights to come on and for television stations to broadcast your favorite programs.

Baca does many important jobs. The one he does most often is installing electric meters that measure the electric power that comes into a building. The utility company that supplies the electric power uses the meter to tell how much electricity has been used.

The meters that Baca installs in houses and apartments are called *watt-hour meters*. They measure electrical energy in kilowatt-hours. When electric power flows through the meter, a disk turns. Each time the disk goes around, a gear makes the pointer on a dial move to the next number. When electricity is not being used in a building, the disk does not turn. The utility company can tell from the numbers on the dials how much electric power has been used.

Baca also checks electric meters to be sure they are accurate. Sometimes people may think they have been charged too much money for the electricity that they used. When that happens, Baca checks the meter to see whether it is accurate. If necessary, he makes adjustments or repairs.

A meter technician cannot tell by looking at a wire whether it has power in it. Because it is not possible to see or hear electricity, people who work with it must be very careful. If you should touch a live wire with your bare hand, you could be badly shocked, burned, or killed. Meter technicians use tools that have nonconducting materials on the handles. They also wear gloves made of rubber and leather so that they will not be shocked or killed if they should touch a live wire.

Meter technicians spend about four years learning their trade. First, Baca became an apprentice. An *apprentice* learns a trade while he or she works at it. Then, Baca worked with experts who had been meter technicians for a long time. Baca also went to classes for two hours a day, twice a week, for four years. He then had to pass several tests to become licensed. It takes about as long to become a meter technician as it does to earn a college degree. ◆

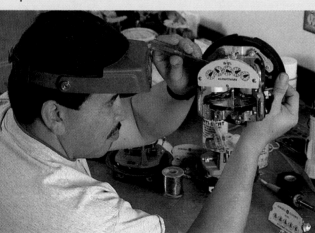

Discover More

For more information about careers for electricians and meter technicians, write to the

National Joint Apprenticeship and
    Training Committee for the
    Electrical Industry (NJATC)
16201 Tradezone Avenue, Suite 105
Upper Marlboro, MD 20772

SUPERCONDUCTING MAGNETS

Something remarkable happens when certain metals are cooled to a point near absolute zero (zero kelvin). At such low temperatures, these metals conduct electricity without resistance. They become *superconductors!*

Power to Spare

Superconductors provide a huge advantage over ordinary electric wire. The wire in household circuits carries 15 amps safely, while wasting about 1.8 watts per meter of wire. A superconducting wire can carry 2000 amps for an infinite distance with no wasted power! Widespread use of superconductors could save billions of dollars in electric costs.

The drawback is that today's working superconductors must be kept at temperatures around 4 K (-269°C). The superconductors are bathed in liquid helium, requiring complicated and expensive refrigeration. Scientists are working on materials that superconduct at around 90 K. Liquid nitrogen is colder than that and is

less expensive than liquid helium. So, liquid nitrogen can by used to cool the new superconductors.

Magnet levitating over superconducting disk

Supermagnets

Superconductors are at the core of a new generation of supermagnets. With electric coils made of superconducting material, these magnets can produce fields 200 000 times the earth's magnetic field at the poles. Superconducting magnets provide the power for magnetic resonance imaging (MRI), the medical device that takes pictures of body tissues.

Superconducting magnets are also powering supercold refrigerators. Conventional refrigerators cool by compressing and expanding gas. Instead of gas, the magnetic refrigerator uses a crystal that heats and cools when passed through a magnetic field. Superconducting magnets help this superfridge to reach temperatures near absolute zero.

Instead of propellers, this ship uses jets of electrified sea water, pushed out by superconducting magnets.

This combination of a superconducting magnet and a conventional magnet is the strongest magnet ever made.

Take a Supermagnetic Cruise

Engineers in Japan have already tested trains and ships powered by superconducting magnets. The *maglev* (magnetic levitation) train rises by the force of mutually repelling superconducting magnets. The train floats on a magnetic field and is pulled forward by magnets along its sides. Without the friction of wheels or rails, the maglev train reaches speeds of 640 km/h, twice as fast as Japan's bullet trains.

Most applications of superconducting magnets remain experimental, since the magnets still require bulky and expensive helium refrigeration. Scientists are working to overcome these obstacles. ◆

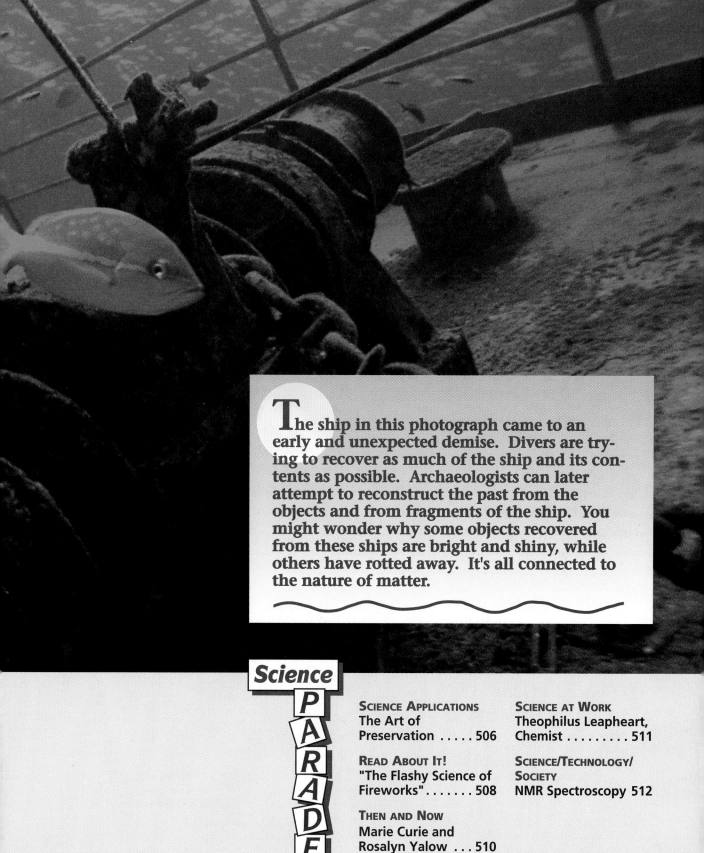

The ship in this photograph came to an early and unexpected demise. Divers are trying to recover as much of the ship and its contents as possible. Archaeologists can later attempt to reconstruct the past from the objects and from fragments of the ship. You might wonder why some objects recovered from these ships are bright and shiny, while others have rotted away. It's all connected to the nature of matter.

Science PARADE

*S*tone may look like it's "solid as a rock," but it isn't. Stone is made of tiny particles that are mostly empty space. The stone in a statue ages, just as humans do, but the particles within the stone stay the same. Science has long sought to understand the world's mysteries by grasping the structure of matter.

THIS STATUE OF A YOUNG MAN MAY BE AN AUTHENTIC GREEK SCULPTURE, CREATED ALMOST 2500 YEARS AGO, OR IT MAY BE A $9-MILLION FAKE! THE STATUE'S TRUE HISTORY IS LOCKED IN THE STRUCTURE OF ITS MARBLE SURFACE. ITS VALUE AS ART DEPENDS ON OUR UNDERSTANDING OF THE STRUCTURE OF MATTER.

Kouros, Unknown artist, 530-520 B.C., Thasian? marble, 68". Collection of the J.Paul Getty Museum, Malibu, California

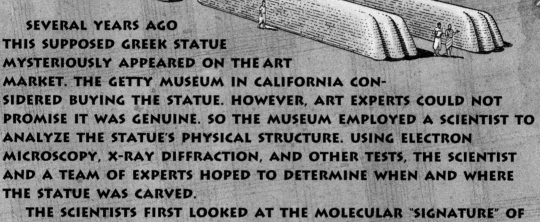

SEVERAL YEARS AGO
THIS SUPPOSED GREEK STATUE
MYSTERIOUSLY APPEARED ON THE ART
MARKET. THE GETTY MUSEUM IN CALIFORNIA CON-
SIDERED BUYING THE STATUE. HOWEVER, ART EXPERTS COULD NOT
PROMISE IT WAS GENUINE. SO THE MUSEUM EMPLOYED A SCIENTIST TO
ANALYZE THE STATUE'S PHYSICAL STRUCTURE. USING ELECTRON
MICROSCOPY, X-RAY DIFFRACTION, AND OTHER TESTS, THE SCIENTIST
AND A TEAM OF EXPERTS HOPED TO DETERMINE WHEN AND WHERE
THE STATUE WAS CARVED.

THE SCIENTISTS FIRST LOOKED AT THE MOLECULAR "SIGNATURE" OF
THE UNWEATHERED STONE IN ORDER TO DISCOVER ITS ORIGIN. THEN
THE MOLECULAR STRUCTURE OF THE STONE'S WEATHERED SURFACE
WAS ANALYZED. SCIENTISTS HAVE USED SIMILAR TECHNIQUES TO ESTI-
MATE THE AGE OF THE GREAT SPHINX IN EGYPT. COULD THE SCIEN-
TISTS TELL THE ART WORLD IF AN ANCIENT SCULPTOR SHAPED THE
GREEK STATUE, OR IF A FORGER HAD MADE IT JUST A FEW YEARS AGO?

AFTER ANALYZING THE CHALKY-LOOKING SURFACE OF THE STATUE,
THE SCIENTISTS DETERMINED THAT THE STATUE WAS GENUINE. THEY
THOUGHT THAT THIS WEATHERING PROCESS COULD NOT BE FAKED.
THE MUSEUM BOUGHT THE STATUE.

BUT WAIT! IN THE MEANTIME A CHEMIST
RE-CREATED THE MARBLE'S CHALKY
COATING IN A LABORATORY. THE
STATUE'S OWNERS AGAIN QUESTIONED
ITS AUTHENTICITY. THE $9-MILLION
QUESTION IS STILL UNANSWERED,
WAITING FOR SCIENTISTS TO DISCOVER
MORE ABOUT THE STRUCTURE OF MATTER.

For Your Journal

- *What do you think matter is made of?*
- *How could most of matter be composed of empty space?*
- *How could understanding the nature of matter help you in your everyday life?*

Atomic Theory

Objectives

Use *a time line to describe the development of atomic theory.*

Compare *different atomic theories.*

Why is a diamond so hard? Why does a "superball" bounce so high? How can we make matter behave the way we want it to? People have tried to understand the composition of matter for many centuries. People in future centuries will probably ask similar questions.

Imagine that you are the captain of a starship in the twenty-third century. At your fingertips you have an amazing amount of technology such as transporters, machines that make any food you want, and engines that can go faster than the speed of light. You know exactly which buttons to push and what commands to give, but you wonder how all these inventions came about. You know that these inventions change or interact with matter, but you wonder, "How did scientists learn about matter and what it can do?"

You are so puzzled by this question that you decide to find some answers. You go to the ship's holodeck and have the computer simulate ancient times. You ask it to take you back in time to observe some of the people who first theorized about matter.

Start at the Beginning

The holodeck program takes you back hundreds of years in time. You look around. You are in a large marble room with several men who are having a heated discussion.

You decide to get closer and to listen to the conversation. "We atomists believe that all matter is made of small particles

called atoms," says one man. He continues to speak. "All matter is composed of four atoms: fire, water, earth, and air."

"What do these atoms look like?" asks a second man.

The first man explains that the atoms of fire are jagged, and that is the reason that fire burns when it is touched. The atoms of water are round, so they flow smoothly. The atoms of earth are solid and stable and are, therefore, probably like little cubes. Air atoms are very light.

Round atoms? Cubed atoms? This wasn't the way you learned it. You know there is something wrong with the ancient Greek philosopher's explanation of the composition of matter. You ask yourself, "What data did they have to back up their statements? Aha! That's the problem—no experimental data!"

You try to put yourself in their shoes. It is very difficult for anyone to understand things that are too small to see or to measure. Yet the ancient Greeks were philosophizing about the nature of matter—something they couldn't possibly see. While many of their ideas were correct, they had no evidence—they were not being scientific.

You can try the next activity to get an idea of what it is like to investigate something you can't see.

ACTIVITY

How can you determine the shape of an unknown object?

MATERIALS
unknown object; rectangular cardboard, 40 cm × 60 cm; a marble

PROCEDURE

1. One member of the group will get the unknown object from the teacher. Without showing the object to the other members of the group, the student should place the object on a table and cover it with the cardboard. The other students should not look under the cardboard.

2. While the first student holds the cardboard, a second student should *gently* roll a marble under the cardboard to hit the object.

3. Use a sheet of paper to keep a record of the bounce-off angle. By rolling the marble from different directions, try to determine the shape of the object.

APPLICATION

1. Form a group conclusion about the object's shape. What led you to this conclusion?

2. How is this activity similar to the philosophizing of the ancient Greeks? How is it different?

 ASK YOURSELF

Why would it have been impossible for the ancient Greeks to create a scientific atomic theory?

Science Hits the Atomic Theory

You decide to go back to the holodeck and ask the computer to simulate a later period. The setting is now an English laboratory in the 1800s. A man is seated at a desk writing something entitled "Dalton's Atomic Theory." Ah . . . this must be John Dalton; you remember reading about him. You peek over his shoulder to read what is on the paper. Dalton's theory starts out with the same idea the ancient Greeks had: "All matter is made up of atoms."

Unlike the Greek philosophers, however, Dalton believed that atoms may be distinguished by their mass. Furthermore, he stated that all atoms of the same element, such as oxygen, have the same mass. Atoms of different elements have different masses. In the process of developing his theory, Dalton had developed several experiments to determine the masses of atoms.

"How could he weigh an atom?" you ask. "Maybe I should try it." The holodeck computer breaks in and suggests that there is an experiment that would tell you what you want to know. That experiment is the Investigation on page 443.

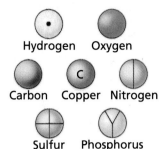

Figure 16–1. John Dalton proposed some of the first symbols for the elements. Some of his original symbols are shown here.

 ASK YOURSELF

Why would weighing an atom be important to the development of an atomic theory?

It's What's Inside That Counts

Dalton's experiments for finding the masses of atoms gave scientists a way to study the elements. He was not concerned with the composition of the atoms.

You and everyone else in the twenty-third century know that atoms are composed of a nucleus and electrons. When did scientists find out about the parts of the atom? You ask the holodeck computer to advance you to the time when the first subatomic particles were discovered. Space seems to whirl about you. Suddenly, you are in a British lecture hall filled with people. It's 1897. A man at the front of the hall is using a cathode-ray tube to perform some experiments. You think to yourself, "What an antique!" The last time you saw a cathode-ray tube was in a museum of the twentieth century. There you learned that the screens of antique television sets were basically cathode-ray tubes.

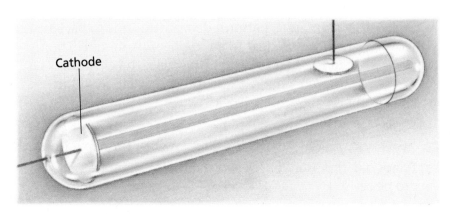

Figure 16–2. A cathode-ray tube similar to this one was used by Sir Joseph Thomson in his experiments.

Cathode

The lecturer's name, Sir Joseph Thomson, is on the board, along with a diagram of the cathode-ray tube he is using in his experiment. The tube has had some of the air removed, and it contains two electrodes. An electric current flows from the negatively charged electrode to the positively charged electrode. The lights are dimmed. As the experiment begins, you see a ray of yellow-green light glow between the electrodes.

Thomson tells the audience that the ray produced between the electrodes is the same regardless of the kind of gas in the tube. This indicates that all matter must contain the substance that produces the ray. What is amazing, Thomson tells the audience, is that the ray behaves as though it were a stream of particles. He briefly discusses other experiments that support his findings. Thomson also explains that this ray is made up of particles that make up the atom and that the particles are charged. To prove this, he brings two electrically charged plates close to the cathode-ray tube. The audience gasps! The ray bends toward the positively charged plate. Since opposite particles attract, Thomson has proven that the ray is made of negatively charged particles.

Figure 16–3. Setup for Thomson's experiments

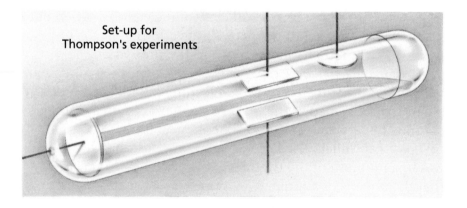

Set-up for
Thompson's experiments

Thomson continues to perform demonstrations to back up his discovery. The point is clear to all the people present. Thomson has discovered the first subatomic particle, the electron. The **electron** is a negatively charged subatomic particle found in almost all atoms.

 **ASK YOURSELF**

What things did Thomson tell his audience that caused Dalton's theory to be revised?

A Positive Step

You wonder when the other particles inside the atom were discovered, specifically the protons and neutrons that everyone knows make up the nucleus of every atom. "Computer?" "Yes, Captain." "Can you transport me to the next major discovery of a subatomic particle?" "Yes, Captain. Just one moment."

You appear in the dimly lit laboratory of a scientist you recognize as Ernest Rutherford. The year is 1911; the place is England. Rutherford's famous gold-foil experiment is underway. Wow! You learned about this at the Academy.

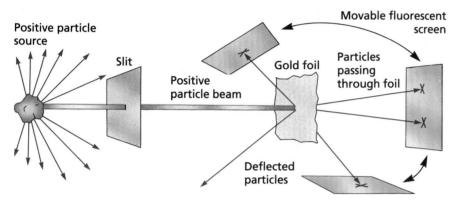

Figure 16–4. Shown here is a diagram of Rutherford's gold-foil experiment. This experiment provided the basis for modern atomic theory.

Rutherford is aiming alpha particles at a very thin piece of gold foil. *Alpha particles* are positively charged particles that are emitted from certain radioactive materials. A screen surrounds the apparatus. The screen glows when it is hit by the alpha particles. The effect is similar to the glow of a luminous watch at night. The lights are dimmed, and the box holding the radioactive material is opened.

Several students are hard at work measuring the angles of the alpha particles that are scattered around the screen. The hypothesis they are testing states that since the atom is mostly empty space, the particles should go right through the gold foil—except for the few that would be deflected by attraction to the electrons.

To their surprise, some of the alpha particles bounce back! "It was almost as if you fired a 15-inch shell into a piece of tissue paper and it came back and hit you," says Dr. Rutherford. After some discussion, the scientists present agree that if some positive particles bounced back they must have come close to other positive particles, since like charges repel. From the number of particles that bounced back, Rutherford concludes that

atoms have a very small but dense nucleus that is positively charged. Today, we call the positive subatomic particles within the dense nucleus **protons.**

OK. This entire process is beginning to make sense. You have seen scientists do experiments that confirmed the presence of electrons and protons. They could do these experiments because the particles were charged. You wonder how they are going to find out about the neutral subatomic particles that you know are also in the nucleus. You ask the computer for a summary of the experiments leading to the discovery of these particles.

According to the computer, the scientists of the early twentieth century knew that there was still something missing: protons alone could not account for the density of the nucleus. Hydrogen was known to be made of one electron and one proton. Helium contained two electrons and two protons. Therefore, the ratio of the mass of a helium atom to the mass of a hydrogen atom should, in theory, be 2:1. But the experimental ratio came out to 4:1.

In 1932 Sir James Chadwick discovered the reason. He performed an experiment similar to Rutherford's. He bombarded a thin sheet of beryllium with alpha particles. Unidentified, high-energy radiation was given off. Chadwick continued to experiment, eventually proving that the radiation was made up of electrically neutral particles that weighed about as much as the proton. Chadwick had discovered the neutron. **Neutrons** are subatomic particles found in the nucleus and have no charge—they are neutral. Now the 4:1 ratio of the mass of helium to the mass of hydrogen could be explained. The nucleus of hydrogen has one proton and no neutrons, but the nucleus of helium has two protons and two neutrons.

 ASK YOURSELF

How did scientists know there was another particle in the nucleus other than the proton?

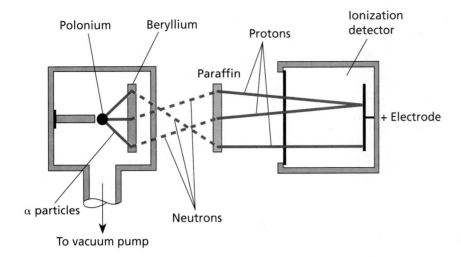

Figure 16–5. Chadwick's experimental apparatus

There'll Be Some Changes Made

You ask the computer to take you back into the early twentieth century. You remember from your student days that the atomic theory continued to be revised throughout this century, and you want to see some of the highlights. The various atomic models developed during the early twentieth century are shown in Figure 16–6.

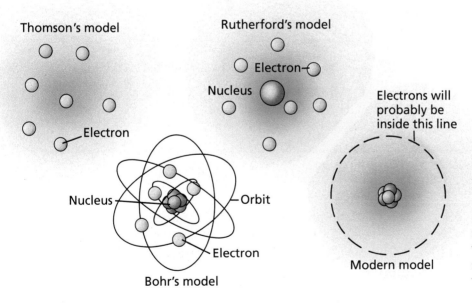

Figure 16–6. Shown here are several different types of atomic models. By studying the changes in these models, you can see how the atomic theory has evolved.

You find yourself back in Dr. Rutherford's lab. Niels Bohr, a Danish scientist working with Rutherford, is discussing his idea that the electrons circle the nucleus in orbits called **energy levels**. (You recall that he won the Nobel Prize for his work on atomic structure in 1922.)

A jump to 1926 and you find yourself in a small seminar room where Werner Heisenberg, a German physicist, is writing equations on the board. He has mathematically demonstrated that both the motion and the exact position of an electron can never be known precisely at the same time. As a result, Heisenberg proposes that there are regions, called *orbitals,* where electrons are most likely to be. Heisenberg and the scientists who worked with him thought that electrons are found around the nucleus in a cloud.

You continue your journey through the "primitive" twentieth century. Eventually, you write what you learned about the atom in the ship's log. "Ship's log: star date 2254.7. I understand much better how hard it is to formulate an atomic theory and why it has been modified so often. I will proceed to summarize my findings . . . ."

 BY *Writing* _____

Summarize your ideas about atomic theories in your journal. Include how experimentation has helped advance the ideas of science.

▶ **ASK YOURSELF**

Why do you think the neutron was among the last subatomic particles to be discovered?

SECTION 1 *REVIEW AND APPLICATION*

Reading Critically

1. Who first used the word *atom?*

2. Describe Rutherford's gold-foil experiment.

Thinking Critically

3. Organize what you have learned about the evolution of atomic theory into a table, a chart, a timeline, or a graph. Use as many diagrams as possible.

4. Suppose that you were trying to explain why it rains. What kind of explanation would you have used if you were an ancient Greek atomist? What kind of explanation would you give if you lived in the early part of the twentieth century? Compare and contrast both theories.

The Modern Atom

It is very hard to visualize particles that are so small you cannot see them. Visual models such as those in Figure 16-6 can help. How can you improve on the modern model in this figure?

Objectives

List the three major types of atomic particles and their properties.

Calculate the number of electrons, protons, and neutrons in a given atom, using the atomic number and atomic mass.

It All Works Together

Back at the starship, you decide to get a bite to eat in your quarters. You are still trying to fit modern atomic theory and its historical development together. How can you visualize an electron cloud? Ah! You have it. You ask the food replicator to make you some cotton candy. You realize that the electron cloud is like a ball of cotton candy. It is not very dense on the outside, and if you touch it, it gives a bit. As you go toward the center, it gets more dense. It has a solid, dense center (the paper cone it is attached to). This center would be similar to the nucleus. "That's it!" you think to yourself. I shouldn't be thinking of electrons as solid balls orbiting a nucleus. That was Bohr's theory, which was outdated even in the twentieth century."

After finishing the cotton candy, you feel you have placed all the concepts in their proper place. You dictate the following summary into the ship's log.

An atom is the smallest unit of matter that retains the properties of matter. The modern model of the atom has electrons swirling around the nucleus in a large region, rather than orbiting in a fixed pattern. The region in which the electrons are found is called the *electron cloud.* As you move toward the center of the atom, chances of encountering a particle increase.

According to modern atomic theory, an atom is composed of three types of particles: protons, neutrons, and electrons. Protons are particles with a positive charge. Neutrons have no charge. The protons and neutrons make up the nucleus. Electrons have a negative charge. The negative charge of an electron is equal in strength to the positive charge of a proton. When the number of electrons equals the number of protons in an atom, the atom has no charge: it is a neutral atom.

It is very difficult to experiment with atoms since they are so small. It would take about 1×10^{18} atoms to make a period at the end of this sentence. Moreover, the nucleus of an atom is even smaller. The computer says that if the diameter of an atom were enlarged to the length of a football field, the nucleus would be slightly smaller than a pea!

 ASK YOURSELF

Match each of the following models with the scientist whose atomic theory it best represents: a swarm of bees; a tennis ball; a ball tied with a string and whirled above your head. Explain your choices.

It's All Relative

DISCOVER BY *Calculating* _____

Suppose you knew the mass of three of your friends—25 kg, 50 kg, and 75 kg. You can make a set of relative masses by assigning your 50-kg friend a relative mass of 1 fmu (friendly mass unit). The friend that has a mass of 75 kg would have a relative mass of $1\frac{1}{2}$, or 1.5, fmu. Every 50 kg would be 1 fmu. What would your 25-kg friend's mass be in fmu? ✏

The mass of an atom is also measured using relative units called *atomic mass units* (amu). Protons and neutrons each have about the same mass—nearly 1 amu. The **atomic mass** of an atom is equal to the sum of the relative masses of all its protons and neutrons. An atom's **mass number** is its atomic mass rounded to the nearest whole number. It indicates the total number of protons and neutrons in the atom.

Although the electrons take up most of the space in an atom, they contribute very little to the mass of the atom. In fact, it would take nearly 2000 electrons to equal the mass of one proton. Because electrons have so little mass, they can usually be ignored when calculating the mass of an atom.

▶ **ASK YOURSELF**

How do scientists determine the atomic mass of an element?

ID Please

Whether you live in an apartment, a house, or a boat at the marina, your home is identified by a number. This guarantees that you get your mail and that friends can find where you live. Your address is unique.

Figure 16–7. A street address is similar to an element's atomic number.

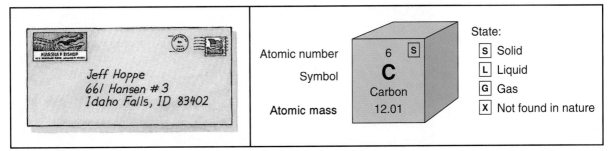

Atoms of different elements can also be identified by a number—the atomic number. The **atomic number** of an element is the number of protons in its nucleus. Each element has a unique atomic number because each has a unique number of protons. The atomic number for oxygen is 8; for calcium, 20; for gold, 79. Knowing the mass number and the atomic number of an element can help you find the number of electrons, protons, and neutrons in a neutral atom.

Look at the calculation below. If you subtract the atomic number from the mass number, the result is the number of neutrons in the atom. Since the number of protons is equal to the number of electrons in a neutral atom, you know the complete particle composition of the atom. For an atom of carbon:

mass number	= 12 = number of protons + number of neutrons	
− atomic number	= 6 = number of protons	
number of neutrons	= 6 = number of neutrons	

 ASK YOURSELF

Element Y is composed of neutral atoms. It has a mass number of 45 and an atomic number of 21. How many electrons, protons, and neutrons does element Y have?

Do the Neutron Dance

You are still dictating your notes in the ship's log when Dr. Shawan calls you on the intercom to come to sick bay. "On my way," you respond. You think to yourself, "It's probably about Ensign Roark."

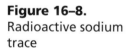
Figure 16–8.
Radioactive sodium trace

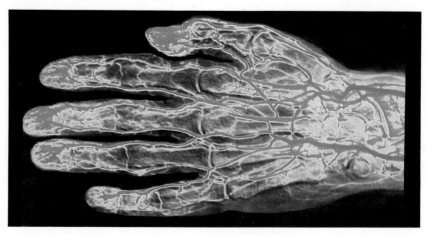

In the last week, Ensign Roark has noticed his toes turning blue and becoming numb after exercising. This condition could be—OH NO!—the rare Malsupian virus or simply poor blood circulation. Dr. Shawan was planning to check the ensign's circulation first.

Since normal blood contains approximately one percent sodium, Dr. Shawan injected a radioactive form of sodium and monitored its progress through the ensign's legs. Look at the graphic of a similar test in Figure 16–8.

Dr. Shawan explains that atoms of the same element vary in mass, just as one apple may have a different mass than another apple, even though it is from the same tree. In atoms, the difference in mass is due to a difference in the number of neutrons. Although the atoms of an element always have the same number of protons, the number of neutrons may vary. Atoms of the same element that have a different number of neutrons are called **isotopes.** Each element has a fixed number of naturally occurring isotopes; most elements have two or more. Some isotopes are radioactive, and their position may be traced using an instrument that detects radioactivity, such as a Geiger counter.

Figure 16–9. Isotopes of carbon

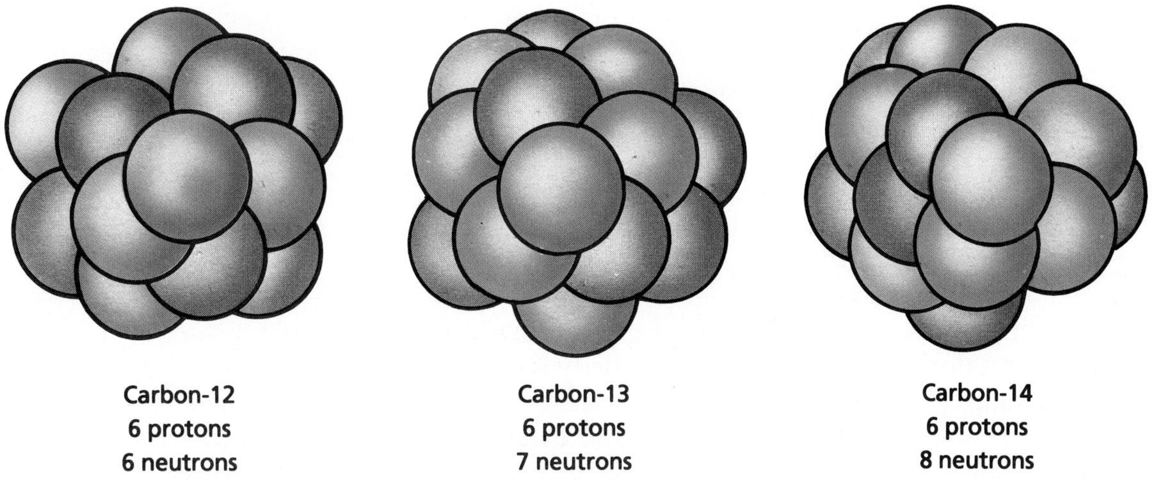

Carbon-12	Carbon-13	Carbon-14
6 protons	6 protons	6 protons
6 neutrons	7 neutrons	8 neutrons

Isotopes are indicated by writing the element's name followed by the mass number. For instance, sodium-23 is the common form of sodium in human blood. Sodium-23 has 11 protons and 12 neutrons. Sodium-24 is a radioactive isotope that is used in medical traces. How many more neutrons does sodium-24 have than sodium-23?

Table 16-1	Common Isotopes and Their Natural Abundance		
Isotope	**Abundance (%)**	**Mass (amu)**
Hydrogen-1	99.98	1.008
Hydrogen-2	0.02	2.014
Carbon-12	98.89	12.000
Carbon-13	1.11	13.003
Oxygen-16	99.76	15.995
Oxygen-17	0.04	16.999
Oxygen-18	0.20	17.999

As Dalton worked on his atomic theory, he assumed that all the atoms of a particular element had the same mass. We now know that atoms of the same element may have a different number of neutrons. This means that atoms of the same element could have slightly different masses. When scientists talk about atomic mass, they are referring to the *average atomic mass*. As the term implies, this is an average mass of all the isotopes for that element.

After examining the results of the test, Dr. Shawan realizes that Ensign Roark has a blockage in the artery to one leg. Dr. Shawan now knows how to treat the problem. Fortunately, the solution to this problem was fairly simple. Thank goodness it was not the dreaded Malsupian virus!

 **ASK YOURSELF**

What is an isotope?

SECTION 2 *REVIEW AND APPLICATION*

Reading Critically

1. Describe and define the subatomic parts of the atom.

2. A neutral atom has a mass number of 28 and an atomic number of 14. How many protons, neutrons, and electrons does it have? Construct a graphic similar to the one on page 440 that shows how you arrived at your answer.

Thinking Critically

3. Electricity is a flow of electrons. Electrons are part of all substances, yet you are not electrically shocked by all the objects around you. Explain.

4. Two isotopes of chlorine are chlorine-35 and chlorine-37. What can you specify about the atomic particles of these two isotopes?

INVESTIGATION

Determining Relative Mass

▶ MATERIALS
- balance ● beads (5) ● medium washers (5) ● small washers (5)

▼ PROCEDURE

1. Make a table like the one shown.
2. Determine the mass of one bead and each type of washer. Record the data in your table.
3. Compile 2.00 g of beads, remove them from the balance, and count them. Record the number of beads, and compare your results with those of three other groups.
4. Repeat step 3 for each set of washers. Record the results in your table.
5. Determine the mass of all five beads. Then calculate the average mass of one bead by dividing the mass by 5. Record your results in the table.
6. Using the method in step 5, calculate the average mass of each type of washer.
7. To determine the standard relative mass, choose the smallest of the average masses you calculated. Then divide the average mass of each group by the standard. For example, if your average masses were:

Washer 1	Washer 2	Bead
0.50 g	0.11 g	0.01 g

the smallest mass would be 0.01 g, so the bead would be your standard. You would divide the mass of each object by the mass of the bead. For Washer 1 you would divide 0.50 g by 0.01 g, which equals 50.

This is the relative mass of Washer 1. It is called the *relative mass* because its measurement is relative to a standard. Record the relative masses of the objects in the table.

8. Without counting, use the balance to determine one-half of the relative mass for each object. For example, if you use the standard, its relative mass is 1.0 g. One-half of the relative mass would be 0.5 g. Now count the number of each object. Record your results.

TABLE 1: CALCULATION OF RELATIVE MASS

Object	Mass of One	Number of Objects in 2 g	Average Mass	Relative Mass	Number in $\frac{1}{2}$ Relative Mass
Bead					
Washer #1					
Washer #2					

▶ ANALYSES AND CONCLUSIONS

1. When you counted the objects in step 8, were there as many as you thought there would be? Explain.
2. How is the atomic mass unit similar to the relative mass units you worked with in this Investigation?
3. What are the benefits, in the case of atoms, to using relative masses rather than actual masses?

▶ APPLICATION

If you were a football player, would it help you to know the relative mass of an opposing team? Explain.

✴ Discover More

The atomic mass unit is defined as $\frac{1}{12}$ the mass of the carbon-12 atom. The relative mass of other elements is found by comparing their average atomic mass to that of carbon-12. Why do you think carbon-12 is used as the standard?

The Periodic Table

Objectives

Demonstrate an understanding of how the periodic table is organized.

List the major parts of the periodic table, and know where they are located.

Analyze the benefits that the periodic table provides.

Many new elements have been made or discovered since the twentieth century. How can any one scientist know the properties of all of these elements? How do scientists and engineers know which elements to use?

The Key Is Organization

One of the most important things you do as captain of a starship is to organize data. Organizing information makes it much easier to know when certain jobs need to be done and when those jobs have been completed.

Scientists also organize data. They organize for many of the same reasons that a starship captain would. Scientists use organized data to find the patterns that exist in nature.

Suppose your teacher gave you a bag of objects and asked you to classify them. Normally, you would use common patterns found among the objects in order to classify them. For example, buttons might be classified by color, shape, or how many holes they have. Scientists do much the same thing. Patterns among data are very important to scientists. In order to appreciate how important pattern recognition is in science, try to organize the following information by using a table.

ᴰⁱˢᶜᵒᵛᵉᴿ ᴮʸ Observing _____

Group the elements that have similar characteristics together. Here are the elements you should consider.

Sodium

Sodium is a soft, silvery metal that can be shaped by pounding or pressing on it. It reacts violently with acids, has an atomic number of 11, and an average atomic mass of 22.99. It conducts electricity.

Fluorine

Fluorine is a pale yellow, poisonous gas. It does not conduct electricity. It reacts vigorously with sodium. It has an atomic number of 9 and an average atomic mass of 19.00.

Potassium

A grey solid, potassium can be shaped into wires that conduct electricity. It reacts violently with acids. It has an atomic number of 19 and an atomic mass of 39.10.

Chlorine

A yellow-green gas, chlorine is very poisonous. It does not conduct electricity. It reacts vigorously with sodium. It has an atomic number of 17 and an atomic mass of 35.45.

Magnesium

Magnesium is a shiny metal that conducts electricity. It bubbles in acids. It has an atomic number of 12 and an atomic mass of 24.31.

Copper

A reddish-brown metal, copper conducts electricity. It may be shaped into wires or pounded to make jewelry. It has an atomic number of 29 and an atomic mass of 63.55.

Silver

Silver is a shiny gray solid that conducts electricity. It may be shaped into wires or pounded to make jewelry. It has an atomic number of 47 and an atomic mass of 107.87.

Try to group these elements together by common properties. Make a list of the elements in each of your "families," and compare them with your classmates' lists. How easy was it to make your groupings? What properties were most important?

Back on the starship, you are in a conference with the ship's science officer Mr. Martinez. Apparently a search team on the planet has discovered a new substance. They suspect it might be a new element. Mr. Martinez is trying to decide what properties this element might have. Will it conduct electricity? Will it be radioactive? Will it have useful properties as an insulator? Will it be too reactive in the earth's atmosphere? You tell the science officer that in order to answer his questions he needs to find out what family in the periodic table this new element might belong to. The process will be similar to the grouping activity you just did.

By the late 1800s, many elements had already been discovered. Scientists tried (just as you did) to arrange the elements into a logical pattern. Dmitri Mendeleev, a Russian chemist, proposed an arrangement of the known elements based on their atomic mass. He organized the elements in horizontal rows, with each row containing elements with similar properties. This arrangement of elements also created vertical columns, which Mendeleev called **periods**. His table is called the **periodic table** of the elements.

Figure 16–10. Dmitri Mendeleev fashioned the first periodic table.

The Periodic Table

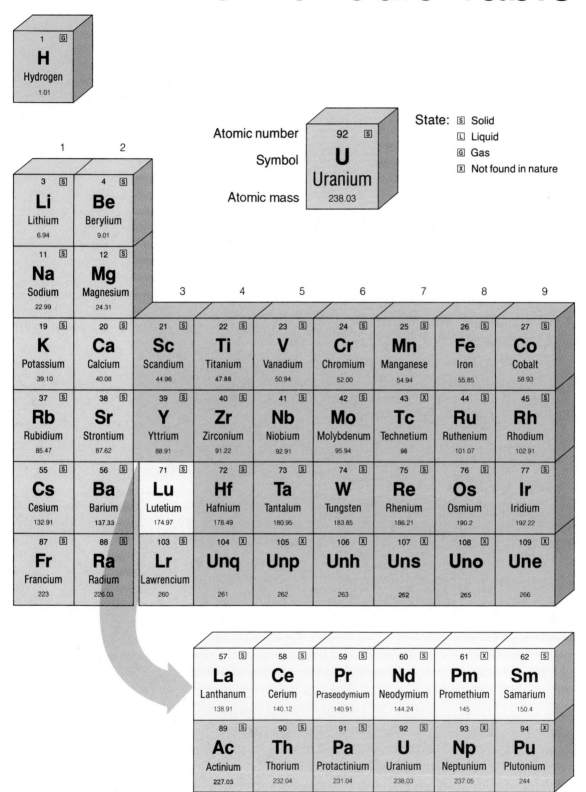

Atomic number — 92
Symbol — **U**
Uranium
Atomic mass — 238.03

State: Ⓢ Solid
Ⓛ Liquid
Ⓖ Gas
☒ Not found in nature

	1	2	3	4	5	6	7	8	9
	H 1 Ⓖ Hydrogen 1.01								
	Li 3 Ⓢ Lithium 6.94	**Be** 4 Ⓢ Berylium 9.01							
	Na 11 Ⓢ Sodium 22.99	**Mg** 12 Ⓢ Magnesium 24.31							
	K 19 Ⓢ Potassium 39.10	**Ca** 20 Ⓢ Calcium 40.08	**Sc** 21 Ⓢ Scandium 44.96	**Ti** 22 Ⓢ Titanium 47.88	**V** 23 Ⓢ Vanadium 50.94	**Cr** 24 Ⓢ Chromium 52.00	**Mn** 25 Ⓢ Manganese 54.94	**Fe** 26 Ⓢ Iron 55.85	**Co** 27 Ⓢ Cobalt 58.93
	Rb 37 Ⓢ Rubidium 85.47	**Sr** 38 Ⓢ Strontium 87.62	**Y** 39 Ⓢ Yttrium 88.91	**Zr** 40 Ⓢ Zirconium 91.22	**Nb** 41 Ⓢ Niobium 92.91	**Mo** 42 Ⓢ Molybdenum 95.94	**Tc** 43 ☒ Technetium 98	**Ru** 44 Ⓢ Ruthenium 101.07	**Rh** 45 Ⓢ Rhodium 102.91
	Cs 55 Ⓢ Cesium 132.91	**Ba** 56 Ⓢ Barium 137.33	**Lu** 71 Ⓢ Lutetium 174.97	**Hf** 72 Ⓢ Hafnium 178.49	**Ta** 73 Ⓢ Tantalum 180.95	**W** 74 Ⓢ Tungsten 183.85	**Re** 75 Ⓢ Rhenium 186.21	**Os** 76 Ⓢ Osmium 190.2	**Ir** 77 Ⓢ Iridium 192.22
	Fr 87 Ⓢ Francium 223	**Ra** 88 Ⓢ Radium 226.03	**Lr** 103 Ⓢ Lawrencium 260	**Unq** 104 ☒ 261	**Unp** 105 ☒ 262	**Unh** 106 ☒ 263	**Uns** 107 ☒ 262	**Uno** 108 ☒ 265	**Une** 109 ☒ 266

La 57 Ⓢ Lanthanum 138.91	**Ce** 58 Ⓢ Cerium 140.12	**Pr** 59 Ⓢ Praseodymium 140.91	**Nd** 60 Ⓢ Neodymium 144.24	**Pm** 61 ☒ Promethium 145	**Sm** 62 Ⓢ Samarium 150.4
Ac 89 Ⓢ Actinium 227.03	**Th** 90 Ⓢ Thorium 232.04	**Pa** 91 Ⓢ Protactinium 231.04	**U** 92 Ⓢ Uranium 238.03	**Np** 93 ☒ Neptunium 237.05	**Pu** 94 ☒ Plutonium 244

										18

Legend:
- Metals
- Transition Metals
- Nonmetals
- Noble gases
- Lanthanide series
- Actinide series

			13	14	15	16	17	2 [G] **He** Helium 4.00

| | | | 5 [S] **B** Boron 10.81 | 6 [S] **C** Carbon 12.01 | 7 [G] **N** Nitrogen 14.01 | 8 [G] **O** Oxygen 16.00 | 9 [G] **F** Fluorine 19.00 | 10 [G] **Ne** Neon 20.18 |

| | | | 13 [S] **Al** Aluminum 26.98 | 14 [S] **Si** Silicon 28.09 | 15 [S] **P** Phosphorus 30.97 | 16 [S] **S** Sulfur 32.07 | 17 [G] **Cl** Chlorine 35.45 | 18 [G] **Ar** Argon 39.95 |

10	11	12						
28 [S] **Ni** Nickel 58.69	29 [S] **Cu** Copper 63.55	30 [S] **Zn** Zinc 65.39	31 [S] **Ga** Gallium 69.72	32 [S] **Ge** Germanium 72.61	33 [S] **As** Arsenic 74.92	34 [S] **Se** Selenium 78.96	35 [L] **Br** Bromine 79.90	36 [G] **Kr** Krypton 83.80
46 [S] **Pd** Palladium 106.42	47 [S] **Ag** Silver 107.87	48 [S] **Cd** Cadmium 112.41	49 [S] **In** Indium 114.82	50 [S] **Sn** Tin 118.71	51 [S] **Sb** Antimony 121.75	52 [S] **Te** Tellurium 127.60	53 [S] **I** Iodine 126.90	54 [G] **Xe** Xenon 131.29
78 [S] **Pt** Platinum 195.08	79 [S] **Au** Gold 196.97	80 [L] **Hg** Mercury 200.59	81 [S] **Tl** Thallium 204.38	82 [S] **Pb** Lead 207.2	83 [S] **Bi** Bismuth 208.98	84 [S] **Po** Polonium 209	85 [S] **At** Astatine 210	86 [G] **Rn** Radon 222

63 [S] **Eu** Europium 151.96	64 [S] **Gd** Gadolinium 157.25	65 [S] **Tb** Terbium 158.93	66 [S] **Dy** Dysprosium 162.50	67 [S] **Ho** Holmium 164.93	68 [S] **Er** Erbium 167.26	69 [S] **Tm** Thulium 168.93	70 [S] **Yb** Ytterbium 173.04
95 [X] **Am** Americium 243	96 [X] **Cm** Curium 247	97 [X] **Bk** Berkelium 247	98 [X] **Cf** Californium 251	99 [X] **Es** Einsteinium 252	100 [X] **Fm** Fermium 257	101 [X] **Md** Mendelevium 258	102 [X] **No** Nobelium 259

A periodic property is one that repeats every so often (periodically). The coming of winter, for example, is a periodic event. Some of the properties of the elements are also periodic. Compare Mendeleev's periodic table with your grouping of similar elements. How are they alike? How are they different? Why do you think the tables are different?

ASK YOURSELF

Why is the periodic table called periodic?

Reading the Periodic Table

Since Mendeleev developed the periodic table, many more elements have been discovered and added. What is surprising is that the modern periodic table is very similar to Mendeleev's original. The major difference, other than the addition of elements, is that the vertical columns on the modern table are like Mendeleev's horizontal rows.

Figure 16–11.
Mendeleev's original manuscript is shown at the left. His findings were first published in the journal shown at the right.

The modern periodic table is a useful summary of many atomic properties. Knowing how to use the periodic table will help you understand many of the properties of elements without having to memorize the information.

Look at any square of the periodic table. In the square, you can see the following information: the symbol for the element, the atomic number, and the average atomic mass. In addition, the state of the element as it is found in nature is shown in the upper right-hand corner.

The current periodic table is arranged in order of atomic number. Scientists have found that this method best illustrates periodic properties.

If you look at the top of the table, between Group 13 and Group 14, you will see the beginning of a zigzag line. To the left of this line are the metals, and to the right are the nonmetals. We will focus on metals and nonmetals in the next chapter.

Elements arranged in vertical columns are known as **groups,** or families. These are similar to the "families" you used when you organized data about the elements. Some of these families have specific names. In Figure 16–12 you see some important groups. All the elements in any of these major groups have similar properties.

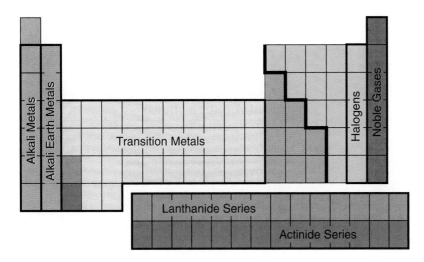

Figure 16–12. Some of the groups and periods are marked on this shell of the periodic table.

Thanks to the holodeck's computer, you have a better understanding of atomic theory and the elements. However, you still have some questions. What do subatomic particles have to do with how matter reacts? Why do some elements combine with others in a particular way? Why are some elements unreactive? There is still much to learn, but you are interrupted by the chief engineer—something is wrong with the engines, and you'd better take a look. Your research will have to wait.

 ASK YOURSELF

What information would you find in one square of the periodic table?

SECTION 3 *REVIEW AND APPLICATION*

Reading Critically

1. Use the periodic table to find the elements whose atomic numbers are 12, 34, and 17. Write the period and the group of each element.

2. Why is the periodic table important to scientists?

Thinking Critically

3. Describe the usefulness of the periodic table.

4. You have discovered a new element and are trying to decide what group it belongs to in the periodic table. What information do you need to accurately place the element?

SKILL Making Tables

▶ **MATERIALS**
- 15 to 20 objects

▼ **PROCEDURE**

1. All tables should be organized in a way similar to the diagram shown here.

A table should have a title that describes the information shown.

Each row should be clearly labeled.

Each column should be clearly labeled.

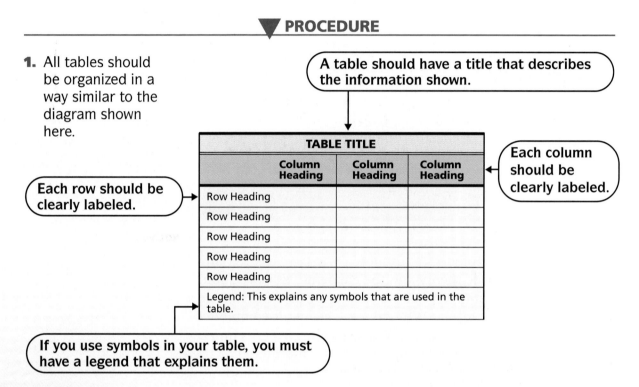

TABLE TITLE			
	Column Heading	Column Heading	Column Heading
Row Heading			
Row Heading			
Row Heading			
Row Heading			
Row Heading			
Legend: This explains any symbols that are used in the table.			

If you use symbols in your table, you must have a legend that explains them.

2. Choose 15 to 20 objects, and write down their names. Decide on ways to group the objects by characteristics they may have in common. For example, objects made out of wood may be one group.

Remember to use only those characteristics that you can observe. Do not take any of the objects apart. You might want to design symbols for the objects you choose for your table.

▶ **APPLICATION**
What characteristics did you use to separate the objects in the table into rows and columns? How did your choice of characteristics help you to classify the objects you were given?

✳ *Using What You Have Learned*
List at least three ways your table is like the periodic table and three ways in which your table is different from the periodic table.

HIGHLIGHTS

The Big Idea

All matter is made up of small particles called atoms. Scientists use models to show their ideas about the structure of an atom and the interaction of its particles. As scientists gain new information from experiments, they revise their models. Scientists today use the electron cloud model. This model, too, may change as new information is discovered.

Scientists organize what they know about matter in the periodic table. Identifying patterns and organizing them into tables helps scientists predict the properties of different kinds of matter.

For Your Journal

Look back at the ideas you wrote in your journal at the beginning of the chapter. Have your ideas changed? Revise your journal entry to show what you have learned. Be sure to include information on how understanding the nature of matter could help you in your everyday life.

Connecting Ideas

This timeline shows the basic development of the atomic theory. Copy the timeline into your journal, and use what you have learned in the chapter to complete it.

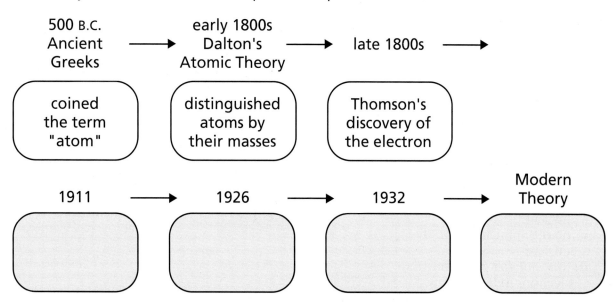

500 B.C.
Ancient Greeks → early 1800s Dalton's Atomic Theory → late 1800s →

coined the term "atom"

distinguished atoms by their masses

Thomson's discovery of the electron

1911 → 1926 → 1932 → Modern Theory

REVIEW

Understanding Vocabulary

1. Explain the meanings of each term or set of terms below.
- **a)** proton (434), neutron (434), electron (432)
- **b)** energy levels (435)
- **c)** atomic mass (439), atomic number (440)
- **d)** isotopes (441)

Understanding Concepts

MULTIPLE CHOICE

2. According to modern atomic theory, which of the following is not a particle in an atom?
- **a)** electron
- **b)** nucleus
- **c)** neutron
- **d)** proton

3. The model of the atom that consisted of a positive center with electrons fastened to its surface was proposed by
- **a)** Rutherford.
- **b)** Thomson.
- **c)** Aristotle.
- **d)** Dalton.

4. The negative charge of an electron is equal in strength to the positive charge of the
- **a)** proton.
- **b)** electron.
- **c)** neutron.
- **d)** mass.

5. The atomic number of an atom is equal to the
- **a)** number of protons in the nucleus.
- **b)** number of electrons in the nucleus.
- **c)** number of neutrons in the nucleus.
- **d)** total number of protons and electrons in the nucleus.

6. The mass number equals the
- **a)** number of protons.
- **b)** number of electrons.
- **c)** total number of protons and neutrons.
- **d)** total number of protons and electrons.

SHORT ANSWER

7. Briefly explain the structure of the modern periodic table.

8. Compare and contrast the modern model of the atom and the Bohr model.

Interpreting Graphics

9. Look at this square of an element in the periodic table. What element is it?

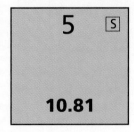

10. Using the periodic table and what you have learned about the elements, find the placement on the periodic table of the following types of elements: gases, metals, nonmetals, synthetic elements, and radioactive elements.

Reviewing Themes

11. *Systems and Structures*
Describe the three types of subatomic particles and explain how they interact.

12. *Systems and Structures*
Using the periodic table, explain how one element differs from the one before it and the one after it.

Thinking Critically

13. Explain why most of the positive particles in Rutherford's experiment passed through the solid gold foil. Why were some particles deflected at an angle? Why did some particles bounce back at the particle source?

14. One periodic property is atomic size, which is compared in terms of the radii (plural of *radius*) of the atoms. The trend is that atomic size decreases as you move along the period, as shown in the diagram below. Explain how this could occur, given that each successive atom in a series has one more proton and one more electron, as well as more neutrons, than the atom before it.

15. Why have all the explanations on the nature of the atom been theories instead of laws or principles?

16. In what way does the nucleus of an atom differ from the electron cloud?

17. If protons and neutrons are present in the nucleus of an atom and are surrounded by a much larger electron cloud, what parts of atoms are most likely to interact when atoms come together?

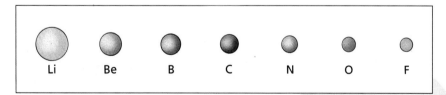

Li   Be   B   C   N   O   F

Discovery Through Reading

von Baeyer, Hans C. "Atom Chasing: Thanks to Sam Hurst, the Chemical Composition of Matter Is Now an Open Book." *Discover* 42(April, 1992). This article focuses on the founder of Atom Sciences, Inc., who has developed a resonance technique for detecting the smallest particles of matter.

METALS AND NONMETALS

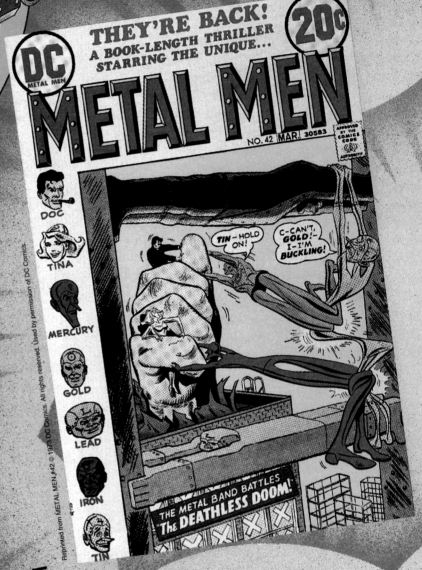

We use metals in our lives every day. Although few people think about the properties of these metals, it is those very properties that make the metals useful. Nonmetallic elements, while more difficult to classify, also have specific properties that enable us to use them.

The unique METAL MEN were comic-book heroes with personalities that reflected the properties of the metals they were made of. The METAL MEN never made it to television cartoons, but in a vintage comic-book adventure they battled an evil toxic-waste monster by using their metallic properties.

There was IRON, the "muscular" metal, who could shape himself into girders and cranes. PLATINUM was an elegant metal who could stretch herself thinner than a human hair. The pushover in this crew of heroes was TIN, a soft metal who was always "buckling" under pressure. TIN just knew he would be a failure. The temperamental metal was MERCURY, who could feel himself expanding as he got angry.

The METAL MEN could not match today's high-tech cartoon heroes, but their comic-book story shows how science fiction anticipates the future. The METAL MEN battled "Chemo," a monster made of discarded toxic waste. Chemo's harmful chemicals corroded buildings and melted railroad tracks as if they were wax. The METAL MEN may be out of date, but they fought scientific problems that scientists and citizens still face today.

For Your Journal

- How can you tell the difference between a metal and a nonmetal?
- How do you use metals in your daily life?
- How are nonmetals useful to you?

Characteristics of Metals and Nonmetals

Objectives

List *the general properties of metals.*

Name *five metallic elements, and* **describe** *their uses.*

Compare and contrast *the properties of metals and nonmetals.*

Some elements are found in nature as beautiful crystals; others occur as gases you breathe every day. Still others exist only under laboratory conditions. They range from very active to totally nonreactive, and they have a wide variety of colors, shapes, and smells. Some elements even produce characteristic colors when they ignite, as they do when fireworks explode.

You know that the periodic table is divided into metals and nonmetals. How do chemists decide whether an element is a metal or a nonmetal?

Figure 17–1. Properties of elements are as varied as the colors of a fireworks display.

From Lead to Gold

In ancient times metals were very important for making tools and weapons. So important, in fact, that historians have given certain periods of history the names of the metals that were used during the period, such as the Iron Age and the Bronze Age.

Knowledge of metals increased throughout the ages, from prehistoric times up to the Middle Ages, where we will take a

Figure 17–2. Alchemists worked in laboratories just as modern scientists do. They even used some of the same kinds of equipment.

short stop to look at some research. The science of chemistry was being born. Early experimenters were called *alchemists*; the subject they studied was called *alchemy*—a word that comes from an old name for Egypt. The term *chemist* is derived from the original *alchemist*.

Alchemists had two main objectives: to find the elixir of perpetual youth and to convert ordinary metals to gold. They dreamed of "striking it rich" by finding a way to convert copper, tin, or iron into beautiful, shiny, valuable gold. You, too, can experiment in alchemy. Try changing copper into gold.

ACTIVITY

Can you change copper to gold?

MATERIALS
safety goggles; laboratory apron; spatula; zinc dust; beaker, 400 mL (2); sodium hydroxide solution, 6 M; hot plate; forceps; copper penny, pre-1982; tongs; paper towels; laboratory burner

CAUTION: Zinc dust can ignite. Sodium hydroxide solution can cause burns. Be sure to wear safety goggles and a laboratory apron for this activity.

PROCEDURE

1. Use the spatula to put about 5 g of zinc dust into a beaker. Add enough sodium hydroxide solution to cover

the zinc, and gently warm it on the hot plate until the mixture is steaming.

2. Using forceps, place the copper penny in the mixture. Use the tongs to remove the beaker from the hot plate.

3. Observe the penny. Record any changes in your journal.

4. Use the forceps to remove the penny from the solution. Swish the penny in a beaker of clean water, and dry it with a paper towel.

5. Hold the penny with the forceps, and insert it into the outer part of the burner flame for several seconds until you observe a change. Immediately place the penny back into the beaker of water. What has happened?

APPLICATION

1. Cut the penny, and describe what the inside looks like. Explain what you observe.

2. Did you turn copper to gold? Explain your answer.

Did you make gold? How do you know? What metal coats your token? To find out, you can do what scientists do every day—research.

DISCOVER BY Researching

Use the card catalog in the library to find at least three references that could help you decide what happened in your experiment. Three subjects that may be of help are alchemy, copper, and gold. Other subjects you may want to check are transmutation and alloys. Use your card catalog information to start digging for your answer. Write your findings in your journal. ✏

► ASK YOURSELF

Why did alchemists try to change other metals into gold?

Luster or Lackluster

Your experiment as an alchemist has taught you that you can't change one metal into another. Why do you think that is so? To formulate an answer to this question, let's look at the periodic table (Figure 17–3). You can see that most of the elements are metals, but what makes an element a metal? Remember the *METAL MEN*? Each had different characteristics, but they did have some things in common. Metals are defined by their properties.

Metals are good conductors of electricity and heat. They can be shaped into sheets and formed into wires. All metals have luster. Yet while some shine, others tarnish easily and become dull looking.

Many of these properties are due to the arrangement of the atoms of metals as shown in Figure 17–4. Atoms of metallic elements are arranged in rows stacked on top of one another. The outer electrons in a metal atom are not held closely by the nucleus. The result is one electron cloud covering all the nuclei in a metal, rather than separate ones covering each nucleus. You could think of metal atoms as being similar to glass beads covered by a thick oil. The coating of oil would represent the outer electron cloud.

This electron arrangement is responsible for the excellent heat conductivity

Figure 17–3. The periodic table

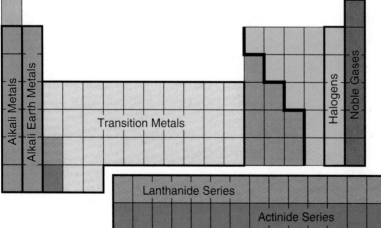

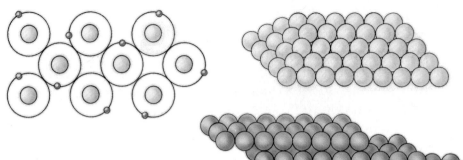

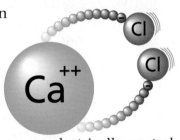

Figure 17–4. Metal atoms are arranged in sheets. The electrons of these metal atoms form a cloud around all of the atoms in the metal.

of metals. The electrons can transfer heat as they move about randomly in the metal. The properties attributed to the *METAL MEN* were in part due to their electron arrangements.

Since metals have loose outer electrons, they tend to lose electrons easily, forming positively charged particles. Remember that atoms are electrically neutral when the number of protons and the number of electrons are the same. If an atom loses an electron, it has an extra positive charge. A neutral substance that loses or gains electrons is called an **ion.** Look at Figure 17–5 to see how metals form ions.

Figure 17–5. Metals form positive ions easily.

The properties of nonmetals are harder to describe than the properties of metals because nonmetals don't have very many properties in common. They are very different from one another. Nonmetals do, however, have a characteristic that helps us identify them: poor conductivity of electricity and heat.

How elements form ions also gives us a clue to their identity. You know that metals form positive ions. When nonmetals form ions, however, they tend to form negative ions. The outer electrons in many nonmetals are held tightly by the nucleus, so they are not easily lost. Therefore, nonmetals tend to gain electrons rather than lose them.

 ASK YOURSELF

Why are metals easier to describe than nonmetals?

A Pinch of Zinc, a Scoop of Copper

What was your "gold" token really made of? If you did your research well, you found out that the copper and zinc made brass. Mixtures of two or more metals, or mixtures of metals and nonmetals, are called **alloys.**

Figure 17–6. Aluminum is a lightweight metal. Why would it be useful in the manufacture of aircraft bodies?

Alloys have dozens of uses. Suppose you wanted to make a very light form of steel for airplane wings. It would have to be strong enough to resist heavy winds but light enough for the plane to fly. Steel is an alloy of iron and other substances. What light, shiny metal that you have in your kitchen could you combine with iron to make a durable but light alloy?

Sometimes people need metals that have very specific properties. They may have a use for a very strong metal or a metal that is a better conductor of heat than any of the metallic elements alone. Mixing metals and other elements can sometimes produce these specific properties. For example, brass, the mixture of copper and zinc that you made, is harder and far more durable than either copper or zinc alone.

Think about the pots and pans in your kitchen. You may have some pans made of iron, some of aluminum, some of steel. You may even have some that are made of glass! The properties of the elements in these materials enable you to cook food efficiently. What would happen if you had a pan made of tin?

"I'm *TIN.* I'm softer than zinc but harder than lead. I turn to powder at 200°C. Cooking with me could be a problem."

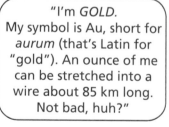

"My name is Tina, and my platinum properties allow me to stretch myself into a strand over 85 km long."

"I'm *GOLD.* My symbol is Au, short for *aurum* (that's Latin for "gold"). An ounce of me can be stretched into a wire about 85 km long. Not bad, huh?"

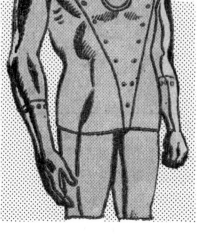

Gold is a very soft metal. For this reason it cannot be used alone to make durable items, such as jewelry. Therefore, gold jewelry is made using an alloy. For instance, copper is combined with gold to make yellow gold. Platinum or silver is added to gold to make white gold.

 **ASK YOURSELF**

Why are alloys important?

Metallic Families

Metals tend to lose electrons to form positive ions. Group 1 metals tend to form +1 ions, while Group 2 metals tend to form +2 ions. The metals in the other groups may form more than one positive ion. That is, they can form +1, +2, or even +3 ions.

The properties of the metals depend on how easily they lose electrons. The easier it is for them to lose electrons, the more reactive they are. In order to analyze the reactivity of metal families, your teacher will demonstrate how they react in water or acid solutions. Make careful observations, and write them in your journal.

DISCOVER BY Observing

Your teacher will place the following metals in water: sodium (Na), calcium (Ca), magnesium (Mg), copper (Cu), and zinc (Zn). What happens? Your teacher will place the metals that didn't react in water into a weak acid solution. Acids are more reactive with metals than water is. Record your observations.

Was the reaction in acid the same as the reaction with water? Describe the similarities and differences.

Now, place the metals in order of reactivity, with the most reactive metal at the top of the list. Then, look up the position of the metals in the periodic table. How does their position in the table relate to their reactivity? 🖉

Potassium (K) wasn't used in the activity because it reacts so violently with water that it is safer to show you a photograph (Figure 17–7). Where would you place potassium in your scheme?

Now that you have organized the metals in order of reactivity, can you see any particular trend? In general, the reactivity of metal families increases as you move from the metals on the top right on the periodic table to the bottom left-hand corner. The alkali metals are the most reactive of the metals, with francium—the last member of the family—being the most reactive.

The transition metals do not follow this trend exactly; there are some exceptions. Which element(s) from the activity did not follow the trend?

Rust Protection   How can the concept of reactivity be used? You may be using it right now to protect your bike. Let's discover how.

Figure 17–7. Alkali metals are very reactive in water. If potassium is placed in water, it bursts into flame.

Discover by Doing

Place an iron nail in a small dish, and cover the nail with water. Take another iron nail, and wrap a strip of zinc tightly around one end. Place this second nail in another dish of water. Leave the nails in the water for several days, and observe the changes. What happened to the nail with the zinc strip? Why? 🖉

Reactivity becomes very important when metals or alloys are used in making things, especially in construction. Many metals react with oxygen in the air to form oxides, or "rusts." This weakens the metal. Iron oxidizes easily; gold and copper do not. Sometimes, to avoid oxidation, one metal is coated with another metal that is more reactive. The more reactive metal will oxidize and form a "crust" that protects the metal underneath. For example, galvanized nails are made of iron (Fe) coated with zinc. The zinc is more reactive than the iron and reacts with oxygen first, thus protecting the iron underneath.

ASK YOURSELF

Describe one way in which the reaction of oxygen with metals could be an advantage and one way in which it could be a disadvantage.

To Beam or Not to Beam

Recall that there is a zigzag line on the periodic table that indicates the boundary between the metals and the nonmetals. The elements that border the line cannot be classified as either metals or nonmetals. These elements are known as metalloids. **Metalloids** are substances that exhibit some but not all of the properties of metals.

Many metalloids are important because they can be used as semiconductors. A *semiconductor* has properties that are in between those of metals and nonmetals. Sometimes they can conduct electricity, and under other circumstances they can act as insulators. Selenium is an example of a semiconductor. It is useful in photoelectric cells such as photographic light meters.

Figure 17–8. Silicon is used in many manufactured goods such as glass, cookware, fiberglass car bodies, and silicon chips. What is one way in which you use silicon every day?

Silicon is a very useful element. It can be used in combination with other elements to make items ranging from glass to silicon chips used in electronics. Silicon chips are used to make wafers for radios, televisions, computers, and video games. How can one element do such different things? Silicon is an excellent semiconductor even if combined with small amounts of other elements.

 ASK YOURSELF

Why are semiconductors useful?

SECTION 1 *REVIEW AND APPLICATION*

Reading Critically

1. What are the differences between the Group 1 and Group 2 metals?
2. Why are metalloids useful?

Thinking Critically

3. Why can't scientists change one metal into another?
4. The chemical activity of metals increases with an increase in atomic number. Why do you think this is so? (Hint: Use your atomic structure models.)

INVESTIGATION

Testing Ions for Flame Colors

▶ MATERIALS

- laboratory apron • safety goggles • wax pencils • test tubes (7) • test tube rack • test solutions (7) • laboratory burner • wire loops • HCl, 1 M

CAUTION: Put on an apron and goggles before starting this experiment. Slight spattering may occur.

▼ PROCEDURE

1. You will be observing the flame colors created by different compounds dissolved in water. Make a table like the one shown to record your observations.

TABLE 1: FLAME TESTS	
Compound	**Color Observed**

2. Your teacher will list the compounds on the chalkboard. Copy the names of the compounds into your table.
3. Use a wax pencil to label the test tubes A through G. Place them in the test tube rack in order. Place a small sample of each compound in the corresponding test tube and moisten with a few drops of HCl.
4. **CAUTION: Flames are hot and can burn you. Do not get your hands too close.** Light your burner. You may wish to refer to page 632 for instructions on adjusting the burner flame.
5. Insert the wire loop into compound A. Hold the loop in the flame. Record the color you see.
6. Repeat step 5 for each of the compounds in the test tubes.

▶ ANALYSES AND CONCLUSIONS

1. You tested different compounds that contained the same element. For instance, LiCl, NaCl, and KCl all contain chlorine. Do all of these compounds have the same flame color? Explain.
2. The elements in each compound are present as ions. In your table, add the names of the ions in each compound and identify their specific colors.
3. What do you think will happen if you mix two of the compounds? Explain. After you have answered the question, try it. Were the results what you expected? Why? Why not?

▶ APPLICATION

Salt substitute usually contains some compound that can be used to replace sodium chloride (NaCl). Do a flame test on some of this substitute and, without looking at the label, determine what metal is present. Look at the label when you are done to see if you were correct.

✳ Discover More

Choose some household compounds and test them. Determine from their flame colors which metals they contain.

Nonmetals

The comic book characters in the beginning of the chapter had only metallic properties. Could we have nonmetal people? *IODINE* would be purple since that is the color of iodine crystals. Iodine easily sublimes (converts from a solid directly into a gas) into a purple vapor. *IODINE* could turn into a purple gas and move through keyholes and under doors. Imagine the possibilities! What other nonmetal can you personify? What properties would be interesting? In order to find out, you will need to study the properties of nonmetals further.

Objectives

Compare the noble gases with other elements in the periodic table.

Diagram the electron configuration for elements in Periods 1 and 2 in the periodic table.

Explain what is meant by diatomic, and give two examples of diatomic elements.

It's a Gas

Drop two balls made of different material on the pavement, and chances are that they won't bounce to the same height. One of them may rebound only a few centimeters, while the other may bounce two meters into the air. The composition of the "dud" is responsible for the small bounce.

The noble gases are the "duds" of the periodic table. The noble gases, Group 18, were once called *inert gases* because it was believed they would not combine with other elements. In other words, they were duds. Scientists have discovered, however, that some of these elements *will* react—but only under high temperatures and pressures.

Just as the composition of a ball is responsible for its bounce, the extraordinary lack of reactivity of the noble gases is due to electron configuration. The noble gases have a peculiar arrangement of electrons that provides them a nonreactivity that other elements do not have.

ASK YOURSELF

How are the noble gases different from other elements?

Electrons, Go to Your Rooms

In order to understand why the noble gases are relatively inert, you need to understand more about electron configurations. Think of it this way. When you go to the music store to buy a new CD, the CDs are arranged in a specific pattern. If you're looking for a specific kind of music, you know which section of the store it will be in. Suppose you want a CD of flute quartets. Where do you go to find it? Would you head for rhythm and blues? Of course not! You would go to the classical section. The music store is like the electron cloud that surrounds the atom. Music stores are divided into sections, while electron clouds are divided into energy levels. Furthermore, just as you wouldn't find a classical selection in R&B, you wouldn't find an electron out of its energy level.

Energy levels can be thought of as compartments in which the electrons move. The energy level closest to the nucleus is the smallest and can hold only two electrons. The farther away the compartment is from the nucleus, the bigger it is and the more electrons it can hold. The second energy level, for instance, can hold a maximum of eight electrons, while the third energy level can hold 18 electrons. Look at the diagram show-

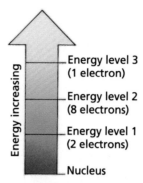

Energy level 3
(1 electron)

Energy level 2
(8 electrons)

Energy level 1
(2 electrons)

Nucleus

Energy increasing

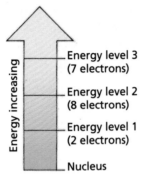

Energy level 3
(7 electrons)

Energy level 2
(8 electrons)

Energy level 1
(2 electrons)

Nucleus

Energy increasing

Figure 17–9. The electron configurations of sodium (top) and chlorine (bottom) can be shown as energy ladders. The farther away the electrons are from the nucleus, the more energy they have.

ing the electrons for sodium and chlorine. How do the levels change as the number of electrons increases?

The Higher You Climb

The farther the electrons are from the nucleus, the more energy they have. Why is this? Remember that the nucleus is positively charged and the electrons are negatively charged. You know that opposite charges attract, so it makes sense that protons and electrons are attracted to each other. It also makes sense that the closer the electrons are to the nucleus, the more easily they are held there.

In order to move electrons from a location close to the nucleus to one farther away, some energy is required. This energy is stored by the electron as potential energy. The farther away from the nucleus the electron is, the more potential energy it has. An electron in the third energy level has more potential energy than one in the first energy level.

Just as you can climb to the top of a tall skyscraper if you have enough energy, electrons can move from level to level if they have enough energy. Electrons fill the energy levels starting from the lowest one and moving to the levels with more energy.

Figure 17–10. Helium, which is lighter than air, is often used to fill balloons.

Figure 17–11. The hydrogen that was used to fill the *Hindenburg* was very reactive. Today, dirigibles are filled with helium.

A Stable Octet Is Not Eight Singing Horses!

The stability of the noble gases is due to the number of electrons in their outer energy levels. All of these elements except helium have eight electrons in their outer energy level. This seems to be a stable configuration and, in fact, is called a *stable octet*. Helium has only two electrons, but it has a filled first energy level, which also gives it great stability.

When elements react, they lose, gain, or share electrons to achieve an electron arrangement like one of the noble gases. Here are some examples. Sodium forms Na^+, a positive ion, by losing one electron. This gives the sodium ion the same electron arrangement as argon, a noble gas. Chlorine, a nonmetal, forms a negative ion, Cl^-, by gaining one electron. This gives the chlorine ion the same electron arrangement as argon. Both elements form stable octets as their electron arrangements change.

Stability in Your Life

The noble gases have many practical uses. You know about helium balloons, but did you know that helium is also used in blimps? Since helium is less dense than air, it makes sense to use it in blimps and other airships. But this has not always been the case. Once, hydrogen was used in giant airships called zeppelins. Hydrogen is also less dense than air, so the zeppelins floated. Hydrogen, however, has one significant characteristic that helium doesn't have. It explodes! The passenger ship *Hindenburg* exploded in 1937 when its hydrogen ignited. Since that time, nonreactive helium has been used to fill airships.

There are other noble gases that you use or see used every day. Neon, as you have probably guessed, is used in neon signs. It produces a red color when an electric current passes through it. Argon is used to fill regular light bulbs. It provides an inert atmosphere for longer-lived bulbs. Xenon and radon are rarer and, therefore, have fewer applications. Xenon is used in photographic flashbulbs for high-speed photography, while radon is used to treat cancer.

Radon, however, has a downside. Radon is produced when radium undergoes radioactive decay. (You will learn more about this type of decay in the next chapter.) This radon is released from soil and rocks and can leak into houses through the cracks in the foundation or basement. Usually any radon is diluted by natural circulation and open windows. However, in new, energy-efficient buildings radon can build up to dangerous levels. If inhaled in large enough quantities, radon can cause lung cancer.

ASK YOURSELF

Why do metals tend to form positive ions?

Nonmetal Families

Besides the noble gases, only one other nonmetal family has a commonly used specific name. The halogens, Group 17, are among the most reactive of the nonmetals. Recall that the most reactive metals, the alkali metals, are on the opposite side of the periodic table. In their uncombined form, the halogens are poisonous and highly corrosive. All the halogens exist naturally as molecules composed of two atoms, or **diatomic molecules**. For example, chlorine gas is Cl_2, and fluorine gas is F_2.

No doubt you would like to test the reactivity of nonmetals in the same way as you did for the metals. Unfortunately, many of them are toxic and very dangerous, so we will tell you about their reactivity.

I'm Always Hungry!   The halogen fluorine is the most reactive of all the nonmetals. In fact, it is so reactive that it is often called the *Tyrannosaurus rex* of elements. That means that

Figure 17–12. One fluorine compound can be "breathed" and is being used experimentally for deep-sea dives. The mouse shown here is being submerged in the compound. It "breathes" in the liquid and gets plenty of oxygen. After the compound is drained from its lungs, the mouse is fine!

it will "eat" (combine with) almost anything. Fluorine compounds have many uses. One fluorine compound, Teflon, is used to make nonstick surfaces on items ranging from cookware to adhesive bandages. Another fluorine compound is used as a blood substitute in emergencies. Yet another fluorine compound can be breathed, as shown in Figure 17–12.

The reactivity of the nonmetals increases as you move toward fluorine. That is, as you move from the lower left to the upper right of the periodic table, the reactivity of the nonmetals increases. Don't forget, though, that the noble gases don't fit the trend.

What Phase Are You?

Have you noticed that the nonmetals described have all been gases? This is not true of all nonmetals. Bromine, another halogen, is a red liquid. Some bromine compounds are used in agriculture for insect control. The element iodine is a purple crystalline solid. The liquid antiseptic with which you might be familiar is a tincture of iodine, that is, iodine crystals dissolved in alcohol.

In the nitrogen family, Group 15, only nitrogen is a gas. In fact, it is a diatomic gas, N_2. All the other elements in this family are solids.

Figure 17–13. Iodine crystals change into a gas, or sublime, at room temperature.

Figure 17–14. White phosphorus, when exposed to air, bursts into flame.

What's the Same but Different?

Twins. Have you ever known a set of twins? Even though they have the same birthday and often look just alike, they are different. They may have different like and dislikes, or they may excel at different things. Some nonmetals are like twins. They are the same but different. These nonmetals exist in two or more molecular forms called **allotropes**.

Phosphorus, a member of the nitrogen family, exists in two major forms—red and white. Red phosphorus is fairly stable. White phosphorus, however, reacts violently with oxygen in the air, so it must be stored under water.

Figure 17–15. Which twin is white phosphorus? How do you know?

The differences in the properties of red and white phosphorus are due to the arrangement of atoms in their molecules. White phosphorus exists as molecules of four atoms, whereas red phosphorus molecules are large disorganized clumps of atoms.

Other families also have allotropes. Oxygen, for example, exists in two forms. The oxygen you breathe is diatomic oxygen (O_2). Oxygen can also be found in a triatomic state (O_3) known as *ozone*.

In the Ozone   Diatomic oxygen is converted to ozone by the sun's ultraviolet rays. The ozone forms a layer high in the atmosphere that screens out or absorbs most of the ultraviolet rays from the sun. However, ozone is a pungent and toxic gas and, when found close to the earth, is considered a pollutant.

Ultraviolet rays cause sunburn, and long-term exposure to them may cause skin cancer. The ozone layer is therefore a very important defense against the sun's harmful rays. Recently, scientists have found holes in the ozone layer. These holes, which seem to open and close at random over different parts of the earth, might be caused by certain gaseous molecules that reach the upper atmosphere and break down O_3 into O_2. One of the culprits might be a class of compounds known as chlorofluorocarbons, or CFCs. (You may have heard them called fluorocarbons.) Read the following article to find out more about CFCs.

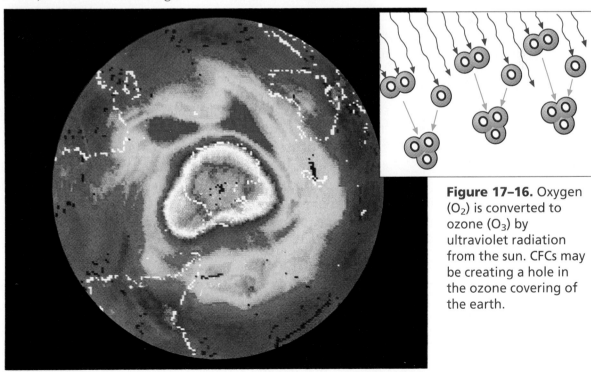

Figure 17–16. Oxygen (O_2) is converted to ozone (O_3) by ultraviolet radiation from the sun. CFCs may be creating a hole in the ozone covering of the earth.

THE TROUBLE WITH CFCs

from Odyssey

They're at the heart of refrigerators and air conditioners. They're useful as solvents to clean electronic parts. They help make insulation for your house, containers for your hamburger, and nice soft cushions for your family-room couch. They're the chlorofluorocarbons (CFCs for short), a type of chemical that has become almost indispensable in the modern world. (CFCs can be used as propellants in spray cans. Although this use is now illegal in the United States, CFC propellants are still used in other countries.)

Prized for years because they seemed to be nontoxic and inert—that is,

they don't react easily with other substances—these apparently harmless chemicals have turned out to be wolves in sheep's clothing.

When a car's air conditioner leaks, or when a foam hamburger container is crushed, chlorofluorocarbons are released into the atmosphere. Some of the CFCs eventually make their way up to the stratosphere.

In the stratosphere, sunlight breaks up the CFC, releasing chlorine from it. The chlorine destroys ozone (O_3) by turning it into regular oxygen (O_2).

When there is less ozone in the stratosphere, more damaging ultraviolet light from the Sun can reach Earth's surface. An increase in ultraviolet light could have serious consequences. It is bad for plants, including food crops. It can hurt some

sea creatures. It can cause skin cancer in humans.

To avoid these serious health and environmental problems, scientists agree, we must act to keep ozone from being depleted any more. We must cut back on our use of CFCs, either by recycling some CFCs or by finding new chemicals to replace them. We don't have a lot of time to waste. The CFCs already in the stratosphere will affect the ozone layer for more than a century.

In September 1987 an international panel worked out an agreement to limit future use of chlorofluorocarbons. Production of these chemicals was halted at the 1986 level; by 1999 it must be cut in half. The U.S. Environmental Protection Agency is calling for even stricter cutbacks. We're taking our first steps in the right direction.

DISCOVER BY Writing

Look up another recent article that talks about the holes in the ozone layer. Write a summary of the article in your journal. Be sure to include the possible causes of and solutions to the problem that the article suggests. ✎

Carbon Has Many Disguises Look at the photograph shown here. What do you think these "things" have in common?

Carbon black

All of the substances in the figure are allotropes of carbon. Carbon black, one common allotrope, is very similar to charcoal. It exists as disorganized atoms of carbon. A second allotrope of carbon is something you use every day. It is called graphite, and it is used in pencil "leads." Its atoms are arranged in sheets, which makes pure graphite very slick. The third allotrope has carbon atoms in a very rigid crystal structure; it is a diamond. What useful things could *GRAPHITE MAN* do? (He could give the word *graffiti* new meaning!) What useful properties would *DIAMOND WOMAN* have?

Graphite

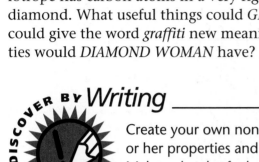

DISCOVER BY *Writing*

Create your own nonmetal person. Describe his or her properties and special abilities in detail. Make a sketch of what your character would look like. ✐

Diamond

What Do All These Things Have in Common? Did you have a hard time at the beginning of the section listing the properties of nonmetals? Now you can understand why. Nonmetals are as varied in properties as they are in color, appearance, and physical state. They generally do not conduct electricity or heat, and they tend to form negative ions by gaining electrons. Aside from that, these elements have exciting and varied properties that scientists continue to explore and utilize.

▶ **ASK YOURSELF**

Explain why allotropes can have different properties.

SECTION 2 *REVIEW AND APPLICATION*

Reading Critically
1. How is chemical activity of the nonmetals organized in the periodic table?
2. Distinguish between metals and nonmetals based on their properties.

Thinking Critically
3. Element *X* is a gas that has a mass number of 16. It has 6 electrons in its outer energy level. Which element is it?
4. Why are the noble gases particularly useful?
5. Metals tend to form positive ions while nonmetals tend to form negative ions. In what ways are these tendencies useful?

SKILL Drawing Conclusions

▶ **MATERIALS**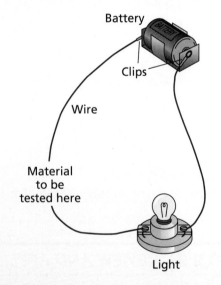
- safety goggles
- aluminum • carbon • copper • lead • sulfur • hammer • anvil
- light bulb and socket • battery and battery holder • wire • wire clips

▼ **PROCEDURE**

1. Copy the table shown.

TABLE 1: TESTS FOR METALS AND NONMETALS			
Element	Shaped by Hammering	Electric Conductor	Luster
Aluminum			
Carbon			
Copper			
Lead			
Sulfur			

2. **CAUTION: Put on safety goggles and leave them on throughout this entire activity.** Test each of the elements listed to determine whether it is a metal or a nonmetal. Remember that all of the characteristics listed must be present if the element is a metal.

3. Test to see if the shape of the element can be changed by hammering.

4. To test for electric conduction, set up an electric circuit like the one shown. A metal will conduct electricity, and the bulb will light.

5. Observe the element to determine whether it is shiny. If it is, it has luster.

▶ **APPLICATION**

Use the results of your experiment to draw conclusions about the elements you tested. Which elements are metals? Which are nonmetals? In what ways does your data support your conclusions?

✳ *Using What You Have Learned*

Think of a problem, and write a description of it. Exchange problems with a classmate. Design an experiment that will help you collect data to solve the problem, complete the experiment, and draw conclusions to solve the problem.

The Big Idea

Have you ever watched a large building under construction? It is hard to guess the final appearance of that building when only the steel girders are up. Yet this framework is essential and allows the building to take the shape the architect designed. It will determine whether the building is massive and solid or airy and covered in glass.

In this chapter, you have looked into a different sort of framework, a different structure. You have learned how the atomic framework, or the electron configuration, affects the properties of a substance. Science seeks to discover the links between the structure of matter and its properties. By dividing the elements into metal and nonmetal categories and looking at their atomic structure, you have gained some understanding of their reactivities.

For Your Journal

Look back at the ideas you wrote in your journal at the beginning of the chapter. How have your ideas changed? Revise your journal entry to show what you have learned. Be sure to include information on how different elements are useful in your daily life.

Connecting Ideas

Below are some photographs of elements. Can you tell by looking at them whether they are metals or nonmetals? Do the names of the elements help you figure it out? Look at the photographs, and make a table that shows what you know about each element. Then write a summary statement that tells how you can differentiate between metals and nonmetals.

Calcium

Mercury

Sulfur

Bromine

Understanding Vocabulary

1. For each set of terms, explain the relationship in their meanings.
   a) alchemist (457), chemist (457)
   b) metals (458), alloys (459)
   c) metalloids (462), semiconductors (462)
   d) noble gases (466), inert gases (466)
   e) halogens (470), diatomic molecules (470)
   f) oxygen (473), ozone (473)

Understanding Concepts

MULTIPLE CHOICE

2. Elements that have some properties of metals and some properties of nonmetals are
   a) metalloids.
   b) alloys.
   c) semiconductors.
   d) ions.

3. Because they are naturally found as two-atom molecules, nitrogen, oxygen, and fluorine are examples of
   a) more stable molecules.
   b) diatomic molecules.
   c) ionic molecules.
   d) less stable molecules.

4. Substances that are good conductors of electricity and heat are
   a) metals.
   b) nonmetals.
   c) noble gases.
   d) inert gases.

5. Which element is *not* an example of a halogen?
   a) fluorine
   b) chlorine
   c) iron
   d) iodine

SHORT ANSWER

6. What are some characteristics of nonmetals?

7. Explain why noble gases are so stable.

8. What are some properties of metals?

9. Explain why it is unwise to mix water with alkali metals.

10. Why are some cooking pots made of copper but not tin?

Interpreting Graphics

Refer to the periodic table in this chapter to answer the following questions.

11. How many different types of elements are found on the periodic table?

12. Where are the metals located on the periodic table?

13. Which elements are part of the halogen family?

14. What pattern of chemical activity is reflected in the periodic table?

Reviewing Themes

15. *Systems and Structures*
How do chemists decide whether an element is a metal or a nonmetal?

16. *Systems and Structures*
Discuss how you might use metals and nonmetals in your daily life.

Thinking Critically

17. What are some advantages of using alloys?

18. Why is helium rather than hydrogen used to fill blimps?

19. You are an engineer designing a new airplane. You know you want to use an aluminum alloy. What properties would you want in this alloy?

20. How are metals held together? How does this arrangement affect the properties of metals?

21. Explain why carbon is considered a metalloid. Give specific examples to support your explanation.

22. How are the Group 1 metals similar to the halogens? How are they different?

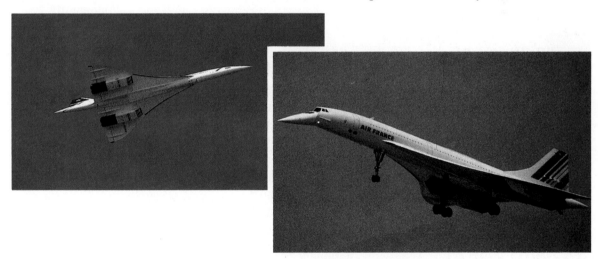

Discovery Through Reading

Berger, Melvin. *Our Atomic World.* New York: Franklin Watts, 1990. This book provides a clear, interesting discussion of atoms in our environment.

CHAPTER 18

RADIOACTIVITY

A PINPRICK OF BRILLIANT LIGHT PUNCTURED THE DARKNESS, SPURTED UPWARD IN A FLAMING JET, THEN SPILLED INTO A DAZZLING CLOCHE OF FIRE THAT BLEACHED THE DESERT TO A GHASTLY WHITE....

*T*he power of the atom has haunted scientists for years. Once this power was discovered and unleashed as a weapon, science has endeavored to find a way to harness the power for the good of humanity. The puzzle of nuclear reactions and the radiation they cause contin-ues to be studied.

For a fraction of a second the light in that bell-shaped fire mass was greater than any ever produced before on earth. Its intensity was such that it could have been seen from another planet. The temperature at its center was four times that at the center of the sun and more than 10,000 times that at the sun's surface. The pressure, caving in the ground beneath, was over 100 billion atmospheres, the most ever to occur at the earth's surface. The radioactivity emitted was equal to one million times that of the world's total radium supply.

No living thing touched by that raging furnace survived. Within a millisecond the fireball had struck the ground, flattening out at its base and acquiring a skirt of molten black dust that boiled and billowed in all directions. Within twenty-five milliseconds the fireball had expanded to a point where the Washington Monument would have been enveloped. . . . At 2000 feet, still hurtling through the atmosphere, the seething ball turned reddish yellow, then a dull blood-red. It churned and belched forth smoking flame in an elemental fury. . . . At 15,000 feet the fireball cleaved the overcast in a bubble of orange that shifted to a darkening pink. Now, with its flattened top, it resembled a giant mushroom trailed by a stalk of radioactive dust. . . . The air had ionized around it and crowned it with a lustrous purple halo. As the cloud finally settled, its chimney-shaped column of dust drifted northward and a violet afterglow tinged the heavens above Trinity.

from *Day of Trinity*
by Lasing Lamont

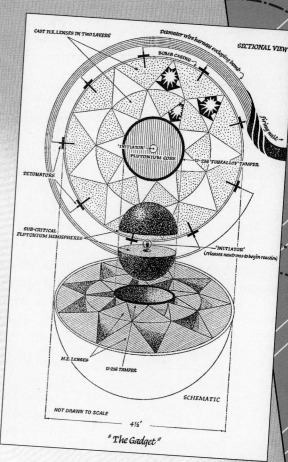

Trinity. The site of the world's first atomic explosion, with a bomb made of uranium and plutonium. The explosion melted the desert sand into pearls of glass. Desert animals were vaporized, leaving only their carbonized shadows on the desert floor. A radioactive cloud spread north across the New Mexico mountains and settled silently into the valleys.

In the fierce light of Trinity, scientists knew they had opened a new age. From that day, science was compelled to explore the mysteries of the atom, discovering its awesome capacity for destruction and its elusive promise of inexhaustible power.

For Your Journal

- *What do you think happens inside an atom when nuclear reactions take place?*
- *What are the effects of radiation?*
- *How could understanding the energy trapped inside the nuclei of atoms affect your life?*

Radioactive Elements

Objectives

Explain what is meant by radioactivity.

Differentiate among the three types of radioactive decay.

List two ways in which radioactivity can be detected.

Antoine-Henri Becquerel was a French physicist who was interested in the absorption of light. Becquerel's research in the 1890s centered around fluorescent compounds. He wanted to know if fluorescence and X-rays were related.

In 1896 Becquerel was studying the effect of sunlight on certain minerals. He stored a sample of uranium wrapped tightly in black paper in his desk drawer. In the same drawer there was an undeveloped photographic plate. A few days later, after he had developed several photographic plates, including the one he thought had not been exposed, he found an image on the unexposed plate. The image was the same shape as the uranium sample! Becquerel concluded that something in the uranium had come through the black paper and exposed the plate.

Does this account of such a small observation sound unimportant, even boring? If so, consider that the discovery of radioactivity has led to increased production of electricity, nuclear medicine, and the construction of the atomic bomb.

Figure 18–1. Becquerel discovered radioactivity while studying how certain minerals glow after being exposed to sunlight. The rock shown here contains willemite and calcite. When exposed to ultraviolet light, the willemite glows red and the calcite glows green.

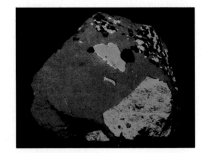

What's in a Name?

You probably already have some ideas about nuclear energy. Think about the terms you would use to describe radioactivity and nuclear energy. Chances are that you would use more negative terms than positive ones. You might remember hearing about the Chernobyl disaster or the accident at Three Mile Island. You may know something about the problem of the disposal of radioactive waste. The idea of radiation might send you running in the opposite direction!

Studying the Unknown

Remember the marble shooting activity in Chapter 16? It wasn't easy to determine the unknown object's shape. It's hard to gather information about an object you can't see or touch. Scientists spend years working on such puzzles. What kind of person would spend a lifetime trying to find out about the unknown? About 100 years ago, there lived such a person; her name was Manya Sklodowska. The world knows her as Marie—Marie Curie.

Marie Curie was born in Warsaw, Poland, in 1867. By 1898, when she was 31, she was already a well-known scientist. Marie was a student of Becquerel and was immediately interested in his discovery of radioactivity. Marie and her husband Pierre were investigating a mineral called *pitchblende*. The Curies observed that pitchblende gave off a strange form of energy, similar to that given off by uranium.

Figure 18–2. Marie Curie (left) and her family

After two years of work, the Curies announced that they had discovered two new elements in pitchblende. The Curies named one of the new elements *radium,* because it gave off rays. They named the other *polonium,* after Marie Curie's beloved homeland, Poland.

In order to learn more about radioactivity, elements were isolated from pitchblende and then purified. By 1902 Madame Curie was able to produce a nearly pure sample of radium. She had many questions: What made radium radioactive? Why are radioactive elements unstable?

Decaying Atoms Madame Curie knew that certain elements were unstable and that they spontaneously broke apart and gave off particles or high-energy radiation. That is, they were **radioactive.** Madame Curie's hypothesis was that this instability came from the nucleus of these atoms. She believed that as particles or energy are released from the nucleus, the atom becomes more stable. This process is known as *radioactive decay.*

Why are some elements less stable than others? Let's see how well you can hypothesize about the answer.

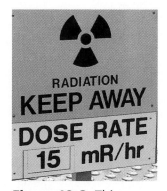

Figure 18–3. This symbol is used to label radioactive substances or areas where radiation is present.

Look at the graph below. It shows the number of neutrons and protons for several stable (nonradioactive) isotopes. The shaded area represents a belt of stability. The solid line shows when the number of neutrons is equal to the number of protons. Notice that many of the smaller nuclei are stable if the number of protons equals the number of neutrons. Why do you think this is so? (Hint: What do you know about like charges?) Write your hypothesis in your journal. As the atomic number gets larger than 20, the belt of stability drifts away from the solid line and seems to favor more neutrons than protons. Why do you think the trend changes? Write your ideas in your journal. ✎

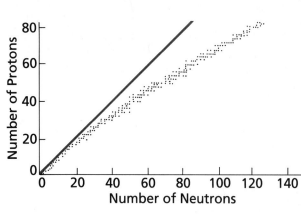

Why is carbon-12 stable and carbon-14 radioactive? The full explanation of what makes an isotope radioactive is too complicated to cover here, but you might have come close to the answer in your hypothesis. Protons, as you know, repel each other because they have like charges. These forces would tend to push the nucleus apart if they were all that was at work. But, besides these repulsions, there are forces of attraction between particles called *quarks* that make up protons and neutrons. These forces tend to hold the nucleus together and make it stable. If there are not enough neutrons, the repulsive forces between protons are greater than the attractive forces between the quarks that make up protons and neutrons. When forces of repulsion are greater than forces of attraction, the nucleus is unstable. As the nuclei become larger, more neutrons are needed in order to maintain a greater number of attractive forces than repulsive forces. That is why the graph shows the belt of stability dropping below the line.

Unstable radioactive atoms lie somewhere away from the belt of stability shown on the graph. In order to become stable, some of these atoms release particles from the nucleus in the form of radiation. You will find out more about radiation in the next section.

ASK YOURSELF

Why are some atoms radioactive while others are not?

Jittery Isotopes

You know that radioactive substances give off radiation. You may even know that some kinds of radiation are more dangerous than others, but do you know why?

Three major types of radiation are given off by radioactive elements: alpha particles, beta particles, and gamma rays. Radiation occurs when a radioactive nucleus decays. When the nucleus decays, it shoots out or emits particles or energy waves or both. This flow of particles or waves is called *radiation.*

When an atom undergoes **alpha decay,** it emits an alpha particle, which is made up of two protons and two neutrons. An example of alpha decay is shown below.

Beta decay occurs when a beta particle shoots out of the nucleus at high speed. A beta particle is formed when a neutron is converted into a proton and an electron. The electron shoots out of the nucleus much like a cannonball shoots out of a cannon when the gunpowder inside it explodes.

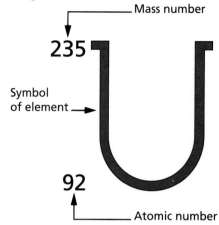

Figure 18–4.
Radioactive isotopes are written as shown here.

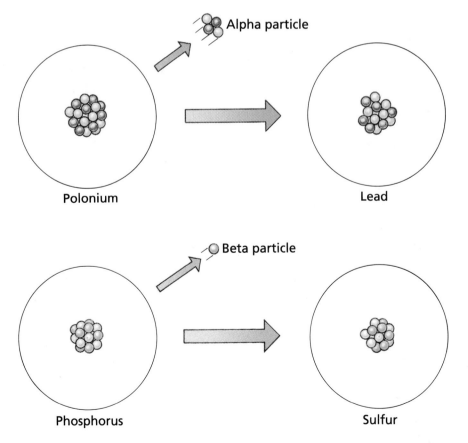

Figure 18–5. Alpha decay is the result of the expulsion of a helium nucleus from the nucleus of an atom. Polonium becomes lead as a result of alpha decay.

Beta decay is the result of the expulsion of an electron from the nucleus of an atom. Radioactive phosphorus undergoes beta decay to form sulfur.

Gamma decay, which does not involve a particle, is the emission of high-energy radiation similar to X-rays. This radiation has no mass and no charge. How can radiation that has no particles make an atom more stable? This process is similar to the one in which an electron drops to a lower energy level. When an atom gives off energy, it moves to a lower energy level, and lower energy levels are more stable than higher ones. Gamma rays can penetrate tissue more deeply than alpha or beta particles, so this type of radioactive decay is considered the most dangerous.

When an atom undergoes radioactive decay, it is changed in atomic number, mass number, or both. Remember that the atomic number of an element is the number of protons in the nucleus. If the atomic number changes, the atom changes into another element. These changes are called *nuclear reactions*. Nuclear reactions occur in an element until a stable, nonradioactive atom is formed. Sometimes an element undergoes many types of decay before it becomes stable. An example is the decomposition of uranium to lead, shown in Figure 18–6.

In contrast, a polonium atom can become nonradioactive by emitting a single alpha particle. When this happens, an atom of lead is formed.

Figure 18–6. Uranium decays into many different elements before it forms lead as shown in this diagram.

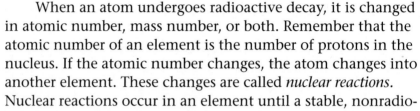

ASK YOURSELF

What can happen to an atom when a nuclear reaction occurs?

The Downside of Radiation

All types of radiation given off during radioactive decay have the power to go through objects. On April 26, 1986, the world learned more about this power than it ever wanted to know. The Chernobyl disaster was the world's introduction to the dangers of nuclear power production. The following article, written one year after the accident, is a firsthand account of what happened.

"Chernobyl—One Year After" from *National Geographic*

The air smelled of scorched metal, and to breathe without a mask was to cough. Helicopters swung low on quick bombing runs, dropping sacks of lead, boron carbide, sand, clay, dolomite. Their target was a tangle of machinery and pipe, visible through a gaping hole in a 70-meter-high building.

On the ground moved a veritable army, hastily and desperately assembled. In white garments, with masks and caps, many looked like physicians dressed for the operating room. Army personnel carriers, their armor augmented by slabs of lead, rumbled to and fro on deadly serious taxi duty.

Now and again buses passed, removing the people of whole villages. In the nearby city of Pripyat, where 45,000 people had lived, laundry still hung on clotheslines, and a carnival carousel spun empty in the wind. An elderly woman departed her home carrying only her identity card, her spectacles, and her house key. Though she might never see her house again, she locked the door.

At times some of the people in this frantic scene paused to think of fellow workers—firemen, a doctor, two paramedics, a woman guard—who were rushed to hospitals in the early hours of April 26, 1986.

I describe the nightmare at the Chernobyl Nuclear Power Plant as it was a year ago, on, say, about the second of May. With four working reactors and two more being built, Chernobyl was to be one of the most powerful nuclear power stations in the Soviet Union. At 1:24 on the fateful April morning, one or possibly two explosions blew apart reactor No. 4—the worst reported accident in the history of the harnessed atom.

The blast(s) knocked aside a thousand-ton lid atop the reactor core and ripped open the building's side and roof. Reactor innards were flung into the night. These included several tons of the uranium dioxide fuel and fission products such as cesium 137 and iodine 131, as well as tons of burning graphite. Explosion and heat sent up a five-kilometer plume laden with contaminants.

Why were authorities dumping lead on Chernobyl? To understand, you need to know that all radiation has the ability to go through solid matter. This ability is called *penetrating power*. Alpha particles have the lowest penetrating power; they can be stopped by several sheets of paper. Beta particles have a greater penetrating power, but they can be stopped by a piece of aluminum 0.75 mm thick. That is about as thick as five pieces of aluminum foil. The penetrating power of gamma radiation is the greatest. Thick layers of lead and cement are needed to contain this radiation.

Figure 18–7. The penetrating powers of the three types of radioactive decay are shown here.

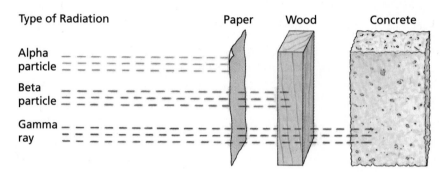

The lead that was dumped on the Chernobyl reactor would help contain the gamma rays—the most dangerous type of radiation. Boron carbide acts in the same way as a control rod does inside a reactor. It absorbs neutrons. The sand and clay dropped by helicopters would act as a shield for the alpha and beta particles. They would also put out some of the fires. Later on, the reactor was encased in concrete and steel to hold the radiation that would be there for thousands of years.

Clouds containing radioactive substances, which were the result of alpha and beta decay, traveled away from Chernobyl. These substances have been detected in the Scandinavian countries, in Greece to the southeast, and even as far away as the west coast of the United States. Milk and fresh produce were contaminated throughout much of Europe. None of these products could be eaten because alpha particles, even though they do not have much penetrating power, will cause radiation damage to internal tissues of living things if they are eaten.

As you can see, radioactivity is filled with potential hazards. Later in this chapter we will also look at how it can be helpful.

 ASK YOURSELF

How is gamma decay different from other types of radioactive decay?

Detecting the Invisible, Hearing the Silent

Many of the initial researchers in radioactivity were not aware of its dangers. Both Marie Curie and her daughter Irene, who continued her mother's research, died of leukemia—a form of cancer believed to be produced by exposure to radiation.

Today, researchers may use any of several devices to detect even small amounts of radiation. The simplest device for detecting radioactivity is the *electroscope*.

Tracks in the Clouds

Usually, however, more sophisticated devices are used to detect radioactivity. One instrument that is commonly used is the *cloud chamber*. A cloud chamber is a compartment filled with a gas kept at a temperature just below its condensation point. Water and alcohol are two substances often used in cloud chambers. When a radioactive particle passes through the chamber, it forms ions. The ions cause the gas to condense into drops of liquid. These drops attach to the ions the same way that dew condenses on the grass. The ions move very fast, causing the droplets to make lines in the cloud chamber. These lines are called *tracks* because they show where the particles have been, just as animal tracks in the snow show the movements of animals. Try the next activity to find out more about how a cloud chamber works.

Figure 18–8. When an electroscope is charged with static electricity, the leaves of foil inside the container repel each other. When placed near a natural radioactive source, the leaves of the electroscope collapse.

Figure 18–9. A cloud chamber shows the tracks of atomic and subatomic particles as the result of ionizing the gas inside the chamber.

DISCOVER BY Doing

Sprinkle some iron filings on a thin piece of cardboard about the size of a sheet of paper. Bring a bar magnet close to the bottom of the cardboard piece but do not let it touch the cardboard. Move the magnet underneath the cardboard. What happens? How is this similar to the way ionizing radiation affects a cloud chamber?

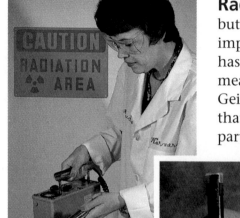

Figure 18–10. A Geiger counter (left) is a portable device used to detect radiation. A dosimeter (right) is a personal radiation device that can be worn or put into a pocket.

Radiation Counts   Detecting radiation is important, but measuring the amount of radiation present is also important. This measurement may be critical if a person has been exposed to radiation. One instrument that can measure the level of radiation is the **Geiger counter.** A Geiger counter contains a compartment filled with gas that forms ions when radiation passes through the compartment. The ions trigger a flow of electric current through a circuit, which is connected to a meter. The meter displays the amount of current produced, and the current is proportional to the level of radiation.

Some people work in places where there could be hazardous levels of radiation. They need to have a radiation detection device with them at all times. Devices, smaller than the Geiger counter, such as the dosimeter, have been developed for these people.

ASK YOURSELF

Why is it important to know whether you have been exposed to radiation?

SECTION 1   REVIEW AND APPLICATION

Reading Critically

1. Under what circumstance can alpha radiation be harmful?
2. How can scientists study radioactive particles?

Thinking Critically

3. Element X is produced when uranium emits an alpha particle. What is the atomic number of the new element? What is its atomic mass? What is its symbol?
4. Explain why gamma radiation is the most dangerous type of radiation for humans.

490   Chapter 18   Radioactivity

Using Radiation

Have you ever put tincture of iodine on a cut? Did you know that this chemical is poisonous? You know it prevents infection. Why? In small amounts, the chemical will kill the microorganisms that might infect your cut, but it won't harm you.

The effect of iodine is similar to the effects of radiation. Despite the dangers of radioactivity, there are many useful applications. Many common elements have radioactive isotopes that are used in medicine and scientific research.

Objectives

State *three uses for radioisotopes.*

Explain *what half-life is.*

Compare *and* **contrast** *methods of radiochemical dating.*

Atomic Tailors

What would happen if you shot a marble at a hollow glass ball? You guessed it, the ball would shatter. What would happen if the marbles were made of clay? With the right amount of energy, the clay might stick to the glass ball, forming a new shape. This is the principle used to create new elements in a particle accelerator.

Particle accelerators move particles at high speeds and smash them into atoms of other elements. If two nuclei collide and remain together, a new type of atom is formed. All the elements heavier than uranium have been created by smashing atoms together under conditions of intense heat and pressure. These elements are all radioactive.

Figure 18–11. A particle accelerator may be either linear or circular. Shown here is the main ring of the Fermi Accelerator in Batavia, Illinois.

DISCOVER BY *Writing*

Suppose you are an alchemist of the Middle Ages. You are brought forward in time and are taken to visit a particle accelerator. Has science achieved what you have tried to do? Write a short story that examines how you, as an alchemist, view the creation of new elements in the particle accelerator. ✏

In Chapter 16, Ensign Roark—one of the ship's crew members—was sick. Do you remember the test that was done to check his leg? (See page 441 to refresh your memory.) The radioactive isotope that was used was sodium-24. This isotope is uncommon in nature. So where does it come from?

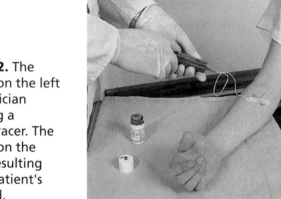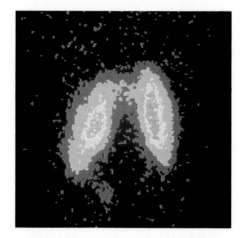

Figure 18–12. The photograph on the left shows a physician administering a radioactive tracer. The photograph on the right is the resulting scan of the patient's thyroid gland.

Scientists can make radioactive isotopes. They use particle accelerators to change a stable nucleus into an unstable, radioactive one. These manufactured radioisotopes have many practical uses, especially in medicine. For example, radiation therapy for the treatment of cancer uses X-rays and gamma rays produced by cobalt-60 or cesium-137. Both are manufactured radioisotopes. Natural and manufactured radioisotopes are also used as *tracers*. Tracers are radioactive chemicals that follow certain reactions inside living organisms. The major advantages of using radioisotopes as tracers are that they can be given in very small amounts and they can be directed to specific places in the body. The sodium-24 injected into Ensign Roark's leg is an example of a tracer. Why do you think these chemicals are called tracers?

Figure 18–13. The mushrooms shown here were grown under the same conditions and were harvested at the same time. The mushrooms on the left were irradiated.

Certain radioisotopes can be used to treat food. As a result of this treatment, called *irradiation,* the food can be stored for a long time without refrigeration. How safe is irradiated food? Read the following article to get some ideas.

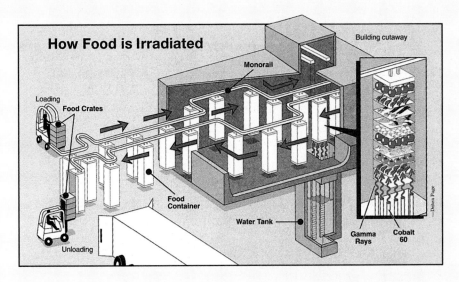

How Food is Irradiated

Loading
Food Crates
Monorail
Building cutaway
Food Container
Water Tank
Gamma Rays
Cobalt 60
Unloading

Crates of food are loaded into containers that pass over rods of cobalt-60 that bathe the food with gamma rays. The rods are stored in water when they are not in use.

How Safe is Irradiated Food?

from *Current Science*

If you were told that the food sold in your local grocery store had been zapped with gamma radiation, would you be worried? Many Americans are, now that irradiated food is available for the first time in the United States.

But are the fears of these people realistic? Is irradiated food really dangerous to human health? Is it radioactive?

To answer these questions, it's helpful to know that the idea of food irradiation was hatched many years ago in an effort to improve food quality. Exposing fruits, vegetables, and meats to gamma radiation kills organisms (molds, insects, bacteria, parasites) that spoil food or cause food poisoning. About 6.5 million Americans contract food poisoning every year, and some elderly people and young children die from the poisoning.

Is Food Radioactive?

Contrary to popular belief, irradiated food is not *radioactive*—it doesn't give off radiation. When food is irradiated, invisible gamma rays pass through the food (see drawing). The gamma radiation does not make food radioactive, just as microwaves passing through food in a microwave oven do not make food radioactive.

However, exposure to gamma rays does rob food of some vitamins. But so do other forms of food preservation, including freezing and canning. The U.S. Food and Drug Administration (FDA) has concluded that irradiation does not destroy enough vitamins to jeopardize public health.

The biggest fear about irradiation is that it might make harmful chemical changes in food. Numerous studies have found that irradiated food contains tiny quantities of *radiolytic products (RPs)* These are created when ionizing energy splits food molecules (fats, carbohydrates, proteins).

Some scientists say that some RPs are *mutagenic*—they have the ability to change DNA. Also, excess exposure to mutagens may cause cancer. However, many years of testing have failed to find cancer in any laboratory animals fed large amounts of irradiated food.

Risky Business

Still, some people worry about RPs—they want to be absolutely certain that eating irradiated food is risk free. But not many foods are truly risk free. Many foods contain chemicals that are identical or highly similar to the RPs in irradiated foods, and eating too much fatty food may lead to heart disease and cancer.

Reviewing all the scientific evidence, the FDA has decided that the risk of eating irradiated food is acceptable. From here on, all irradiated foods will carry a special label so that your choice is clear—and now it is up to you.

ASK YOURSELF

How are radioisotopes useful in medicine?

Atoms that Keep on Ticking

When doctors use radioactive isotopes, they are limited to those that give off small amounts of radiation and decompose into nonradioactive elements in a short time. But how do scientists know how fast the radioisotopes will decompose?

Scientists have found that each radioactive element breaks down at a specific rate. The time required for one half of a sample to decay is known as the *half-life* of that substance. The half-life is constant regardless of what you do to the element. The table shows the half-lives of several radioisotopes.

Table 18-1	Half-Lives of Some Common Radioisotopes		
Isotope	**Type of Decay**		**Half-Life**
$^{238}_{92}U$ (uranium)	alpha decay		4.5 billion years
$^{14}_{6}C$ (carbon)	beta decay		5730 years
$^{3}_{1}H$ (hydrogen)	beta decay		12.26 years
$^{32}_{15}P$ (phosphorus)	beta decay		14.3 days
$^{131}_{53}I$ (iodine)	beta decay		8.1 days

Figure 18–14. All living organisms contain carbon-14. Knowing the half-life of this element allows scientists to determine the age of organic remains

Some of the radioisotopes released during the Chernobyl accident were uranium-238, which has a half-life of 4.5 billion years, and cesium-137, with a half-life of 30 years. The land around Chernobyl cannot be used for years due to the long lifetime of the radioactive wastes that spilled out of the reactor.

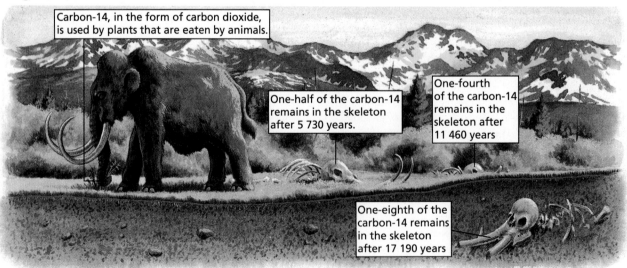

Carbon-14, in the form of carbon dioxide, is used by plants that are eaten by animals.

One-half of the carbon-14 remains in the skeleton after 5 730 years.

One-fourth of the carbon-14 remains in the skeleton after 11 460 years

One-eighth of the carbon-14 remains in the skeleton after 17 190 years

Scientists have put the idea of half-life to work in dating artifacts. Various isotopes can be used in this process, but the most common is carbon-14. Carbon-14 is used because it is found in all living things. Since carbon-14 exists in the atmosphere, plants absorb it in the form of carbon dioxide. Animals get carbon-14 by eating plants or by eating animals that have eaten plants. So, the amount of carbon-14 in a living organism stays constant as long as the organism is alive. When the plant or animal dies, carbon-14 keeps on disintegrating, but it cannot be replaced. By knowing how much carbon-14 was in an organism while it was alive and how much remains in the object being tested, scientists can estimate the age of the material. This method is called *radiochemical dating.*

Other radioactive elements can be used for dating nonliving materials. (Remember that *nonliving* means that the material has *never* been alive.) Rubidium-87 is often used because it undergoes beta decay to form strontium-87. Since strontium-87 is a stable isotope, the amount of strontium-87 in a sample can tell scientists how old the sample is. Rubidium dating was used to determine the age of lunar rocks brought back as part of the Apollo 15 mission.

Figure 18–15. The rocks gathered from the moon were dated using rubidium-87. Why didn't scientists use carbon-14 to date the rocks?

 ASK YOURSELF

Why is carbon-14 used to date only once living material?

SECTION 2   *REVIEW AND APPLICATION*

Reading Critically

1. Why are radioactive tracers useful in medicine?
2. What safety measures must scientists take to make irradiated food safe to eat?

Thinking Critically

3. Carbon is a part of all living tissue, which should make it ideal for use as a medical tracer. Why would doctors not use a radioisotope of carbon for medical purposes?
4. Oxygen-15 has a half-life of 2.0 minutes. Using a line graph, illustrate the decay process of 20 g of oxygen-15. Label the vertical axis "Mass" and the horizontal axis "Time." Choose an appropriate scale for each axis. Calculate the amount of oxygen-15 remaining every 2.0 minutes.

SKILL Communicating Using a Graph

▶ **MATERIALS**
- 100 pennies ● shoe box with lid ● graph paper

▼ **PROCEDURE**

1. Line graphs are a good way to organize some kinds of data, such as the half-life of a radioactive element. Look at the graph of the half-life of radium. How long is radium's half-life? If you started with one gram of radium, about how much would there be after 3200 years?

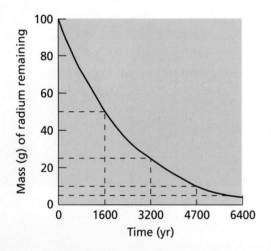

2. Suppose you wanted to make a graph of the half-life of a radioactive element you had just discovered. First you would need to gather data about the element. Then you could make a graph similar to this one. You can use simple materials to make a model that will provide the data you need for your graph.

3. Place the pennies in the box, all "heads up." Each penny represents an atom of a radioactive element.

4. Put the lid on the box, and shake it for 5 seconds. Take the lid off the box, and remove all the pennies that are "tails up." These pennies represent atoms that have broken down into some other element. Record the number of pennies left in the box.

5. Repeat step 4 four more times. Make a line graph of the results. Remember that the atoms of an element, unlike the pennies, take time to change. In a real half-life investigation, you would need to check on the element after certain time periods.

▶ **APPLICATION**
Compare the graph you made with the graph for radium. How are the two graphs similar?

✳ *Using What You Have Learned*
How else could you organize data about the half-life of a radioactive element? What would be the advantages and the disadvantages of organizing the data this way?

Nuclear Reactions

Remember the comparison between the marble-shooting activity and a particle accelerator? What would happen if you did it again, only this time you had a cluster of real marbles at the center of a ring and shot at it with a well-aimed marble? The marbles would fly apart.

Bombarding the nucleus of an atom with a neutron can cause the same type of reaction. Instead of a group of marbles separating, however, the nucleus would split into two nuclei. This process is called *nuclear fission*.

To Split or Not to Split

When an atom of uranium-235 is struck by a neutron, energy and three neutrons are given off and krypton-92 and barium-141 are formed, as shown in Figure 18–16. The three neutrons released by this reaction can go on to hit three more atoms of uranium, each of which would release three more neutrons, for a total of nine neutrons! Imagine this occurring over and over again. The reaction would soon be out of control— limited only by the number of available uranium atoms. The uncontrolled nuclear fission of uranium-235 is the basis of the atom bomb.

When a single action causes the same reaction to occur over and over, the reaction is known as a **chain reaction**. A chain reaction is similar to a landslide. In a landslide, a rock falls and rolls down a mountain, striking other rocks in its path. These rocks move, hitting more rocks, and so on. Try the next activity to find out more about how a chain reaction works.

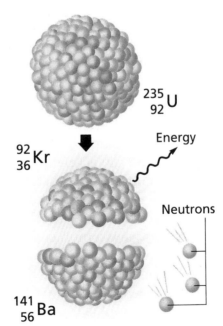

$^{235}_{92}U$

$^{92}_{36}Kr$

Energy

$^{141}_{56}Ba$

Neutrons

Figure 18–16. Uranium-235 undergoes fission to form krypton-92 and barium-141. Three neutrons and energy are given off.

ACTIVITY
What is a chain reaction?

MATERIALS
domino set; small wooden block

PROCEDURE

1. Set up the dominoes as shown. Push the first domino with your finger so that all the dominoes fall. If the chain reaction does not work, investigate the cause.
2. Once the triangular arrangement works, try the following variations: (a) place a small wooden block in the middle of the triangle of dominoes; (b) place some of the dominoes on their sides and some on end; (c) experiment with another design. Record your observations for all variations.
3. The fission reaction of uranium-235 releases three neutrons, which then hit three other atoms of uranium, and so on. Using your dominoes, create a model of this reaction.

APPLICATION

1. Based on the activity, what two things do you think are necessary for a chain reaction to occur?
2. What symbolizes the first neutron in your model of uranium-235? What does each domino represent?

Figure 18–17. An uncontrolled fission reaction results in an explosion. Such a reaction is produced by an atom bomb, which creates the characteristic mushroom cloud shown here.

What can be done to stop a chain reaction? In the activity, the reaction was limited by the number of dominoes. Similarly, fission can be controlled by limiting the number of neutrons available for the reaction. This is what is done in a nuclear reactor, where cadmium rods are inserted between the rods of radioactive fuel. The cadmium rods absorb neutrons, so the number of available neutrons can be limited to one neutron per fission reaction. When there is more than one neutron, the reaction can accelerate at a dangerously rapid rate. If there are no neutrons, the reaction stops.

One problem with fission reactors, however, is that the waste products they produce are radioactive. Disposing of our daily garbage is a problem. Imagine how hard it is to find a dump site for radioactive wastes!

Another problem we must face is the possibility of a nuclear accident like the one at Chernobyl. Are we ready to cope with such a problem? As petroleum and natural gas become scarcer, should we turn to nuclear fission for electricity? The answers are not easy.

 ASK YOURSELF

Why must a fission reaction be controlled?

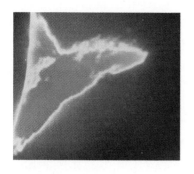

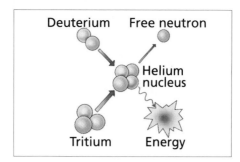

Fusion

You may have heard someone say that the sun is our greatest source of energy. Have you ever wondered where the energy comes from? It comes from nuclear reactions, but not from fission. The sun's energy is the result of nuclei combining and giving off huge amounts of energy. This process is called *fusion*. Figure 18–18 shows you an example of a fusion reaction.

Fusion reactions produce much more energy per gram of fuel than fission reactions. Even though fusion reactors might depend on fission reactors for the creation of fuel, scientists predict that fusion reactors would generate much less radioactive waste and would be safer than fission reactors. So, what's the catch? Fusion can occur only in a special state of matter called a *plasma*. A **plasma** is a low-density, super-heated gas in which the electrons and nuclei are free to move at random. That means the electrons and nuclei are not found together as they are in atoms.

Once a fusion reaction starts, the reaction releases enough energy, in the form of heat, to keep itself going as long as there is fuel. One problem is the very high temperature required for such a reaction to occur—close to 10 million degrees Celsius. For this temperature to be achieved, a way to contain the hot material will have to be found. Researchers are experimenting with ways in which to contain the material and to produce the temperature necessary for fusion to take place.

DISCOVER BY *Researching* _____

A search is underway to understand what is known as "cold fusion." Go to the library and find out what cold fusion is and what has been done in this area. In your journal, record three sources you used to gather information. Share what you learn with your classmates. ✐

Figure 18–19. Radium was once sold as an over-the-counter medication for the treatment of general aches and pains.

► **ASK YOURSELF**

Would controlled fusion be a better energy source than fission? Explain your answer.

How Much Is too Much?

As you can see, radiation is a double-edged sword. It can be useful in medicine, industry, and energy production, but it is lethal if not managed correctly.

If radiation can kill you, it must be incredibly dangerous. Is any amount of radiation safe? If so, how much exposure is okay? Is a very small exposure a problem?

The fact is, you are exposed to radiation every day of your life. No, don't run for cover. You won't glow in the dark—that only happens in science fiction. The radiation you are exposed to is natural. It comes from the sun and other stars, from soil, and even from building materials. This natural amount of radiation is called *background radiation*.

To measure the exposure of living things to radiation, a unit called the *rem* is used. A rem is a unit of absorbed radiation. Exposure to radiation is usually measured in millirems, or one thousandth of a rem. Table 18–2 shows the average radiation exposure for the average person during a year.

Table 18-2 **Average Annual Radiation Exposure in the United States**	
Source	**Millirems**
Natural sources	
From outside the body	
Cosmic radiation	50.0
The earth	47.0
Building materials	3.0
From inside the body	
Inhalation of air	5.0
Elements found naturally in the body	21.0
Manufactured sources	
Medical	
Diagnostic X-rays	50.0
Radiotherapy, radioisotopes	10.0
Internal diagnostics	1.0
Atomic energy industry, laboratories	0.2
Luminous watch dials, television tubes, microwaves, radioactive industrial wastes, and so on	2.0
Radioactive fallout	4.0
Total exposure	193.2

How much radiation is too much? Experts disagree on the maximum number of rems to which a human can be safely exposed. A generally accepted figure is 5 rems (5000 millirems) per year. How much radiation were the people close to Chernobyl exposed to? One estimate suggests that the 24 000 people closest to the reactor received more than 45 rems of radiation after the reactor exploded. That is nine times the safe dose for an entire year!

All types of radioactive decay are potentially dangerous because they cause atoms in living things to form ions. These ions can interrupt the normal processes of living things. Gamma and beta radiation can penetrate deep into a living organism and cause extensive tissue damage. The heavy alpha particles cannot penetrate the skin. However, alpha particles may enter the system through ingestion of contaminated food, giving internal tissue a full dose of this radiation.

Did you know that smokers are exposing themselves to small amounts of radiation every day? The phosphate fertilizers used to grow tobacco are rich in uranium and some of its decay products. The smoke particles that are inhaled by the smoker are rich in lead-210. Lead-210 decays into bismuth-210 and polonium-210. These radioisotopes continue to build up in certain internal organs and expose them to alpha and beta particles. Lead-210 has a half-life of 20.4 years. So you see, it is not just the nicotine and tar that are bad for you.

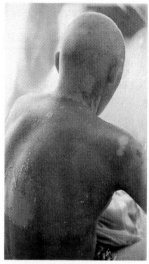

Figure 18–20. Exposure to radiation may cause severe burns.

 ASK YOURSELF

If alpha particles cannot go through your skin, why are they a dangerous source of radiation?

SECTION 3 REVIEW AND APPLICATION

Reading Critically

1. What is a chain reaction? Give an example of a reaction that simulates a chain reaction.

2. Why do some scientists believe that fusion is the answer to the energy needs of tomorrow?

Thinking Critically

3. Why does your dentist give you a lead apron when your teeth are X-rayed?

4. Are both fusion and fission chain reactions? Explain why or why not. You may wish to draw a diagram to support your explanation.

INVESTIGATION

Making a Spinthariscope

▶ MATERIALS

- matchbox, 2.5 cm × 3.5 cm ● clear nail polish ● zinc sulfide powder
- thick needle ● radioactive source ● shellac ● hand lens

▼ PROCEDURE

CAUTION: Treat all radioactive material as extremely hazardous. Avoid touching the material or spreading it on the table. Wash your hands thoroughly when you have finished.

1. Remove the inner box from the matchbox. Brush one of the outer ends of the inner box with clear nail polish, and sprinkle it with a thin layer of zinc sulfide powder. Allow the nail polish to dry.

2. Push the needle through the zinc sulfide coated end of the box, as shown in the figure. If the needle does not seem secure, you can fasten it in place by taping it to the inside of the box.

3. Carefully follow these instructions to attach a tiny piece of the radioactive material to the eye of the needle: Use the box as a holder to avoid touching the needle or the radioactive source. Place several drops of shellac on the radioisotope itself, and carefully

mix the two together with the eye of the needle. Most of the material should adhere to the eye. You have made a spinthariscope (spihn THAWR uh skohp).

4. Push the inside box partially back into the matchbox. Look at the zinc sulfide layer with the hand lens. The matchbox cover can be pushed forward or backward to get the best view with the hand lens. For best results, use the spinthariscope in a dark room.

5. Return all your materials to your teacher for proper disposal.

▶ ANALYSES AND CONCLUSIONS

1. Describe what you observe.

2. What do you suppose is causing it?

3. If the radioactive material you are using was from the luminous dial of an old watch, it probably contains tiny amounts of uranium. How long do you think your spinthariscope will work?

▶ APPLICATION

Do you think the amount of uranium found on watch dials would be hazardous to the wearer? Explain.

✳ *Discover More*

Many older watches had radium numbers on the dials so they would glow in the dark. The people who painted the numbers on would dampen the brushes in their mouths. Go to the library and find out what effect this had on the workers.

HIGHLIGHTS

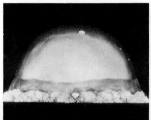

The Big Idea

Within every atom, inter-actions between matter and energy go on constantly. These interactions eventually form atoms with greater stability.

Some naturally occurring elements, such as uranium, are not stable. To achieve stability, they give off energy in the form of alpha, beta, or gamma radiation. As radiation is given off, the nucleus of the element changes, producing a more stable atom.

By altering the stability of naturally occurring atoms, scientists can create new radioactive elements. As with their natural counterparts, these elements emit alpha, beta, or gamma radiation until they become stable again. Returning to a nonradioactive state may require thousands of years.

For Your Journal

Look back at the ideas you wrote in your journal at the beginning of the chapter. How have your ideas about radiation changed? Revise your journal entry to show what you have learned. Don't forget to add what you have learned about how radioactivity is used in medicine.

Connecting Ideas

Substances may change in several ways. They may undergo nuclear change, chemical change, or physical change. Below is the beginning of a concept map. Copy it into your journal, and add what you know about how atoms and molecules change to complete the map. Be sure to show how the types of change are related.

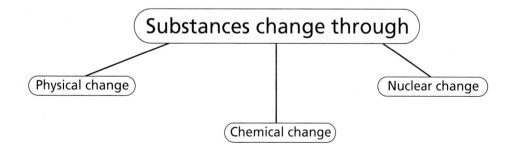

Substances change through

Physical change   Chemical change   Nuclear change

REVIEW

Understanding Vocabulary

1. For each set of terms, explain the similarities and differences in their meanings.
   a) radioactive (485), half-life (496)
   b) alpha decay (487), beta decay (487), gamma decay (488)
   c) chain reaction (499), plasma (501)
   d) Geiger counter (492), dosimeter (492)

Understanding Concepts

MULTIPLE CHOICE

2. Which of the following is not a use of radioactive isotopes?
   a) used as tracers in the human body
   b) used in the production of paints and other commercial products to increase durability
   c) used in medicine to treat many forms of cancer
   d) used to treat food so that it can be stored for long periods of time without refrigeration

3. In a fusion reaction
   a) large atoms split into smaller atoms.
   b) dangerous waste products are produced.
   c) cadmium rods are used to absorb neutrons.
   d) the nuclei must be at very high temperatures.

4. What type of radiation is given off in the following transformation?
   $$^{226}_{88}\text{Ra} \rightarrow {}^{222}_{86}\text{Rn}$$
   a) alpha radiation
   b) beta radiation
   c) gamma radiation
   d) none of the above

5. Which of the following is not an advantage of nuclear fission?

a) Fuel is inexpensive and easily obtained.
b) Fission could replace natural gas and petroleum as an energy source.
c) Large quantities of energy are produced.
d) Fission can be controlled by limiting the number of neutrons available for the reaction.

SHORT ANSWER

6. Discuss the pros and cons of the irradiation of food.

7. Scientists opened a new age when the first atom bomb was detonated. Make a list of technological advances and problems that have occurred in the nearly 50 years since their experiment.

Interpreting Graphics

8. Look at this line graph that contains data about carbon-14. What percentage of the carbon-14 breaks down after approximately 2865 years? after 5730 years?

9. What percentage of the original material remains after each half-life?

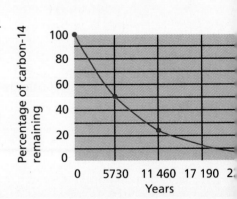

Reviewing Themes

10. *Changes Over Time*
In nature, radioactive atoms become stable over time by releasing particles or rays. Explain this phenomenon in terms of alpha, beta, and gamma radiation. Cite specific examples in your explanation.

11. *Systems and Structures*
Radioactive decay occurs in specific patterns and at specific rates called half-lives. The half-life of carbon-14 is 5730 years. Discuss how carbon-14 is used to date once-living material.

Thinking Critically

12. Compare and contrast nuclear fission and nuclear fusion.

13. List some ways in which people who work with radioactive materials can protect themselves. Explain why this is an important thing to do.

14. The notation for a radioactive isotope of Uranium is $^{238}_{92}$U. Explain what both the upper and lower numbers mean. Why do you think radioactive isotopes have a special kind of notation?

15. Ernest Rutherford used the apparatus on the right to identify radioactive particles. Beams emitted by radioactive materials would travel through the charged plates and hit the photographic plate. By developing the film, Rutherford could find information about the type of particle emitted and its charge. Using the figure shown, where do you think each of the three types of radiation should appear on the photographic plate?

16. If an aluminum sheet is bombarded by alpha particles, neutrons and other particles are given off. These particles continue to appear after the radioactive source is taken away. What can you conclude from this experiment?

17. Two atoms, $^{22}_{11}$Na and $^{26}_{13}$Al, undergo an unusual type of radioactive decay not covered in this chapter. They emit protons. What effect does this have on the stability of the products? Why does this type of decay occur?

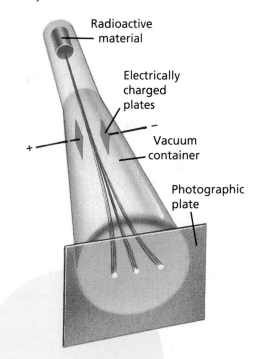

Radioactive material

Electrically charged plates

Vacuum container

Photographic plate

+

−

Discovery Through Reading

Miller, Peter. "A Comeback for Nuclear Power? Our Electric Future." *National Geographic* 180 (August 1991): 60–89. This article presents information on the usefulness and practicality of nuclear power.

Science
PARADE

The Art of Preservation

Gold bars, silver goblets, glittering jewels—who hasn't read or dreamed about the adventure of finding a sunken ship loaded with treasures? Sunken ships sometimes do contain treasures of gold and silver. They also contain valuable information that can tell scientists many things about the time in history when the ship sailed. For example, the materials that were used for trade and commerce can often be identified. The tools used to build the ship and to navigate it are often found. Sometimes cooking implements and eating utensils can provide information about the foods that were carried on long voyages.

How Do Elements Behave?

The greatest challenge to archaeologists and divers is to restore and preserve the ancient objects they uncover. To do this, they must have scientific knowledge of the properties of metals and nonmetals.

A famous treasure hunter once said, "Gold shines forever." Notice that the luster of the gold objects in the diver's hands is bright, even though they have been on the ocean floor for hundreds of years. Gold is one of the most stable metals. It does not tarnish, and it resists corrosion by most acids. Therefore,

gold is not difficult to restore.

Silver, however, reacts very differently to sea water. The diver in the photograph is also holding silver coins stuck together by chemical reactions. Besides the outer crust, which can be chipped off, there is a blackened crust covering each coin. This crust is silver sulfide—formed by the reaction of the silver with the dissolved sulfur in sea water.

Silver sulfide can be cleaned off silver objects by a process known as *electrolytic reduction*. This process

X-ray of the sword (right) shows that all the metal has dissolved and that only the crust remains.

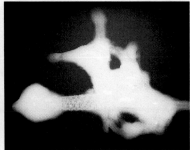

involves dipping the pieces of silver into a solution of zinc and caustic soda (sodium hydroxide) through which an electric current is passed. The sulfur separates from the silver and reacts with the silver in the solution.

Drawing of the Atocha, *which sank in 1622*

other cases, an encrusted object can be X-rayed. The resulting X-ray shows the original shape of the metal object underneath.

Wood is another material that disintegrates easily once recovered from the ocean floor. If the wood is allowed to dry in the air, it will crack and warp. To avoid this damage, wood from ancient ships is submerged in several freshwater baths to remove all traces of salt and silt. If the wooden object is large, such as a ship, it is placed under sprinklers.

Gone, but Not Forgotten

Iron is probably one of the most difficult metals to restore. Iron rusts very easily, combining readily with oxygen in the water to form iron oxide. Moreover, rusted objects tend to disintegrate when exposed to air.

Most iron objects completely rust away on the ocean floor. Before this happens, though, sand and shells may form crusts on the objects' surfaces. Even as the objects continue to rust, a crust remains as a permanent record of the original shapes.

In some cases, the crust that formed around a long-dissolved iron object can be used as a mold. The crust can be cut in half, and any remaining iron oxide washed out. Then the crust can be filled with plaster or a rubber compound to form a cast of the original object. In

Restore with Care!

After the wooden object is washed, any material that is stuck to its surface is carefully chiseled away. Then a fluid such as alcohol or polyethylene glycol is injected into or painted onto the wood. This fluid must be added to the wood to reinforce its structure and prevent warping and cracking.

The work of the archaeologist, diver, and restorer is always challenging. A knowledge of how elements, especially the metals, behave is essential. It is also necessary to work with the utmost care in order not to destroy valuable clues to the past. ◆

Archaeologist examining a timber from a sunken ship

Diver holding treasure from the Atocha

The Flashy Science of

FIREWORKS

from *Current Science*

WHOOOMPH! Sssssssss—POP! "Ooooh, Aaaaah." Pause. Then, *BOOM! BADA-BOOOOOOOOM!*

Don't you just love fireworks? The dazzling colors, the intricate patterns, all that really great noise? If so, you'll probably be one of the many millions of Americans watching fireworks displays this July 4.

For centuries, fireworks makers passed on their secret formulas only to family members. Many of the formulas remain secret today. But research over the last few years has revealed much about *pyrotechnics,* the science of fireworks. Here's some of what scientists now know about pyrotechnics.

Oxygen Fuels the Fireworks

Each firework shell contains a fuel that burns with explosive force and an *oxidizing agent.* The type of oxidizing agents used in fireworks are chemicals that readily give off oxygen. The burning of a substance—a process called *combustion*—requires the presence of oxygen or an oxidizing agent. In short, oxygen fuels the fuel.

Combustion in a firework shell occurs when the oxidizing agent (usually potassium perchlorate or ammonium perchlorate) ignites and releases oxygen. The sudden spurt of oxygen allows the fuel to burn. Fireworks fuels burn rapidly, producing sudden blasts of white light.

By adding other chemicals to the fuel-oxidizer mixture, a fireworks designer can add various colors to the fireworks flare. Barium chlorate, for instance, produces a green color. Strontium carbonate makes red.

Blue presents the toughest challenge for pyrotechnics experts. Getting a rich blue color requires a precise mixing of chemicals. Says fireworks expert Dr. John Conkling, "I pay close attention to flame colors. If a decent blue color appears, I am always impressed."

How Fireworks Work

1. Technician or computer ignites fuse.
2. Short fuse ignites black powder, sending shell skyward.
3. Delay fuse ignites first chamber, creating colorful blast.
4. Explosion of first chamber ignites delay fuse to second chamber, which explodes, usually with different color.
5. Delay fuse ignites flash and sound mixture—BANG!

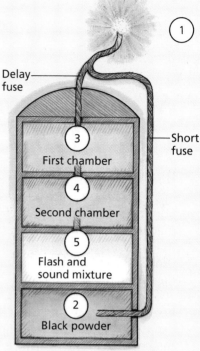

1

Delay fuse

Short fuse

3
First chamber

4
Second chamber

5
Flash and sound mixture

2
Black powder

Fireworks shell

Color + Timing =

Color alone, however, doesn't make a great fireworks show. More and more, fireworks designers launch fireworks in rhythm to particular pieces of music. Here's how.

Each shell contains two or more *fuses*, cords or wires that burn or melt at precise speeds. To launch a shell, a worker lights a rapidly burning fuse at the base of the shell. This fuse ignites a packet of explosive black powder, which sends the shell aloft at 390 feet (117 meters) per second.

The fuse also ignites a slower burning fuse. This fuse allows the shell to reach a safe height before exploding. By carefully timing the lighting of the fuses, workers can make, say, a red and blue shell explode at just the right moment during "The Star-Spangled Banner."

"We think of ourselves as artists," says Butch Grucci, president of Fireworks by Grucci, a prominent family-operated outfit. "The sky is our canvas." ◆

509

Marie Sklodowska Curie (1867-1934)

Born in Warsaw, Poland, Marie Sklodowska left her native country to obtain an education. She went to Paris, where she studied mathematics, physics, and chemistry. While in Paris, she met a French physicist, Pierre Curie. They eventually married and worked together.

Shortly after her marriage, Marie Curie began working with radioactive elements. In fact, she was the first to use the word *radioactivity*. Together, she and Pierre discovered that the radiation coming from uranium ore was much more than the amount the uranium alone could account for. They hypothesized that there must be other radioactive material present in the ore. From tons of uranium ore they isolated two new radioactive elements, which they named *polonium* and *radium*. For their discoveries, they received the 1903 Nobel Prize in physics.

In 1911 Marie Curie was awarded a second Nobel Prize, this time in chemistry. Aided by her daughter Irène, Marie worked on the application of X-rays in medicine. After Marie's death, Irène and her husband, Frédéric Joliot-Curie, continued research on radioactive materials. ◆

Rosalyn Yalow (1921-)

What is a physicist doing in a Veterans Administration hospital working with physicians? Earning a Nobel Prize in medicine is the answer that Rosalyn Yalow might give you. Born in the Bronx, New York, Yalow graduated with honors from Hunter College in Manhattan. Following graduation she was accepted as a graduate student in the College of Engineering at the University of Illinois. While studying for her degree, Yalow became interested in the measurement of radioactive substances, a skill she later put to use as a medical physicist.

In 1947 Yalow converted a closet at the veterans' hospital where she worked into a radioisotope laboratory. Soon after this, she met Solomon Berson, a physician, with whom she would work for more than 22 years. In 1977 Yalow and Berson won a Nobel Prize in medicine for a technique that they developed. The technique, called *radioimmunoassay,* is a test that uses radioactivity to identify and measure small traces of substances in the blood.

Today Yalow travels throughout the world to lecture. She has received many awards and prizes from universities, medical societies, and associations and has also been awarded the National Medal of Science. ◆

Theophilus Leapheart, Chemist

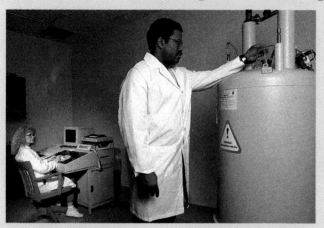

Each year chemists make thousands of new chemical compounds. These compounds can be very complex molecules. How do chemists know the structures of the compounds they make? Theophilus Leapheart, a chemist from Midland, Michigan, uses special instruments to find out the structures of new chemical compounds and medicines he makes.

Leapheart works as part of a team of scientists. "Chemistry is very team oriented," he explains. "No one can put the whole puzzle together alone. Chemists, engineers, mathematicians, and physicists all need to work together to solve problems and to develop new products. To be successful, you have to know how to work with others."

One of the tools that Leapheart uses to find out the structure of a compound is called a *nuclear magnetic resonance* (NMR) instrument. When Leapheart makes a new compound, he takes a small sample of it and dissolves it in a solvent. He then puts the sample into the NMR's strong magnetic field. The magnetic field causes the molecules to line up, just like the earth's magnetic field causes a compass needle to line up.

Next, Leapheart passes radio waves through the sample. This causes the molecules to wobble. When the radio waves are shut off, the molecules try to line up again with the magnetic field. As they do, they give off energy. The frequency of the energy depends on the kind of molecule. These frequencies show on a graph called a *spectrum.*

To a trained scientist, the NMR spectrum of a compound is like its signature. No two compounds have exactly the same NMR spectrum. A spectrum can tell what atoms are present in the compound and how many of each are present. Leapheart uses computers to run the NMR machine. The computers tune the instrument, much like you would tune a television to get a clear picture. The computer also collects the data and stores it for later use.

To prepare for his career, Leapheart took many science and mathematics classes. However, he also thinks that English classes are important. "Writing is very important," Leapheart adds. "In any setting, you must be able to communicate what you have done, how you did it, and how someone else can follow through on your work after you are gone." ◆

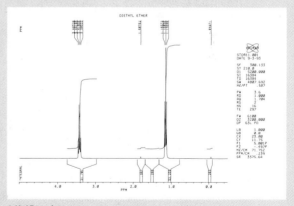

NMR printout

Discover More

For more information about NMR and careers in chemistry, write to the

Office of Pre-High School Chemistry
American Chemical Society
1155 16th Street, N.W.
Washington, DC 20036

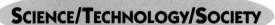
NMR Spectroscopy

Nuclear magnetic resonance (NMR) spectroscopy has become a powerful tool in chemistry and medicine. Chemists use it to determine the structure of complex substances. NMR is also used by physicians to make pictures of the soft tissues in the body and to help diagnose some diseases.

NMR spectroscope

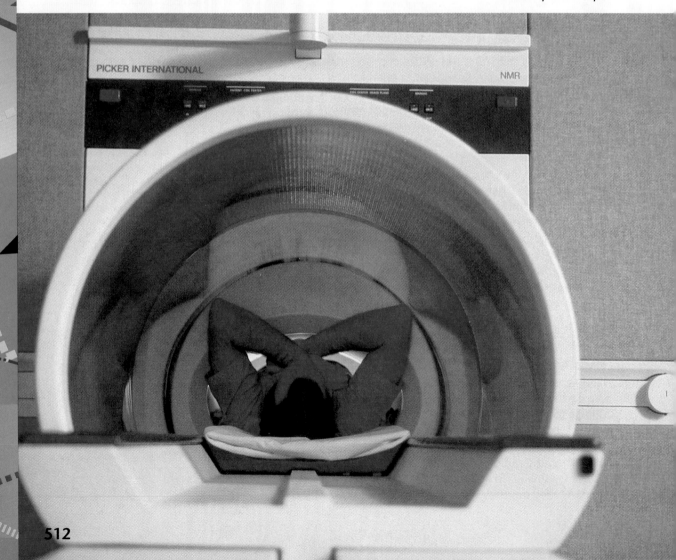

Nuclei— Line Up!

NMR uses isotopes of elements that have magnetic nuclei. Hydrogen-1, carbon-13, and phosphorus-31 are often used. If a chemist is trying to find the exact structure of a substance, a sample of the substance is placed in a strong magnetic field. This field is 30 000 to 240 000 times greater than the magnetic field of the earth. The field causes the nuclei of the unknown sample to line up. This happens for the same reason that a compass needle points to the magnetic north pole.

Sending Out a Signal

When the nuclei are lined up, radio waves are sent through the sample. These waves make the nuclei wobble in much the same way that a rotating top does. After the radio waves are discontinued, the nuclei try to line up again. As they line up, they send out radio waves like the ones that they received. These waves can be detected by a computer. The computer can take information from the waves and make a graph that shows chemists such as Leapheart where the atoms are located in a molecule.

Many important molecules in living things contain phosphorus-31. For this reason,

NMR can be used to study how cells and organs function. The use of NMR makes it possible to learn about the structure of molecules in living organisms without killing the subject or even performing surgery!

An Image— a Diagnosis

The application of NMR in medicine is known as Magnetic Resonance Imaging (MRI). MRI gives physicians a two-dimensional picture of the human body, but it does not use or produce any harmful radiation. Therefore, it provides physicians with a safe tool for diagnosing illnesses.

MRI can also be used to study what is happening in living tissue. The image on this page shows the inside of a human brain. The colors indicate different areas of the brain and give important information on the chemicals that make up the tissue of the human brain. You can see how using MRI enables physicians to study what is happening in living systems. Additionally, physicians can study what happens to tissue or body systems when they are under stress, under the influence of drugs, or diseased. ◆

MRI of the human brain

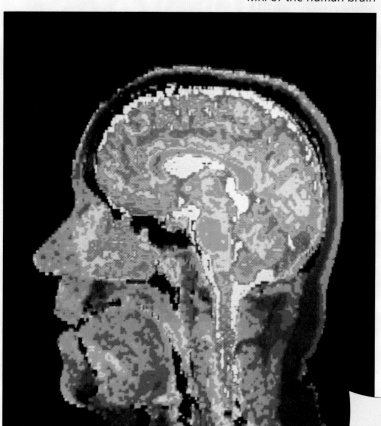

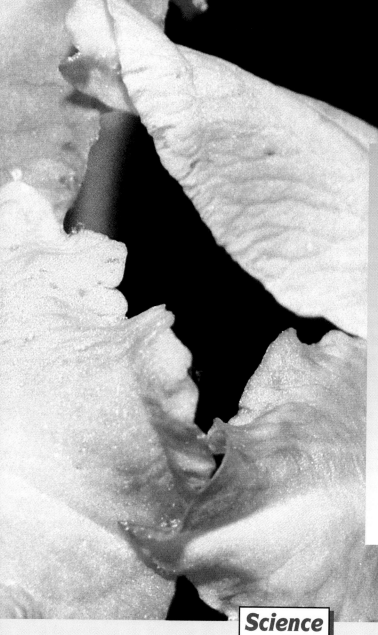

Swishing its tail back and forth, the house cat crouches, waiting for the right moment to pounce on its prey. Success! Frog in mouth, the cat begins to prance away. Suddenly, the cat drops the frog and begins foaming at the mouth. The cat has just discovered that nature, the frog in this example, sometimes uses chemistry to discourage predators. Nature also uses chemicals to attract. Plants use chemistry to invite insects, such as the bee in this photograph, to pollinate their flowers. Both animals and people have a variety of uses for chemistry, or changes in matter.

Science **PARADE**

INTERACTIONS OF MATTER

In the early 1800s, a person needed only careful observation to know what he or she was eating. "Tell me what you eat, and I shall tell you what you are," wrote the Frenchman J. Brillat-Savarin in his classic book on eating, *The Physiology of Taste*. Today, however, you need to be a scientist to know what holds your food together.

If you have ever eaten Italian dressing, you probably know that water and oil do not stay mixed. You may not know, however, that at one time people had to knead margarine or else a watery liquid seeped out of the vegetable shortening. Today, chemical additives hold margarine's liquid and oil

Cooking is the art of combining ingredients to make foods look and taste good. In other words, good food results from skillfully controlling the interaction of matter.

together. The same chemicals, called *emulsifiers,* also keep oil from floating to the top of peanut butter. But you are not likely to find a label that reads "Emulsified Peanut Butter!"

If you've ever drunk chocolate milk, you've probably drunk red algae with it. To keep cocoa suspended in milk, manufacturers must use an additive called a *stabilizer.* One of the most common stabilizers is derived from red algae that grow off the coast of Ireland. The same algal extract is added to those "lean" hamburgers sold at fast food restaurants. Algae burgers! Now you may think twice when you order lunch!

For Your Journal

✎ What are mixtures and compounds? How are they formed?

✎ What is a solution? What kinds of solutions do you use everyday?

✎ You use colloids all the time? What do you think colloids are and how might you use them?

Types of Matter: A Review

Objectives

Distinguish *between homogeneous and heterogeneous mixtures, and give examples of each.*

Propose *a method by which you could separate the parts of a mixture.*

Have you ever tried to wash a stain from a shirt? Have you ever made salad dressing or fruit gelatin? Maybe you have helped paint the house or some furniture. What do detergent, salad dressing, and paint have in common? They are all matter, but are they alike in other ways?

DISCOVER BY *Doing*

Examine the different substances your teacher gives you. Briefly describe each in your journal. Decide which ones could be made of two or more substances. Explain your choices. ✐

Mixtures

A mixture is a combination of two or more types of matter, each of which keeps its own characteristics. This means that if you separate a mixture, the parts are just the same as when they were put together. Because the parts of a mixture keep their properties, they can be separated by physical means, such as filtering or evaporation. Go back and look at your ideas from the activity. Which of the substances you observed was obviously a mixture? How would you separate the mixture into its components?

In the next activity, you will be given two liquids. One is colorless, and the other is blue. Do you think they are mixtures or pure substances? How could you find out?

DISCOVER BY *Doing*

CAUTION: Put on safety goggles and a laboratory apron, and leave them on for the entire activity. Pour a small amount of each liquid into a shallow glass dish, and warm each dish gently on a hot plate. What do you observe? Are the liquids pure substances or mixtures? Explain what happened that led you to this conclusion. ✐

There are two types of mixtures: homogeneous and heterogeneous. A homogeneous mixture is uniform; it has the same composition through out. When you look at a homogeneous mixture, you might not be able to tell it is a mixture. Which of the mixtures in the activity were homogeneous?

A heterogeneous mixture is not uniform throughout. Instead it is made up of different substances that can be easily seen. Vegetable soup is an example of a heterogeneous mixture.

▶ **ASK YOURSELF**

What is the difference between a homogeneous and a heterogeneous mixture?

Figure 19–1. You encounter many types of heterogeneous mixtures every day, such as this tossed salad.

Cooking the Chemist's Way

Meet Chef Pierre. He boasts that he is the master of all edible matter. He can convert eggs, flour, sugar, chocolate, and a few other choice ingredients into a delicious heterogeneous

Figure 19–2. This is how chlorine and sodium look when they are pure elements.

mixture known as a chocolate chip cookie. He can turn cream, broth, and seasonings into the most delicate of sauces. But Chef Pierre can make only mixtures. Even with all his skill, he cannot make a pure compound. His friend Kimmie, a chemist, has asked the chef to visit her lab and watch her make a compound that he uses every day.

Kimmie is very secretive about what she will cook up. She asks Chef Pierre to stand back and watch. "I will be making a compound," she says. A **compound** is a pure substance composed of elements that have been combined *chemically.* The compound has totally new properties that are different from those of the original materials.

Kimmie opens a gas cylinder that is connected to a flask and allows a yellow-green gas to flow into the flask. She covers the flask so the gas will not escape. "This is chlorine gas, and it is very poisonous. That is why we are doing this under the fume hood in the lab." Chef Pierre would rather be in his safe kitchen, but he is interested in this compound Kimmie is making.

Kimmie cuts a piece of a soft metal that has been stored in a liquid. The metal has a grayish sheen. Kimmie tells Chef Pierre that the metal is sodium. "Sodium catches on fire in water; therefore, I have to store it in oil or kerosene."

"Now, we are going to combine these two elements to form a compound that I am sure you have on your kitchen shelf, Pierre." "*Cordon bleu!* That is impossible," exclaims Pierre. "I would not have such dangerous substances in my kitchen."

Figure 19–3. This is how sodium and chlorine react with each other.

A Flash and a Bang Kimmie places the piece of sodium on a spoon with a very long handle. She opens the flask containing chlorine and drops the sodium into it, capping it immediately. Chef Pierre is amazed. The piece of sodium ignites with a bright flame. It seems as though the whole flask is on fire. Even though he is not close to it, Chef Pierre knows the reaction is giving off a lot of heat. The green gas is also disappearing. Whatever was inside the flask is turning colorless!

Kimmie opens the flask under the hood to get rid of any leftover chlorine. She asks Pierre to come closer and look inside the flask. "What do you see? Look closely."

Pierre notices that there are small, colorless crystals around the inside of the flask. "Are these little crystals in my kitchen?" he asks.

Kimmie laughs. "The compound that you see is made of sodium and chlorine. I call it sodium chloride; you, Pierre, call it table salt."

"*Fondue!* Is it possible that substances can change so much when they combine?" Kimmie explains that chemists perform even more amazing feats every day.

Breaking Up Is Hard to Do   Chef Pierre looks worried. Will his table salt break down into its component elements and burst into flame? Kimmie assures him that while mixtures can be separated by physical means such as filtration, evaporation, or distillation, the elements in a compound cannot be separated in this way. In order to separate the elements in a compound, chemical changes involving energy are required.

There is another difference between a compound and a mixture. Substances that form a mixture can be combined in any amounts. For instance, you can add a pinch of sugar to a cup of tea, or you can add several spoonfuls. The result is still sweetened tea. Compounds, however, can be made only by combining elements in definite ratios. A water molecule, for example, is formed by combining two atoms of hydrogen with one atom of oxygen. Water cannot be formed with any other ratio of hydrogen and oxygen. Furthermore, once formed, water cannot be easily reseparated into hydrogen and oxygen.

Figure 19–4. Two atoms of hydrogen and one atom of oxygen form water. Two atoms of hydrogen and two atoms of oxygen form hydrogen peroxide, a common disinfectant.

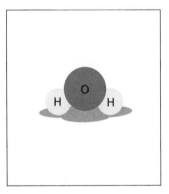  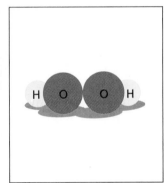 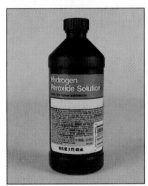

 ASK YOURSELF

How is making a compound different from making a mixture?

SECTION 1   *REVIEW AND APPLICATION*

Reading Critically

1. Give two examples of heterogeneous mixtures that might be found in your home.

2. Compare and contrast mixtures and compounds.

Thinking Critically

3. How would you separate a mixture of sand, sugar, oil, and water?

4. Why do you think the properties of sodium chloride are so different from the properties of the elements that are combined to make it?

SKILL Problem Solving

▶ **MATERIALS**
- plastic tub ● water ● cooking oil, 200 mL ● graduate ● tissues
- liquid soap ● wooden sticks or long pencils ● plastic wrap

▼ **PROCEDURE**

1. Fill a plastic tub half-full of water. Put 200 mL of cooking oil into the water.

2. Using the equipment provided, design a plan to remove the oil from the water. Write down the steps you will use in your cleanup operation. Predict how effective your procedure will be.

3. Carry out your plan. Does any oil remain in the water? How does your result compare with your prediction?

4. Revise your plan. Predict how changes in your plan will affect how much oil is removed from the water.

▶ **APPLICATION**

1. How did you use what you know about mixtures to devise your plans?

2. How accurate were your predictions? What knowledge did you use to revise your plan and to increase the accuracy of your predictions?

3. What are some practical applications of this procedure in the real world? How could your ideas be used to solve environmental problems?

✳ *Using What You Have Learned*

How is this miniature oil spill like an oil spill in the ocean? How is it different?

Solutions

Objectives

Define and ***give*** examples of a solution, a solute, and a solvent.

Describe factors that affect the rate at which a solid dissolves in a liquid.

Interpret a solubility graph to determine whether a solution is saturated, unsaturated, or supersaturated.

Chef Pierre and Kimmie are sitting down to a nice glass of iced tea. Kimmie notices that Chef Pierre is putting three spoonfuls of sugar in his tea! She laughs. "Pierre," she says, "most of that sugar will end up on the bottom of the glass. Trust me! We chemists know these things."

Making Solutions

A soft drink is a mixture of water, flavorings, and carbon dioxide gas. A soft drink, however, looks like a single material; it is homogeneous. A homogeneous mixture is called a **solution**. Every solution has two parts, the solvent and the solute. The **solvent** is the material in which the solute is dissolved. Because there is always more solvent than any other component in a solution, the solvent is the major component. In a soft drink, water is the solvent.

The **solute** is the substance that is dissolved in the solvent. It is the minor component. The gas and flavorings in a soft drink are the solutes.

Although many common solutions are made from solutes added to liquids, solutions can also be combinations of liquids, solids, and gases. Table 19–1 lists examples of solutions.

Table 19–1	**Types of Solutions**	
Example	**Solvent**	**Solute**
Air	Gas	Gas
Soda water	Liquid	Gas
Silver amalgam	Solid	Liquid
Brass	Solid	Solid

▶ **ASK YOURSELF**

Give an example of a solid/liquid solution not mentioned in the section. Tell which component is the solute and which is the solvent.

Figure 19–5. This soda water is a solution of carbon dioxide gas in water.

What's the Solvent?

Chef Pierre is back in his kitchen trying to solve a problem. One of his helpers has spilled an Indian spice called turmeric, and it has left a yellow stain on his white counter top. Your teacher has a similar counter top material with a turmeric stain. See if you can get it clean.

DISCOVER BY Doing

Try to get the stain out. In your journal, keep a careful record of what you do and the results you achieve. ✐

Now you can understand Chef Pierre's problem; turmeric is a hard stain to get out. Chef Pierre calls Kimmie and explains his problem. According to Kimmie, scientists have a rule that might help: Like dissolves like. That is, chemically similar substances dissolve in each other. Nail polish, for example, does not dissolve in water. This is actually a good thing, because if it did, you couldn't wash your hands without the nail polish coming off. To take off the nail polish, you use a chemical that is similar in structure to the nail polish—nail polish remover.

"The principal ingredient in turmeric," says Kimmie, "is an organic compound. Organic compounds contain carbon. You can probably get the stain out if you use another organic compound." The chef asks what organic compound he might have in his kitchen. "Do you have some cooking sherry? Try that."

You cannot use sherry in the classroom, but you can use rubbing alcohol. Alcohol is an organic compound, similar to the turmeric. How well did it work on the stain? Do you think that Pierre will get good results using his cooking sherry?

Water is sometimes called a universal solvent. This is because water can dissolve many different substances. Did water dissolve the turmeric stain? Finding solvents for substances that are not water soluble is important to the cleaning industry. Many dry cleaners have used fluorocarbons or chlorocarbons as stain removers. Since some of these compounds are either toxic or harmful to the ozone layer, they might have to be replaced.

Figure 19–6. Dry cleaners use solvents to remove stains.

ACTIVITY

What are the effects of solvents on tincture of iodine?

MATERIALS
cotton swabs (3), tincture of iodine, water, ethanol

PROCEDURE
1. Use a cotton swab dipped in tincture of iodine to make two small spots on the palm of your hand.
2. Dip a second cotton swab in water, and wash one iodine spot with it. Dip another swab in ethanol and wash the other spot.

APPLICATION
1. What happened when you put water on the iodine spot?
2. What happened when you put ethanol on the iodine spot?
3. Which solvent removed the iodine spot better? Why?

Removing stains is quite a science when you think of it. You have to find the proper solvent to remove a particular stain without bleaching away the color of the stained material or eating through it. Many industrial chemists dedicate long hours to the manufacture of good detergents and cleaners.

 ASK YOURSELF
What is meant by the "like dissolves like" rule?

Figure 19–7. Rock candy is formed when sugar crystalizes out of a saturated solution.

A Little or a Lot

Chef Pierre is making some rock candy to decorate one of his desserts. In order to make the candy, Pierre must dissolve as much sugar as he can in very hot water. He needs a saturated solution. Solutions can be *dilute* (not much solute) or *concentrated* (a lot of solute). A **saturated solution** has the maximum amount of solute that can be dissolved in a given amount of solvent at a given temperature.

When Pierre sees that no more sugar will dissolve after he stirs his mixture, he knows he has a saturated solution. The next step is to allow the solution to cool untouched for a few days.

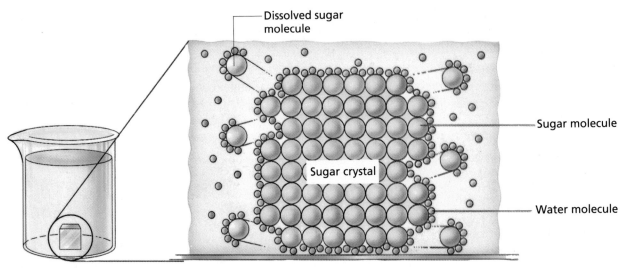

Dissolved sugar molecule

Sugar molecule

Sugar crystal

Water molecule

Figure 19–8. When a sugar cube is put into water, the water molecules surround the sugar molecules, causing them to go into solution, or dissolve.

He places a weighted string in the jar holding the sugar syrup. Sugar crystals will form on the string. They will be large and beautiful and—of course—edible. They will look like diamonds on his award-winning chocolate torte. What makes the sugar come out of solution and form crystals?

A solution is saturated when no more solute seems to dissolve. In reality, however, some solute continues to dissolve. For example, salt added to saturated salt water still dissolves in the

Figure 19–9. The watch glass (left) contains crystals of copper sulfate that were formed after a solution of copper sulfate in water was evaporated. In the salt pond (right), the same principle is used to retrieve salt from sea water.

water. What happens is that for every grain of salt that dissolves, another grain comes back out of solution. The solution has reached equilibrium. Equilibrium in a solution is a state of balance between what is going into solution and what is coming out of solution.

A saturated solution contains as much solute as the solvent can hold at a given temperature. In the case of sugar water, you may wonder how much sugar makes the solution saturated. Look at the graph in Figure 19–10. What usually happens when you increase the temperature of a solution? For most substances the amount of solute that can be dissolved increases. There are a few exceptions to the rule, as you can see with $Ce_2(SO_4)_3$.

The maximum amount of solute that can be dissolved at a specified temperature and pressure is known as **solubility**. Solubility is usually expressed as grams of solute in 100 g of solvent.

Which of the solutes in Figure 19–10 dissolves most easily in hot water? Which dissolves most easily in cold water? Which is the least affected by temperature change? Each curve on the

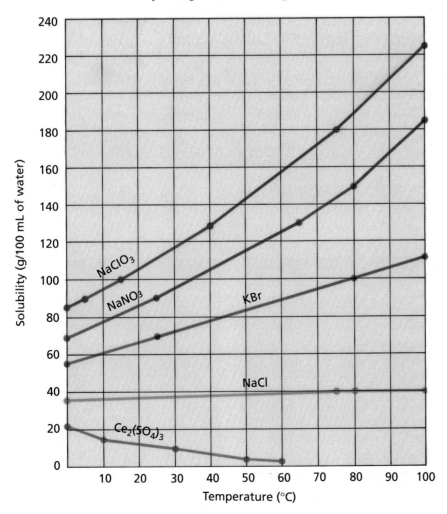

Figure 19–10. This graph shows the solubility (in water) of several substances at different temperatures. Notice that most of the substances have higher solubilities at higher temperatures.

graph shows the maximum amount of solute that can be dissolved at a given temperature. The area below each curve gives the amount of solute in an unsaturated solution. Above each curve is the area indicating that no more solute would dissolve. Solubility is a characteristic property of each substance.

DISCOVER BY *Problem Solving*

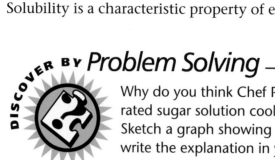

Why do you think Chef Pierre has to let his saturated sugar solution cool in order to get crystals? Sketch a graph showing what happens, and write the explanation in your journal. 🖉

It is sometimes possible to dissolve more solute in a solvent than you could normally. When this happens, the solution is *supersaturated*. Supersaturated solutions are not stable because they are no longer in equilibrium. As Chef Pierre's solution cooled, what happened?

A supersaturated solution is similar to a tightrope walker teetering on a rope before falling into the safety net. It doesn't take much of a push to cause the tightrope walker to fall. Similarly, it doesn't take much to make the solute in a supersaturated solution come out of solution. Often, you can bring the excess solute out of solution by cooling the solvent. Sometimes you can bring it out by stirring the solution, scratching the side of the container, or by adding just one crystal of the solute.

► ASK YOURSELF

Describe a condition under which you would want to have a saturated solution.

Figure 19–11. A supersaturated solution (left) contains more solute than it can hold under normal circumstances. Scratching the side of the beaker will force the solute to come out of solution (right).

Now You See It, Now You Don't

The dissolving of a solute can be affected by many factors. You are probably aware of at least three but have never really thought about them. Think about the answers to these questions; they may help you identify the three factors. Which is easier to dissolve in water—a spoonful of sugar or a sugar cube? When adding sugar to lemonade, what do you do to speed up its dissolving? Does sugar dissolve faster in iced tea or hot tea?

Let's see how many correct answers you have. The speed of dissolving can be increased by powdering the solid or breaking it into small pieces. When you make the pieces smaller, more of the solute is exposed to the solvent. So, a spoonful of sugar dissolves more quickly than a sugar cube.

Stirring also brings more of the solute into contact with the solvent and breaks up clumps of the solute. Therefore, stirring is another way to speed up the rate of solution. You stir your lemonade to dissolve the sugar.

Figure 19–12. The rate of solution can be increased by several methods, including stirring.

As you saw in Figure 19–10, an increase in temperature often increases solubility. It can also make the solute dissolve faster. An exception to the effects just described is the dissolving of a gas in a liquid. If you have ever opened a soda bottle that is warm or has been shaken, you know that the dissolved gas fizzes out of the bottle, spilling some of the liquid. When that happens, it shows that the gas is coming out of solution. Shaking or stirring the liquid causes the gas molecules to escape. Agitation allows more of the bubbles to reach the surface.

Sometimes factories and power plants use water from rivers and lakes to cool their machinery. If this heated water is not cooled before being returned, the river's or lake's water temperature can be increased. This is called *thermal pollution.* Because less gas can stay dissolved in warmer water, the amount of oxygen dissolved in the water is reduced. This can have a drastic effect on the fish and other living things in the water that depend on the dissolved oxygen to stay alive.

Figure 19–13. Factories and power plants sometimes return heated water to lakes and rivers.

DISCOVER BY *Doing*

Pour regular sand into a 500-mL beaker half-filled with water. Describe what happens. Do the same with some "magic sand." Describe what happens. Did you expect this? Can you explain what is so special about magic sand?

Try to separate the regular sand from the water by pouring the water into another beaker. Do the same thing with the magic sand. Describe what you observed. Relate what you observed to what you know about solutions and solubility. ✎

▼ ASK YOURSELF

How can you increase the solubility of a gas in a liquid?

SECTION 2   *REVIEW AND APPLICATION*

Reading Critically

1. Identify the solvent and solute in each of the following solutions:
   a. carbonated water
   b. air
   c. salt water
2. Why do crushing and stirring speed up dissolving?

Thinking Critically

3. Why do you think that ice cubes made with cold water are often cloudy, while ice cubes made with hot water are usually clear?
4. Using Figure 19–10, determine whether the solutions listed below are saturated, unsaturated, or supersaturated.
   a. 40 g of $NaNO_3$ in 100 mL of water at 40°C
   b. 70 g of KBr in 100 mL of water at 20°C
   c. 20 g of NaCl in 50 mL of water at 100°C

Suspensions and Colloids

Chef Pierre's assistant is making salad dressing. He mixes oil, vinegar, and spices for an Italian dressing. As soon as he stops shaking it, the dressing separates into two layers and the parsley and oregano drift down to the bottom. "Chef, I will have to shake vigorously each time I serve the dressing. I have 50 customers to please—I am a chef, not a weight lifter—can you help me?" Chef Pierre proceeds to show his assistant how to prepare a creamy salad dressing that will not separate as easily. He takes the oil and vinegar mixture and puts it in the blender for a few minutes. "*Voilà!*" says Pierre. "That should take care of the problem. This dressing should remain creamy for the duration of the banquet."

Objectives

Distinguish among a suspension, a solution, and a colloid.

Describe and ***give*** examples of different types of colloids.

Suspensions

What happened to the oil and vinegar mixture as Chef Pierre mixed it in the blender? You can try a similar experiment.

DISCOVER BY Doing

Combine a small amount of oil and water in a container and shake it. What happens? Now take the same oil and water, put it in a blender, and whip it for a few seconds. Describe any changes. What has happened? Will the oil and water separate? Observe for several hours or overnight. ✐

What you have made is a suspension. A **suspension** is a heterogeneous mixture in which one of the parts is a liquid. Visible particles will settle out of any suspension. This is often because the two substances are insoluble in each other. That is, one will not dissolve in the other. Remember oil and water do not mix. Is the oil-water mixture that you placed in the blender a solution or a suspension? How do you know?

Figure 19–14. In order to make different colors of paint, pigments are suspended in a liquid.

You have probably used thousands of suspensions during your life. They are everywhere. Suspensions are used frequently as medicines. Liquid medicines that require shaking before they are taken are suspensions. Some paints contain solid pigments in a liquid. These pigments must be finely ground so that the paint does not separate readily. You may have noticed that paint stores have special machines that shake the cans of paint in order to mix the components.

The usefulness of suspensions is demonstrated by the many applications of MCP (microcrystalline polymer) suspensions. MCP suspensions were discovered by American chemist and inventor Dr. Orlando A. Battista. He discovered that he could break down synthetic materials such as nylon and polyester into tiny crystalline structures—MCPs—with the aid of acid.

MCP suspensions can cover hard-to-coat surfaces such as glass and aluminum. When sandwiched between two pieces of glass, MCP suspensions bind the sheets together, producing one of the cheapest forms of safety glass. Suspensions of MCPs in water result in a substance similar to petroleum jelly. These suspensions can be used as greaseless cosmetic bases. The best thing about MCPs is that they are made from waste products such as discarded nylon stockings and plastic bags!

Figure 19–15. The microcrystalline polymer suspension shown here can be made from old plastic bags.

 **ASK YOURSELF**
What is a suspension?

Colloids

Eventually, particles in suspensions settle out. What if we could make the particles small enough so that they would stay in solution? There are mixtures like these; they are called **colloids**. In a colloid, the particles are very, very small. So, even though they do not dissolve, they do not settle out nor can they be filtered out. The reason for this is that colloidal particles are in constant random motion. Because of this motion, the particles do not settle out of the mixture.

Chef Pierre will share his recipe for a particular colloid. Look through Chef Pierre's recipe. Are there any ingredients that will not dissolve in each other? What can you do about it? What is Chef Pierre's colloid normally called?

Chef Pierre's
Colloidal Sandwich Spread

1 egg
1 tablespoon lemon juice
1 teaspoon salt
1/4 teaspoon pepper
1/2 cup oil

Combine first four ingredients in a blender or food processor. Gradually add oil until mixture becomes colloidal.

Figure 19–16. Mayonnaise is an example of an emulsion. The chef shown here is adding oil to the mayonnaise he is making.

Figure 19–17. Gelatin (above) and agar (right) are both examples of gels.

There are probably many colloids at your home right now. Gels, aerosols, and emulsions are all forms of colloids. *Gels* are liquid particles spread out in a solid. Gelatin, jelly, and stick deodorant are gels. *Emulsions* are colloids made of two liquids. You are familiar with many oil-water emulsions such as mayonnaise, hand cream, and milk.

Another type of colloid is an *aerosol.* Aerosols are formed when either solid or liquid particles are suspended in a gas. Fog and smoke are examples of aerosols. A spray can shoots out a colloid of particles finely dispersed in a gas. Chef Pierre uses an aerosol nonstick cooking spray when he cooks.

You might wonder how anyone could tell the difference between a colloid and a true solution. They both seem homogeneous. There is one easily observable way: colloids scatter light that passes through them and solutions do not. You can see this phenomenon in the photographs in Figure 19–18.

Figure 19–18. The photograph at the top shows the light of a slide projector shining through one glass containing a solution and one containing a colloid. You can see the beam of light in the glass containing the colloid. Fog (below) is a natural colloid; it scatters light as the sun's rays pass through it.

▼ **ASK YOURSELF**

How are colloids different from solutions? How are they the same?

SECTION 3 *REVIEW AND APPLICATION*

Reading Critically

1. List five colloids that might be found in your home.
2. What are two differences between a colloid and a suspension?

Thinking Critically

3. How could you determine experimentally if a liquid is a solution, a suspension, or a colloid?
4. What advantage would a medicine in the form of an emulsion have over a medicine that is a nonemulsified suspension?
5. Natural milk contains cream that forms a layer when it rises to the top. The milk you buy at the store does not separate into layers because it has been homogenized. What do you think is invo̶lved in the process of homogenization?

INVESTIGATION

*S*eparating Mixtures

▶ MATERIALS

PART A
- safety goggles ● laboratory apron
- cotton plug ● glass tube (7 cm × 1 cm)
- activated charcoal ● beaker, 150 mL ● clamp or clothespin ● methylene blue solution ● droppers (2) ● copper sulfate solution

PART B ● test tubes (2) ● spatula or scoop
- "mineral ore" ● corks (2) ● mineral oil, 2 mL

PART C ● test tubes (2) ● wax pencil ● unknown liquid ● boiling chip ● two-hole stopper with glass tubing and thermometer ● ring stand ● clamp
- iron ring ● wire gauze ● laboratory burner
- beaker ● ice water

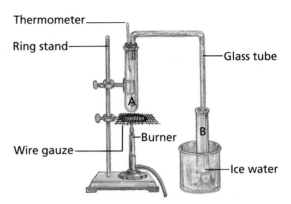

CAUTION: Put on safety goggles and a laboratory apron, and leave them on throughout the investigation.

▼ PROCEDURE

PART A

1. Place a cotton plug in one end of the glass tube. Add enough charcoal through the other end to reach a height of 2.5 cm. Suspend the tube over the beaker.

2. Using a dropper, fill the tube with the methylene blue solution. Collect the filtered liquid in the beaker.

3. Use the same dropper to collect the filtered liquid and put it through the tube again. Repeat until you observe a color change. Record your observations.

4. Rinse all the equipment, and repeat steps 1–3 using copper sulfate solution.

PART B

5. Fill two test tubes halfway with water. Use the spatula to add a pinch of the "mineral ore" to each tube. Stopper and shake.

6. To one of the test tubes, add 2 mL of mineral oil. Stopper and shake. Let the test tubes stand until you see two layers in the test tube with the mineral oil. Compare the test tubes.

PART C

7. Label one test tube "A" and one test tube "B."

8. Fill test tube A one-fourth full of the unknown liquid. Add a boiling chip.

9. Set up your equipment as shown. Be sure the glass tubing is not plugged up in any way.

10. Heat the liquid gently until it boils. Adjust the heat for even boiling.

11. Describe what you see in test tube B. Add a boiling chip, and heat until the test tube is almost dry. What do you observe?

▶ ANALYSES AND CONCLUSIONS

1. Summarize what happened in parts A, B, and C.

2. Why do you use ice water in part C?

3. In part C, did all groups get the same results? If not, explain why.

▶ APPLICATION

How could what you have learned here help you separate a mixture in your kitchen?

✳ *Discover More*

Choose three solutions, and filter them in the ways you learned in the investigation. How does each method affect the solutions?

CHAPTER 19

*H*IGHLIGHTS

The Big Idea

The process of making solutions, suspensions, and colloids could be seen as an exercise in equilibrium. A saturated solution is in a state of dynamic equilibrium. For every solute particle that dissolves, a dissolved particle comes out of solution. This is a balanced state.

Scientists study different examples of equilibrium in order to understand and to control them. Then the equilibrium can be shifted in one direction or another. Understanding how systems interact and how we can affect the stability of them enables us to manipulate reactions.

For Your Journal

How have your ideas about the interactions of matter changed? Revise your journal entry to show your new ideas about mixtures and compounds. Include information on how solutions are used in the everyday world.

Connecting Ideas

This concept map shows how the big ideas of this chapter are related. In your journal, complete the concept map to show how different types of mixtures fit into this arrangement.

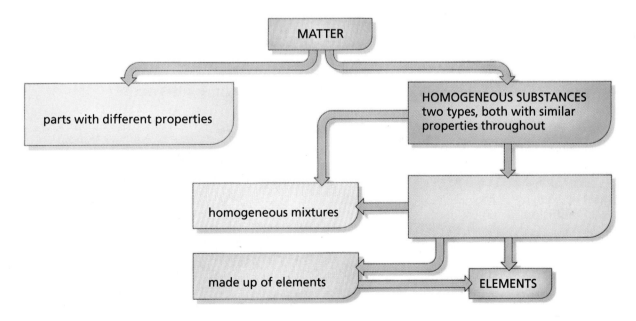

REVIEW

Understanding Vocabulary

1. For each set of terms, explain the similarities and differences in their meanings.
   a) mixture (518), compound (520)
   b) solute (523), solvent (523)
   c) solution (523), saturated solution (526), solubility (527)
   d) suspension (531), colloid (533)

Understanding Concepts

MULTIPLE CHOICE

2. Which of the following is a homogeneous mixture?
   a) tossed salad
   b) soil
   c) lemonade
   d) vegetable soup

3. Suppose you add a teaspoon of salt to a cool salt water solution and stir vigorously until the salt dissolves. The solution you started with was
   a) unsaturated.
   b) supersaturated.
   c) saturated.
   d) heterogeneous.

4. Which of the following affects the solubility of a solute in a solvent?
   a) the volume of the solute
   b) the volume of the solvent
   c) the mass of the solute and solvent
   d) the temperature of the solution

5. A cloud consists of tiny water droplets suspended in the air. Therefore, a cloud is a(n)
   a) aerosol.   b) gel.
   c) emulsion.   d) solution.

6. Suppose your doctor prescribes eardrops for an ear infection. The label on the eardrops says "Shake well before using." This medicine is probably a
   a) solution.   b) suspension.
   c) colloid.   d) gel.

SHORT ANSWER

7. What is the difference between a saturated solution and an unsaturated solution that share these characteristics: the same solvent and solute; the same volume of solvent; the same temperature?

8. Compare and contrast gels, aerosols, and emulsions. Give an example of each.

Interpreting Graphics

9. Study the drawing below. Refer to it to explain how sugar dissolves in warm water.

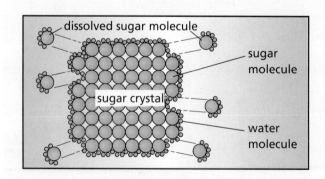

Reviewing Themes

10. *Environmental Interactions*
Compare a saturated sugar solution at a specific temperature with a supersaturated sugar solution at the same temperature in terms of dynamic equilibrium and stability.

11. *Systems and Structures*
Describe a day-to-day situation that shows how dynamic equilibrium leads to stability.

Thinking Critically

12. You accidentally shake up a can of soda that you were planning to drink. What could you do so that the soda will not spray all over when you open the can?

13. How could you determine experimentally whether a solution of a solid in a liquid is saturated, unsaturated, or supersaturated?

14. The salt we use to flavor foods is usually mined from the ground. However, when mined, the salt is often combined with soil and minerals. How could this crude material be treated to yield pure salt? Be specific.

15. Suppose you have a fish aquarium sitting on a window sill in your bedroom. You know that the weather is going to be hot and sunny for the next two weeks. What should you do with the aquarium? Explain why.

16. One method of taking medicine is known as transdermal infusion. The medicine, enclosed in a small, soft package, is absorbed slowly through the skin. Medicines to prevent motion sickness or to treat some coronary conditions are sometimes administered this way. Explain how you think the transdermal infusion systems works. Could all medication be given this way? Explain your answer.

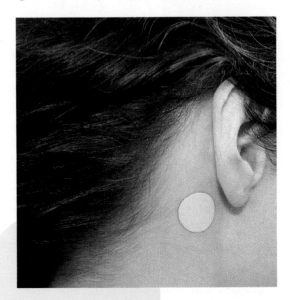

Discovery Through Reading

Soviero, Marcelle. "A Cure For Soggy Sandwiches." *Popular Science* 240 (March 1992): 23. Chitin—a natural substance found in nature—is resistant to solvents. Read this article to find out how this substance is used to prevent foods from becoming soggy.

CHAPTER 20

Bonding and Chemical Reactions

Often science's most intriguing ideas are interestingly wrong. In the early 1800s, scientists believed living things behaved according to the same laws as nonliving substances. The German author Johann W. von Goethe (GUHR tuh) based an entire novel on this fascinating —but scientifically inaccurate— principle. Goethe called the principle "elective affinity."

Listen as the novel's two couples discuss the laws of human interaction:

"Wait a moment and see if I have understood what you are driving at," Charlotte said. "Just as everything relates to itself, so it must have some relation to other things."

"And that relation will be different according to a difference in the elements involved," Eduard eagerly went on. "Some will meet quickly like friends or old acquaintances and combine without any change in either, just as wine mixes with water. But others will remain detached like strangers and refuse to combine in any way—even if they are mechanically mixed with, or rubbed against, each other. Oil and water may be shaken together, but they will immediately separate again."

[The Captain replies:] "Imagine an A so closely connected with a B that the two cannot be separated by any means, not even by force; and imagine a C in the same relation to a D. Now bring the two pairs into contact.

A will fling itself on D, and C on B, without our being able to say which left the other first, or which first combined with the other."

from *Elective Affinities*
by Johann W. von Goethe

Matter combines in certain prescribed ways. It may form compounds or mixtures. Each of these combinations also react in predictable ways.

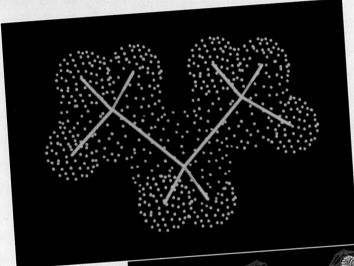

Chemical Bonds

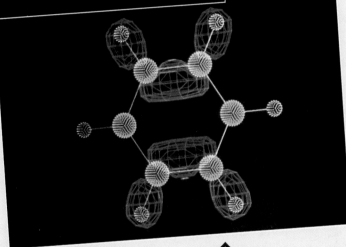

Can you imagine where the plot of Goethe's eighteenth-century soap opera is headed? Would you agree with this version of the scientific law that opposites attract?

For Your Journal

🖉 *What is the difference between an atom and an ion?*

How Atoms Combine

Objectives

List *the types of chemical bonds.*

Differentiate *between ionic and covalent bonds.*

Demonstrate *your knowledge of chemical bonding by identifying the type of bond formed, given the elements involved.*

"The San Antonio Rattlers are ready at the goal line," says the announcer. "Will they pass or let Number 18 run it in?"

The chances to score in a football game depend on the arrangement of the players and how close they are to the goal. Atoms often behave as football players do. You might say atoms pass or hand off electrons to other atoms to form compounds.

In Chapter 17 you learned about the reactivity of elements. There is an important connection between how elements behave and their electron arrangement.

Dance of the Electrons

Juan is using glue to put his kite together. The glue bonds the plastic, wood, and paper pieces together through some "sticky" interaction. The atoms in a compound are held together by similar "sticky" interactions known as chemical bonds. A **chemical bond** is the force that holds the atoms in a compound together. Chemical bonds contain a great deal of energy. When bonds are broken, the energy in them is released. This energy is most often seen in the form of heat, but it can also be released as light or sound. Find out for yourself what happens when bonds are broken by doing the next activity.

DISCOVER BY Doing

Get a 250-mL beaker and put 100 mL of water into it. Add 10 g to 15 g of CaCl₂ (calcium chloride) to the water. Touch the outside of the beaker. What do you observe? What do you think causes the change? ✐

Chemical bonding is a very complex process. So complex, in fact, that it would take years to understand all its different aspects. Since you don't have years to read this chapter, we are going to use a simplified explanation.

Suppose you have a baseball card collection. You exchange cards with friends to complete your collection. What if you have two identical cards? You might trade one of your cards with a friend for a card you don't have.

Atoms do something similar. They exchange electrons to achieve a stable noble gas configuration. That is, atoms trade electrons until they resemble the noble gas closest to them in the periodic table. The electrons involved are the electrons in the outer shell, or the *valence electrons*.

▶ **ASK YOURSELF**

How do atoms form compounds?

Give, Take, or Share

When atoms lose or gain electrons, they are no longer electrically neutral; they become ions. As you know, ions carry an electric charge. Metals, you might remember, lose electrons to form positive ions. Nonmetals may gain electrons to form negatively charged ions. Keep in mind also that opposite charges attract. The bond resulting from the attraction between a positive ion and a negative ion is called an **ionic bond**. This bond is similar to two magnets stuck together. A compound formed by ionic bonds is called an *ionic compound*.

Suppose you wanted to make a compound with potassium and chlorine. Ask yourself these questions: What must potassium do to achieve a noble gas configuration? What must chlorine do to achieve a noble gas configuration? If you look at the periodic table, you'll notice that potassium is closest to argon. Potassium needs to lose one electron to have the same electron arrangement as neon. Chlorine, on the other hand, needs to gain an electron to have the same electron configuration as argon.

Madame, I would be pleased if you would accept this extra electron.

When potassium loses an electron, it becomes an ion with a +1 charge. This is because it now has one more proton, which has a positive charge, than it has electrons, which have a negative charge. Chlorine was neutral but then gained an extra electron. So now chlorine is an ion with a −1 charge. Both the potassium ion and the chloride ion have an electron arrangement like a noble gas. Once the ions are formed, their opposite charges attract so the potassium and chlorine are now held together like two tiny magnets. This attraction forms the ionic bond.

Of course, in order to make some compounds, more than one electron must be transferred. In the formation of magnesium bromide, for example, the charge for magnesium is +2. The charge of each bromide ion is −1. What must magnesium do to gain a stable configuration? What does bromine do to gain a noble gas configuration? To form the compound, a magnesium atom loses two electrons.

When compounds are formed in this manner, the total number of electrons released must equal the electrons taken. The compound is, therefore, electrically neutral, even though the individual atoms are not. The +2 charge of the magnesium ion is balanced by the −1 charges of the two bromides.

▶ **ASK YOURSELF**

How are ionic compounds formed?

Atoms Apply for Loans?

Fluorine gas exists as F_2, or two atoms of fluorine bonded together. Which atom gives or takes an electron? How many electrons must each F atom give or take to achieve a noble gas configuration? None, of course. Puzzled? Both fluorines need one electron to become like the noble gas neon. If each fluorine gives one electron to the other, nothing is changed. How can both atoms gain an electron without losing any? They can share electrons with each other.

Okay, let me get this straight. I'll share one with you, and you'll share one with me. *Right?*

Consider this example. You have a chocolate bar, but your mother says you can't eat it because your brother doesn't have one. You can either save the chocolate, you can buy your brother his own chocolate bar, or you can offer to let your brother have some of your chocolate. The best solution is to share. Most atoms in a similar situation share electrons.

Each fluorine atom needs one electron to complete a stable octet. Transferring one electron from another fluorine atom would create one stable fluoride ion and one fluorine that needed two electrons. This wouldn't help much. Each fluorine atom can, however, share one electron with another fluorine

Figure 20–1. Acrylic glues form covalent bonds between the substances glued together. The man shown here has had his shoes glued to a platform. The glue forms bonds so strong that they are able to support his weight.

Figure 20–2. These computer models show what covalent bonds look like.

atom. Then, each atom has a stable octet. This sharing of electrons is called a **covalent bond.** Compounds made in this way are called *covalent compounds*. The smallest particle of a covalent compound is called a *molecule*.

▼ ASK YOURSELF
How do atoms form covalent bonds?

SECTION 1 *REVIEW AND APPLICATION*

Reading Critically
1. What is a valence electron?
2. What are the differences between ionic and covalent bonds?

Thinking Critically
3. Hydrogen peroxide (H_2O_2) is a covalent compound made of hydrogen and oxygen.

Water (H_2O) is also a covalent compound made of hydrogen and oxygen. Explain how they are different.
4. Use the periodic table to answer this question. How will the following elements combine to form a compound? Indicate whether the compound formed is ionic or covalent.
   a. hydrogen and bromine
   b. magnesium and oxygen
   c. calcium and chlorine

Chemical Formulas

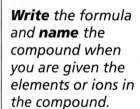

Objectives

Write the formula and **name** the compound when you are given the elements or ions in the compound.

Summarize the crisscross method of formula writing.

Analyze the importance of covalent compounds to the human body.

It was a dark and stormy night. The pair knew their only hope of survival was to find the lost message and get it to central command. They prepared for the treacherous journey to the drop point. Silently they glided through the wet streets, looking in every direction for their enemies. Finally . . . success! They reached the drop and took the message to their contact. The man opened the small container and gazed at the message. "It's in code!" he cried . . . .

Chemical Codes

Have you ever read about spies using secret codes to send messages? Maybe you have even made up your own code to use with a friend. It might seem that formulas are a chemist's secret code, yet anyone can learn to decipher them. A *chemical formula* contains the symbol for each atom in the compound. The formula for an ionic compound gives the simplest ratio of the different ions in the compound. For example, the formula for calcium chloride is $CaCl_2$. The 2 after the Cl, written lower than the symbol for the element, is called a *subscript*. If no subscript is written, as for Ca in this formula, a 1 is understood. By looking at the formula, you know that the crystal contains 2 chlorine

Figure 20–3. The crystal structure of the ionic mineral halite—or table salt—is distinctive and beautiful.

ions for every calcium ion. Compounds made from a metal and a nonmetal are nearly always ionic compounds, while compounds made of combinations of non-metals are covalent. By looking at the formula, therefore, you can tell an ionic compound from a covalent one. Use the periodic table to help you determine which of the following are ionic compounds: NaCl, KF, CO_2, MgO.

All for One Some ionic compounds contain ions made from several nonmetals that are bonded together covalently. These ions form a unit, and they have a single charge. They are called *polyatomic ions*. Table 20–1 lists some common polyatomic ions.

Table 20–1

Common Polyatomic Ions

Ion	Formula	Charge
Ammonium	NH_4	+1
Hydroxide	OH	-1
Sulfate	SO_4	-2
Phosphate	PO_4	-3
Nitrate	NO_3	-1

Charge It! In any compound, the net charge of the compound must equal zero. That is, the number of positive charges must equal the number of negative charges. If you know the charge of the ions in an ionic compound, it is easy to write the correct formula by using the crisscross method. Look at the examples below:

Example 1

$$Mg^{(+2)} + F^{(-1)}$$
$$Mg \quad F_2 \qquad\qquad MgF_2$$

Example 2

$$K^{(+1)} + S^{(-2)}$$
$$K_2 \quad S \qquad\qquad K_2S$$

Example 3

$$Al^{(+3)} + O^{(-2)}$$
$$Al_2 \quad O_3 \qquad\qquad Al_2O_3$$

Example 4

$$Mg^{(+2)} + OH^{(-1)}$$
$$Mg \quad (OH)_2 \qquad\qquad Mg(OH)_2$$

DISCOVER BY *Problem Solving* _____

Look at the examples on page 548, and figure out what the crisscross method is. Write your explanation in your journal. Make sure your explanation works by writing the formulas for compounds made of potassium and bromine, aluminum and sulfur, and ammonium and hydroxide ions. ✐

After you have completed the activity, take a look at your formulas. When you are working with ionic compounds, you should check to be sure that the ratio of elements is the simplest possible. For example, Ca_2O_2 is not the simplest whole number ratio. CaO would be the correct formula.

In contrast to ionic formulas, formulas for covalent compounds tell the actual number of atoms of each kind that make up the molecule. They are never reduced to the simplest whole number ratio. For example, in the compound hydrogen peroxide, H_2O_2, every molecule has two hydrogen atoms and two oxygen atoms.

Consider some simple sugars. Glucose, the sugar made by plants during photosynthesis, has a formula of $C_6H_{12}O_6$. Sucrose, or table sugar, has a formula of $C_{12}H_{22}O_{11} \cdot H_2O$. (Sucrose loses a water molecule as it forms.) If you reduce glucose and sucrose to their simplest whole number ratio, you would get the same compound: CH_2O. This compound is the basic building block for all sugars—how many of them there are in a compound determines the type of sugar.

Floridian

Bostonian

What's Your Name?

The general way to name a compound is to use the name of the first element followed by the root of the second's element name with the suffix *-ide* added. This is similar to calling someone from Florida a Floridian and someone from Boston a Bostonian. Consider two of the formulas you wrote in this section: potassium + bromine (potassium bromide) and aluminum + sulfur (aluminum sulfide).

Sometimes more than a suffix is added to the name. Prefixes can be used to tell you the number of atoms in an element. For example, carbon monoxide is CO (*mono-* means "one"). Carbon monoxide has one oxygen atom. The formula for carbon dioxide is CO_2 (*di-* means "two"), so carbon dioxide has two oxygen atoms.

To get some practice forming and naming compounds, try the next activity.

What would you call him?

Using index cards, make flash cards for each of the elements in the first two periods of the periodic table. Make four cards for each element. With a team of three classmates, shuffle the cards and place them upside down. Have a player draw two cards and tell whether a compound can be made from the elements chosen. If a compound can be named, the player must give the correct formula. ✎

I have two oxygens, and you only have one.

▼ **ASK YOURSELF**

Explain how you balance a chemical formula.

Ionic or Covalent . . . Who Cares?

Why would knowing whether a compound is ionic or covalent be of any interest? For one thing, their properties are different. Different properties make these compounds useful for different purposes. For example, ionic compounds tend to have much higher melting and boiling points than covalent compounds. A second difference is that pure ionic compounds are usually crystalline solids at room temperature, while covalent compounds may be solids, liquids, or gases. If you needed to fill a bicycle tire, would you use an ionic or a covalent compound?

Both ionic and covalent compounds have many uses in daily life. You use thousands of different compounds that you are not even aware of! Let's take a look at a few.

Ionic Compounds in Action Many ionic compounds dissolve easily in water. Your body must keep a precise amount of ions in solution in order for it to function properly. These

Figure 20–4. Ionic compounds have many practical uses.

ions are known as electrolytes. The most important are Na$^+$, K$^+$, Cl$^-$, and Ca^{+2}.

Electrolytes? Doesn't that have something to do with electricity? There is no electricity in your body, you might say. Pinch your arm. The reason you can feel the pinch is because nerve impulses deliver small electric messages to your brain. Without the proper concentration of electrolytes, your nerve impulses cannot travel to your brain. When you sweat due to heavy exercise, you lose some of these ions. Often, athletes use special drinks to keep their electrolytes in balance.

"Oh, what a pretty ring! What kind of stone is that?" Ionic, that's what kind. Ionic compounds are found in some of the minerals in the earth. Many ionic compounds are also very beautiful. You can find out more about crystals by doing the next activity.

ACTIVITY
How do different crystals polarize light?

MATERIALS
wax pencil, microscope slides (8), dropper bottles with solutions (4), light microscope, polarizing film (3 cm^2), tape, scissors

PROCEDURE
1. Using a wax pencil, label each microscope slide with the name of the solution to be used. Add two drops of one solution to the center of a slide. Use another slide to smear the solution across the first slide. Set it aside to dry. Prepare a slide for each solution provided. Allow each slide to dry.
2. Observe each chemical, using the microscope. On your paper, sketch the crystal shapes.
3. After you have looked at all the crystals, place a piece of polarizing film on the microscope stage and fasten it with tape. Cut a circular piece of film to fit over the eyepiece. Place a slide over the film on the stage, and look at the slide. Rotate the eyepiece film to get the best color. Repeat for each slide, and record your results.

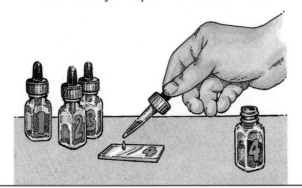

APPLICATION
1. Summarize in a short paragraph what you have observed.
2. The compounds were given to you in aqueous (water) solutions. What type of compounds are these? How do they exist in solution? How do they exist as crystals?

Covalent Compounds to the Rescue

One, two, three, four. One two, three, four. Exercise is good for you. "Breathe," shouts your instructor. "I can't," you cry. "I have a shortage of covalent compounds!"

Covalent compounds make up most of the atmosphere and, therefore, the air we breathe. They also are major components in living cells. Your response to your instructor could mean you couldn't get your breath, or it could mean you didn't have the muscle power to get the job done!

Covalent compounds made of carbon are the basis for fossil fuels. These fuels provide us with a great percentage of the energy we use daily. Without them, at least until new technology is implemented, you might not have lights in your classroom!

So, who cares if a compound is ionic or covalent? You do.

Figure 20–5. Covalent compounds have many practical uses.

 ASK YOURSELF

Name three differences between ionic and covalent compounds.

SECTION 2   REVIEW AND APPLICATION

Reading Critically

1. What is the difference between the formula of an ionic compound and that of a covalent compound?
2. Write the formulas and names for the compounds made from the ions of the following elements:
   a. calcium and fluorine
   b. sodium and sulfur
   c. aluminum and chlorine

Thinking Critically

3. Sodium chloride, NaCl, has a melting point of 801°C, while sucrose, $C_{12}H_{22}O_{11}$, has a melting point of 185°C. Explain why this is so.
4. Covalent compounds are important to humans in many ways. Make a list of how they affect the way your body works, and then write a paragraph explaining how an absence of covalent compounds would affect you.

Chemical Reactions

Write and *balance* chemical equations.

Identify different types of reactions.

Jaime and Minh are walking home after seeing a chemistry show. They saw solutions that changed colors several times and chemicals that ignited when mixed. They even learned how to make a stink bomb.

"You know what I liked the most, Jaime?" says Minh. "The thermite reaction." "Yeah," replies Jaime, "that was a great ending. I liked it too." The chemist-magician had taken the audience outside. She had shown them a setup that held a paper cone filled with aluminum and iron(III) oxide over a sand box. Stuck in the cone was a strip of magnesium that she compared to a candle wick. When she lit the strip, it would burn like a wick and get the reaction started. The chemist lit the magnesium ribbon, which burned very brightly. All of a sudden the materials in the cone caught on fire. The mixture flared up, giving off huge amounts of heat, light, and smoke. Then a glowing molten ball of iron dropped from the cone into the sand below. WOW!

"Can you believe that the iron was between 2000°C and 2400°C?" says Minh. "I wish we could do reactions like that in school." "Are you crazy! That's too dangerous. But I sure would like to make a few stink bombs," adds Jaime. "Really," says Minh, "the best was understanding what was going on. The chemist told us the ingredients that were involved in each of the tricks. She even wrote the chemical reactions on the portable board. I wrote down the equation for the thermite reaction. I'll tell everyone about it in class tomorrow."

Chemical Goings-On

The thermite reaction that Minh and Jaime were talking about is really spectacular. However, it is usually not practical to describe a chemical reaction in a paragraph. Scientists generally describe chemical reactions in a shorter way called a *chemical equation*. A **chemical equation** lists the compounds or elements involved in a reaction and the new compounds or elements that form as a result.

Figure 20–6. The thermite reaction shown here gives off terrific amounts of heat and light.

The chemical equation for the thermite reaction is:

$$2Al + Fe_2O_3 \rightarrow 2Fe + Al_2O_3$$

A chemical equation is like a very condensed recipe. On the left side of the equation is a list of all the materials needed for the recipe. These ingredients are called *reactants*. On the right side of the equation is a list of what is produced by the recipe. The new substances formed are called *products*.

Figure 20–7 Chemical equations are like recipes.

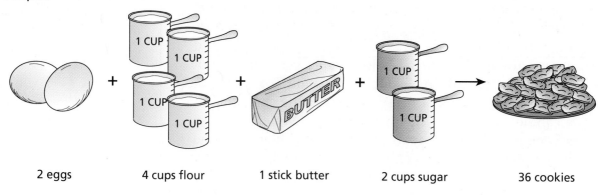

| 2 eggs | 4 cups flour | 1 stick butter | 2 cups sugar | 36 cookies |

If you were making a stew, all the ingredients you added to the stew would still be there at the end. They would only be changed in flavor and appearance. The law of conservation of matter states that in a chemical reaction all matter involved in a reaction is conserved. Atoms are not lost while changing from reactants to products; they are only rearranged.

The arrow in a chemical equation is read "yields," which means turns into. The total mass of reactants used must equal the total mass of products formed. In order for this to be true, numbers are sometimes placed in front of formulas. These numbers are called *coefficients*. In the thermite reaction, the coefficients indicate that two atoms of Al react with one unit of Fe_2O_3 to form two atoms of Fe and one unit of Al_2O_3.

Do not confuse coefficients with subscripts. You can change coefficients in a chemical equation until the number of atoms of each element is balanced. **You cannot balance an equation by changing the subscripts.** If you change the subscripts, you change the composition of the substance to something totally different. Remember there is a great difference between H_2O (water) and H_2O_2 (hydrogen peroxide).

Table 20–2 is a checklist to help you balance a chemical reaction. The formation of hydrochloric acid (HCl) is used as an example.

Table 20–2 Writing Chemical Equations

Procedure	Result
Write the correct formulas for all reactants and products involved in the reaction. Separate each of the substances on the reactant side from each other by a *plus* sign. Separate reactants from products by an arrow. Remember that some gases are diatomic and must be written as such in the reaction.	$H_2 + Cl_2 \rightarrow HCl$
Balance the equation one element at a time. Use only coefficients to balance. Do not change the subscripts you have already written. Look at hydrogen first—there are two hydrogens on the left of the equation but only one on the right. Place the coefficient 2 before HCl.	$H_2 + Cl_2 \rightarrow 2HCl$
When you put the 2 in front of HCl, you have two hydrogens and two chlorines on the right side of the equation. Look back at the left side. Since chlorine is diatomic, there are two chlorines there also. The equation is balanced. Do not be afraid to go back and forth more than once.	
Do a final check. Make sure the chemical formulas are correct and the numbers of all the atoms are balanced.	

ASK YOURSELF

Why are chemical equations useful?

Chemical Recipes

There are so many different types of chemical reactions it would be impossible to learn them all. The study of new reactions and the processes by which they occur is an ongoing area of research for many chemists. Most reactions, however, may be grouped into three basic types: synthesis, decomposition, and replacement reactions.

Synthesis Reactions When you make bread, you mix flour, water, yeast, and other ingredients and come out with a loaf of bread. This would be called a synthesis reaction by chemists. In a **synthesis reaction** two or more substances combine to form one new substance. The formation of water is an example.

$$2H_2 + O_2 \rightarrow 2H_2O$$

Figure 20–8. When iodine is added to zinc, zinc iodide is formed. This is an example of a synthesis reaction.

Decomposition Reactions

Figure 20–9. Decomposition of mercury (II) oxide; fire at fertilizer plant

Decomposition Reactions Sometimes people need to separate the atoms in a compound. Chemically, a **decomposition reaction** involves breaking a compound into two or more substances. For example, mercury(II) oxide can be decomposed into the elements mercury and oxygen gas by heating.

$$2HgO \rightarrow 2Hg + O_2$$

In Figure 20–9, you can see beads of mercury on the side of the test tube after this decomposition has taken place. Another example of a decomposition reaction is shown in the same figure. The burning factory was ignited by the spontaneous decomposition of fertilizer on a ship in the harbor. The factory became an uncontrollable decomposition reaction.

Some metals are purified from ore by decomposition reactions. Aluminum oxide is the ore from which we get aluminum foil. The reaction is shown below.

$$2Al_2O_3 \rightarrow 4Al + 3O_2$$

Figure 20–10. The production of aluminum from ore involves a decomposition reaction caused by electricity instead of heat.

Replacement Reactions   How would you like to make silver from a solution and copper wire? If a piece of copper is placed in a colorless silver nitrate solution, pieces of silver will be deposited on the copper and the solution will gradually turn blue. The blue solution is copper(II) nitrate. The equation for this reaction is:

$$Cu + 2AgNO_3 \rightarrow Cu(NO_3)_2 + 2Ag$$

Sometimes chemical reactions involve changing partners, like in a dance. These types of reactions are known as replacement reactions. During a **replacement reaction,** one atom or group of atoms in a compound is replaced with another atom or group of atoms. In Figure 20–11, copper from the wire replaced the silver in the silver nitrate solution. As the silver comes out of solution, it looks as if it "grows" on the copper wire. As more and more of the copper goes into the solution, the clear solution will become blue. The replacement of copper for silver in the silver nitrate solution is an example of a *single replacement reaction* because only one type of atom or group of atoms is being replaced. If two different types atoms or group of atoms exchange places in a reaction, the reaction is called a *double replacement reaction.*

Below is an example of a double replacement reaction. In this reaction, potassium (K) and lead (Pb) are changing partners in the compounds.

$$2KCl + Pb(NO_3)_2 \rightarrow PbCl_2 + 2KNO_3$$

Sometimes in a replacement reaction one of the compounds formed is not soluble in the reaction mixture. The compound forms a solid that comes out of solution. This solid is called a *precipitate*. This is just one way we know a reaction has taken place.

It is sometimes difficult to tell whether a reaction or chemical change has taken place when two substances are mixed. The formation of a precipitate is a sure sign that a chemical reaction has occurred. Other signs are a change in color, a temperature change, bubbles or a gas being released, or a buildup of the new substance (as in the copper and silver nitrate reaction).

Figure 20–11. When copper wire is placed in a solution of silver nitrate, the silver builds up on the wire. This is an example of a single replacement reaction.

Figure 20–12. In this double replacement reaction, sodium chromate (yellow solution) combines with silver nitrate (clear solution) to form sodium nitrate (clear solution) and silver chromate (pink precipitate).

Figure 20–13.
Electroplating is an example of a replacement reaction.

Replacement reactions are used in electroplating. You might have gold-plated or silver-plated jewelry. These items are not made out of pure gold or silver. Instead, a thin coating of these metals has been stuck on to an inexpensive metal. Beneath your silver-plated earrings might be steel or copper.

Sometimes more than one type of reaction occurs during a single process. Create your own replacement and decomposition reactions by trying the following activity.

DISCOVER BY Doing

 Some fire extinguishers contain CO_2, which puts out fires by smothering them. One of the older types of extinguishers used a compound called *foamite.* You can make foamite by adding two spoonfuls each of baking soda, alum, and powdered gelatin to a large beaker or glass. Stir well. Add about six spoonfuls of vinegar and stir again. What happens? Explain what has occurred using a chemical equation. ✏

ASK YOURSELF

What is the difference between a synthesis and a decomposition reaction?

SECTION 3   REVIEW AND APPLICATION

Reading Critically
1. What information does a chemical equation give you?
2. What are three major types of reactions?

Thinking Critically
3. Look at the equation:

$AgNO_3 + H_2SO_4 \rightarrow Ag_2 SO_4 + HNO_3$

Which of the substances above are ionic? Which are reactants? Which are products? Balance the equation.

4. Hydroxide ions, OH^{-1}, may combine with metallic ions to form compounds. When heated, these metallic hydroxides decompose into metallic oxides and water. Write and balance the equation for the decomposition of calcium hydroxide.
5. Write the balanced equations for the following reactions:
   a. carbon + water → hydrogen gas + carbon monoxide
   b. sulfur + oxygen → sulfur dioxide
   c. copper(II) carbonate ($CuCO_3$) → copper(II) oxide + carbon dioxide

SKILL Drawing Conclusions

Identifying Gases

Several tests can be used in order to identify common gases that may be given off in a reaction. Here are some examples.

1. Oxygen will cause a barely glowing wood splint to burst into flames.
2. Hydrogen will make a popping sound when brought into contact with a flaming splint.
3. Carbon dioxide will put out a flaming splint.

Solubility of Some Solids in Water

KEY: S = soluble, NS = not soluble

zinc chloride—S
potassium chloride—S
sodium nitrate—S
lead(II) iodide—NS
silver chloride—NS
silver nitrate—S
sodium chloride—S
copper(I) nitrate—S

▶ **APPLICATION**

Using the above information and information provided in this chapter, write the complete chemical equations for each experiment listed here. Balance the equations, and identify the type of reaction for each experiment.

Experiment 1

Richard dropped a piece of zinc (Zn) into 10 mL of hydrochloric acid (HCl) in a test tube. He observed a large number of bubbles forming around the piece of zinc and rising to the surface. When he brought a flaming splint close to the mouth of the test tube, he heard a loud pop. After a few minutes, all the zinc had disappeared.

Experiment 2

Liz heated a massed amount of potassium chlorate ($KClO_3$) in a special porcelain dish over very high heat. At the beginning, the solid was composed of white, translucent crystals. After heating for 30 minutes, the solid had changed to a white powder. She held a glowing splint over the dish, and it burst into flame. Liz massed the compound after heating and found it had less mass than before.

Experiment 3

Luis massed a few pieces of copper foil. He placed them in a test tube and began heating the test tube over a burner. After a few seconds, he noticed that the shiny copper began to turn black. Luis let the black substance cool and then massed it. Its mass was greater than the initial mass of copper.

Experiment 4

Leta had 20 mL of sodium chloride (NaCl) solution and 20 mL of silver nitrate ($AgNO_3$) solution in separate beakers. She slowly poured the silver nitrate solution into the sodium chloride. A white solid began to form as the silver nitrate was being added.

✳ Using What You Have Learned

Analyzing information is just as important in everyday life as it is in the laboratory. Write a short paragraph explaining a situation in which you analyzed information to form a conclusion.

INVESTIGATION

Observing Chemical Reactions

▶ **MATERIALS**

- safety goggles
- laboratory apron ● test tubes (8) ● test-tube rack
- labels or wax pencil ● dropper bottles with the following solutions: sodium fluoride, sodium chloride, potassium bromide, potassium iodide
- graduate ● droppers ● calcium nitrate solution ● silver nitrate solution

CAUTION: Put on safety goggles and a laboratory apron, and leave them on throughout this investigation. Silver nitrate solution will stain your skin and your clothes.

▼ **PROCEDURE**

TABLE 1: OBSERVATIONS OF REACTIONS			
Reactants	**Products**	**Balanced equation**	**Observations**

1. Prepare a data table similar to the one shown. Leave plenty of room in each section. As you complete each reaction, fill in the reactant and observation sections.
2. Label four test tubes with the names of the solutions in the dropper bottles.
3. Place 5 mL of each of the solutions into the appropriate test tube.

4. Add 1 mL of calcium nitrate solution, $Ca(NO_3)_2$, to each of the four test tubes. One milliliter is approximately 20 drops. Record your observations.
5. Label four more test tubes as in step 2. Place 5 mL of each of the solutions into the test tubes.
6. Add 1 mL of silver nitrate solution ($AgNO_3$) to each

of the test tubes. Record your observations.
7. Before going farther, complete your data table and have your teacher check it.
8. Obtain an unknown solution from your teacher. Identify the halogen ions in your mixture. Record your procedure, observations, and conclusions.

▶ **ANALYSES AND CONCLUSIONS**

1. Which ion(s) produced a precipitate with calcium nitrate?
2. Which ion(s) produced a precipitate with silver nitrate? Does this test give conclusive evidence for all of the ions? Why or why not?
3. When hydrochloric acid, HCl, is added to limestone, $CaCO_3$, this reaction takes place.

$$CaCO_3 + 2HCl \rightarrow CaCl_2 + CO_2 + H_2O$$

You would observe bubbles fizzing at the spot where the acid touched the stone. Why? Could you use this as a test for an ion? Which ion? Explain.

 Discover More

Use the ion tests you have practiced in this investigation to test various liquids. Make a chart of your findings.

*H*IGHLIGHTS

The Big Idea

Everything in nature moves toward a lower energy state and becomes more stable. A ball rolls down hill. You get tired toward the end of the day and sleep. A hot solution gradually gives off heat and cools. Radioactive elements decay into more stable ones.

What about atoms? Atoms have a unique way of achieving chemical stability: they can exchange or share electrons in order to achieve a noble gas configuration. These movements of electrons are what we call chemical reactions. The reason iron rusts is because the product formed, iron oxide, is less reactive than the initial elements, iron and oxygen. It is amazing that so many different reactions occur spontaneously all around us. The only driving force is the need to achieve stability.

For Your Journal

What is the difference between an atom and an ion? How are atoms held together when matter interacts? Explain how chemical reactions can be written. Look back and review your original answers to these questions. How would you revise your ideas to show what you have learned?

Connecting Ideas

This concept map shows how the big ideas of this chapter are related. Copy the map into your journal, and fill in the blanks to complete it.

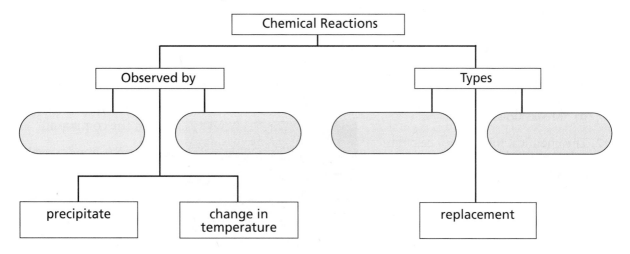

REVIEW

Understanding Vocabulary

1. For each set of terms, compare and contrast their meanings.
   a) chemical bond (542), ionic bond (544), covalent bond (546)
   b) synthesis reaction (555), decomposition reaction (556), replacement reaction (557)
   c) chemical formula (547), chemical equation (553)

Understanding Concepts

MULTIPLE CHOICE

2. $A + B \rightarrow AB$ is an example of a
   a) synthesis reaction.
   b) decomposition reaction.
   c) single replacement reaction.
   d) double replacement reaction.

3. $AB \rightarrow A + B$ is an example of a
   a) synthesis reaction.
   b) decomposition reaction.
   c) single replacement reaction.
   d) double replacement reaction.

4. Which substance contains an ionic bond?
   a) CO   b) CO_2
   c) KCl   d) O_2

5. The smallest particle of a covalent compound is called a(n)
   a) atom.   b) ion.
   c) molecule.   d) reactant.

6. In any chemical equation, the arrow means
   a) "equals."
   b) "is greater than."
   c) "yields."
   d) "breaks down into."

7. The chemical formula for calcium chloride is
   a) CaCl.   b) Ca_2Cl.
   c) $CaCl_2$.   d) Ca_2Cl_2.

8. What is the coefficient needed in front of sodium bicarbonate, $NaHCO_3$, to balance the equation below?

   $$NaHCO_3 \rightarrow Na_2CO_3 + CO_2 + H_2O$$

   a) 1   b) 2
   c) 3   d) 4

SHORT ANSWER

9. What substances are the products in the chemical equation

   $$Cu + H_2O \rightarrow CuO + H_2?$$

10. List the signs that show that a chemical reaction has taken place.

11. Why is the phrase "a molecule of sodium chloride" inaccurate?

Interpreting Graphics

Study the computer-generated model of the compound C_2H_5OH shown below.

12. How many bonds are formed by each carbon atom?

13. How many bonds occur in the compound?

14. Are the bonds covalent or ionic?

Reviewing Themes

15. *Environmental Interactions*
Describe the role of valence electrons in achieving stability through chemical bonding.

16. *Systems and Structures*
Oxygen gas and hydrogen gas react to form water. Use the law of conservation of matter to write a chemical equation for the interaction.

Thinking Critically

17. How can some ionic compounds contain covalent bonds?

18. If two carbon electrodes attached to a battery are placed in a beaker containing a solution of copper(II) chloride, two things happen. A layer of copper coats the electrode at the negative pole of the battery, and bubbles of chlorine gas form at the other electrode. Describe the chemical changes that occur. What must happen for Cu^{+2} to change to copper? Why does copper form only on the electrode at the negative pole of the battery? Predict what would happen if the poles of the battery were reversed.

19. Why is the formula of ammonium sulfate, a fertilizer, written as $(NH_4)_2SO_4$ rather than $N_2H_8SO_4$?

20. Some living organisms produce light through a chemical process called bioluminescence. Use your knowledge of chemical reactions to explain how a chemical reaction could occur that gives off light rather than producing or absorbing heat.

Discovery Through Reading

Uehling, Mark D. "Birth of a Molecule." *Popular Science* 240 (February 1992): 75–77, 88, 90. This article discusses use of computers to design new molecules with specific properties.

CHAPTER 21
ACIDS, BASES, AND pH

I t was the difficult Case of the Dying Lakes. *The patients were dying, but the doctors could not agree on what to do. Scientists knew that some lakes in the northeast United States were dying of acid poisoning, their fish completely gone. However, it took scientists and government officials a long time to diagnose and treat their sick patients.*

The first problem was finding the cause of the lakes' acidification. Many scientists and environmentalists blamed acid rain, the acid-laden rain and snow that fell on northeastern mountain ranges. Standing by a lake in upstate New York, it was easy to see the effects of acid rain. But scientists and officials found it harder to agree on the causes. Was it acidic pollution from power plants and automobiles in the Midwest United States, as many thought? How could this air pollution drift for thousands of miles and affect a handful of lakes in the Northeast? Could scientists pinpoint the sources of acidic pollution and advise governments about the most effective regulations?

While some scientists debated these difficult questions, others tried to give the patients some immediate relief. By the 1980s, up to 10 percent of lakes in some regions had become seriously acidified. Reducing air pollution would take 10 or 20 years. The *Case of the Dying Lakes* called for a short-term solution, something to restore the lakes temporarily until pollution controls took effect.

Scientists knew that the antidote for acid indigestion in humans was an antacid tablet, and to neutralize the acid on a car battery, you poured on a solution of baking soda. So scientists prescribed an antacid for the dying lakes. They poured the equivalent of baking soda into the lakes.

Scientists first attempted to neutralize acidic lakes by dumping carbonates from planes, helicopters, and snowmobiles. But they found that the carbonates didn't prevent fish kills. Acid rains and snowmelts brought acid "pulses" that killed eggs and young fish. So scientists tried a different prescription. They dumped carbonate pellets on the mountainsides around the lake and let the pellets neutralize the acid runoff before it reached the lake.

The acid lakes aren't out of the woods yet. Scientists still are not certain that neutralization and pollution controls will revive the dying lakes. The simple chemistry of acids and bases has created a very complicated environmental problem.

For Your Journal

🖉 What is an acid? What is a base?

🖉 How can you determine if a substance is an acid or a base?

🖉 How do acids and bases affect you every day?

Acids and Bases

Distinguish between acids and bases.

Name some common acids and bases, and list their uses.

Compare the characteristics of strong and weak acids and strong and weak bases.

Hank is watching the lifeguard at the local pool take a sample of the water. The lifeguard adds some drops to the sample that make it turn pale yellow. To another sample of water she adds some red drops, and the solution turns violet. Hank is intrigued by what the lifeguard is doing. "Why do you add those drops to the samples of pool water?" he asks.

"I'm checking to see whether the amount of chlorine in the pool is right and what the pH is," Jessie, the lifeguard, responds. "Oh," says Hank. "I know it's important to have chlorine in the pool, but pH? What's pH?"

Jessie explains that the pool cannot be too acidic or too basic. If it is too basic she must add a substance to reduce the pH. "Wait, whoa, stop," Hank says. "I'm confused. I thought taking care of a pool was easy. What's an acid? What do you mean by basic and pH?"

Jessie smiles. "In order for you to understand pool chemistry, you will need to learn something about acids, bases, and salts. Why don't you come early tomorrow before the pool gets too crowded, and I can teach you."

Figure 21–1. Citric acid got its name because it was first isolated from citrus fruits.

What's in an Acid?

Hank had heard of acids. In fact, he recently saw a horror movie that had an acid pit. It was full of a fuming, corrosive liquid that could "eat" through metal—and anything else. Surely Jessie could not be adding that kind of acid to the pool. Wouldn't that be too dangerous?

What Hank didn't know was that not all acids are as strong as the ones in the movie. In fact, Hank drinks and eats some acids every day. Acids are an important part of carbonated drinks, fruits, vegetables, meats, and even salad dressing!

Look at Table 21–1. Notice that each acid contains hydrogen. An **acid** is a compound that produces hydrogen ions (H^+) in solution. When an acid is dissolved in water, it separates into ions. Hydrochloric acid, for example, forms H^+ and Cl^- (chloride ions).

Table 21–1

Some Common Acids

Name	Chemical Formula
Acetic acid	$HC_2H_3O_2$
Citric acid	$HC_6H_7O_7$
Formic acid	$HCOOH$
Hydrochloric acid	HCl
Nitric acid	HNO_3
Phosphoric acid	H_3PO_4
Sulfuric acid	H_2SO_4

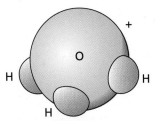

Figure 21–2.
Hydronium ions contain one oxygen atom and three hydrogen atoms. The number of hydronium ions in solution determines the strength of the acid.

What is H^+? If you remove one electron from hydrogen you are left with a proton. Solitary protons do not wander around in solutions. Instead the H^+ combines with a water molecule to form H_3O^+, or a hydronium ion. Any solution that contains H_3O^+ is an acidic solution. For simplicity, the symbol H^+ will be used in this chapter.

 ASK YOURSELF

How can you tell from a chemical formula if a substance might be an acid?

Pucker Up

Jessie had told Hank that the safest way to taste an acid was to taste an unripe apple and compare it to a ripe apple. Hank went to the store and looked for the most unripe apple he could find. On the way home he bit into it. Wow! Was it ever sour!

One property of acids is that they taste sour. Most fruits have some acid content. Oranges and lemons contain citric acid; apples contain malic acid. Unripe fruit tastes sour because it has a greater concentration of acid than ripe fruit does. As the fruit ripens, the acids are converted into sugar.

Another common acid is vinegar. Vinegar is a solution of acetic acid in water. Vinegar and the acids in fruit are weak acids. Hank decided to look for the names of some more acids in the encyclopedia. Table 21–2 lists some weak acids and some strong acids listed in Hank's references. **CAUTION: Strong acids should never be touched or tasted.**

Table 21–2	**Strengths of Some Acids**
Strong Acids	**Weak Acids**
Hydrochloric acid (HCl)	Acetic acid ($HC_2H_3O_2$)
Nitric acid (HNO_3)	Citric acid ($HC_6H_7O_7$)
Sulfuric acid (H_2SO_4)	Carbonic acid (H_2CO_3)
Hydrobromic acid (HBr)	

The strength of an acid is determined by the concentration of hydronium ions in solution. The more hydronium ions the acid produces in solution, the stronger the acid. Strong acids ionize completely in water, producing many hydronium ions. Weak acids do not ionize completely, so there are not as many hydronium ions in solution.

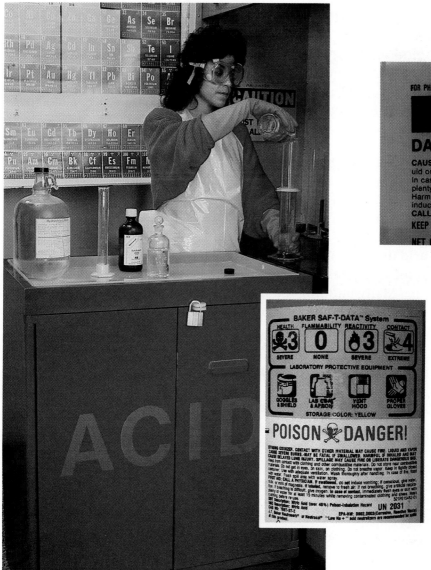

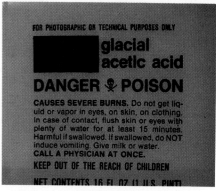

FOR PHOTOGRAPHIC OR TECHNICAL PURPOSES ONLY

glacial acetic acid

DANGER ☠ POISON

CAUSES SEVERE BURNS. Do not get liquid or vapor in eyes, on skin, on clothing. In case of contact, flush skin or eyes with plenty of water for at least 15 minutes. Harmful if swallowed. If swallowed, do NOT induce vomiting. Give milk or water. **CALL A PHYSICIAN AT ONCE.** KEEP OUT OF THE REACH OF CHILDREN

NET CONTENTS 16 FL OZ (1 U.S. PINT)

Figure 21–3. Strong acids must be handled with care. When using acids, you should always wear safety goggles and a laboratory apron and you should work in a well-ventilated area. The label on an acid bottle will provide any special instructions you might need.

DISCOVER BY *Doing*

Use some blue and red litmus paper. What happens when you dip them in vinegar or orange juice? Write the results in your journal. Can you think of any uses for what you have learned? ✐

Strong acids must be handled carefully because they can produce severe burns. Perhaps you have observed how metals react in strong acids. In addition to reacting with metals, acids also react with many other types of compounds. Try the next activity to find out about one kind of reaction.

Figure 21–4. Caves like this one are formed when acid reacts with limestone.

DISCOVER BY *Doing*

Place some baking soda in a glass, and add some vinegar. What happens? Baking soda contains sodium bicarbonate ($NaHCO_3$). What common gas do you think is given off by this reaction? 🖉

A similar reaction to the one in the activity also takes place in nature. In areas where limestone is common, this reaction can cause the formation of caves and sinkholes.

Limestone is mainly calcium carbonate. When carbon dioxide in the air dissolves in rainwater to form a weak acid, the acid dissolves minerals in the cave rock. The dissolved minerals form stalactites and stalagmites.

Many building materials contain carbonates. The same reaction that occurs in caves can occur over time on concrete and marble.

DISCOVER BY *Researching*

Locate an article on acid rain in the library. Find the answers to the following questions about acid rain: What causes it? What are its effects? What are some of the acids found in it? What can be done to prevent it? 🖉

Figure 21–5. Acid rain causes damage to many buildings and to exposed statues and sculptures.

▼ **ASK YOURSELF**

What determines the strength of an acid?

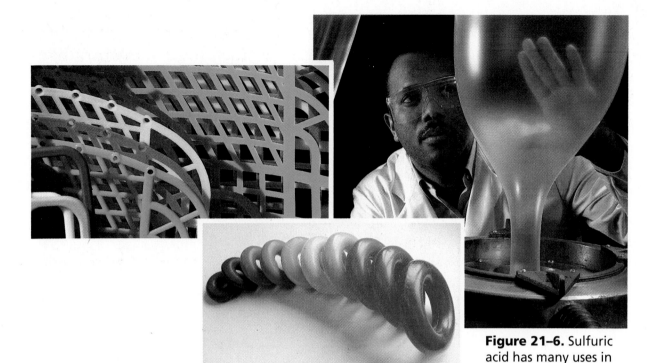

Figure 21–6. Sulfuric acid has many uses in industry. One important use is in the manufacture of plastics.

Uses of Acids

Hank was ready for another chat with Jessie. "I have learned a lot about acids," he says, "but I'm still not sure why it is important to add acid to the pool."

"One reason to add acid to the pool is to prevent deposits of carbonates," says Jessie. "I learned about how acids react with carbonates," says Hank. "Does that mean when you add acid the pool will foam?" "Usually not," Jessie assures him. "But we do use well water for our pool. Well water contains a lot of dissolved carbonates. If we don't do something about it, they will form a crust around the rim of the pool and clog the pipes. So I add just enough hydrochloric acid to react with the carbonates but not enough to harm the swimmers."

Hank went home and decided to look up more information on uses for acids. He found out that acids are very important to many industrial processes. Sulfuric acid, a strong acid, is used to make fertilizers and automobile batteries. It is also an important component in steel production and many other manufacturing processes. Sulfuric acid is used in tanning leather for shoes, in treating the paper

Figure 21–7. Hydrofluoric acid is used to etch fine crystal. Because this acid "eats through" glass, it must be stored in plastic.

Figure 21–8.
Hydrochloric acid is used to clean buildings. Notice the protective gear the worker is wearing. Why do you think this equipment is necessary?

Figure 21–9. One use for nitric acid is to make dyes such as those shown here.

in this book, and even in making the vanilla flavoring for ice cream.

Nitric acid, another strong acid, is used to make explosives, such as nitroglycerin and dynamite. It is also used in the manufacture of ammonium nitrate (NH_4NO_3), an important ingredient in fertilizers.

Then there is the hydrochloric acid that Jessie uses in the pool. It is commonly called *muriatic acid*. Besides balancing the acid content of swimming pools, it is also used to clean concrete.

DISCOVER BY Doing

Test the effects of a common acid on different materials. Get four small paper cups, and fill them about half full of lemon juice. In the first cup, place a small piece of aluminum foil. In the next three cups, place an iron nail, a thumbtack, and a piece of limestone, respectively. Observe and describe the effect of the acid over a period of several days. ✏

ASK YOURSELF

Why are acids important?

The Solution Is Basic

"Jessie, what would you do if the pool was too acidic?" asks Hank while sitting on the edge of the swimming pool splashing his feet. "I would add a base to neutralize the acid." "You mean like first base or home plate?" jokes Hank. "Very funny," Jessie answers. Then she proceeds to explain about bases.

Remember that an acid produces hydronium ions in solution. A **base** produces hydroxide ions (OH^-) in solution in the same way. When bases are added to water, they react to form hydroxide ions. The stronger the base is, the more hydroxide ions are released into solution.

Although many bases contain hydroxide ions, there are some bases that do not. Ammonia gas, for example, is a base. However, it produces hydroxide ions only when it is dissolved in water.

$$NH_3 + H_2O \rightarrow NH_4^+ + OH^-$$

The key to being a base, remember, is the formation of hydroxide ions. Since ammonia forms OH^- in solution, it is a base. Household ammonia is a solution of ammonia gas and water.

Solutions that contain bases are called basic solutions. However, the word *alkaline* is sometimes used instead of *basic*. Strong bases are just as dangerous as strong acids. They can cause serious burns and must be handled with care. **CAUTION: Never touch or taste a strong base.** Find out more about bases by trying the next activity.

Figure 21–10. Strong bases, such as the drain cleaner shown here, are as dangerous as strong acids.

ᴅɪꜱᴄᴏᴠᴇʀ ʙʏ Doing

Take two strips of litmus paper, one red and one blue. Dip each into some household ammonia. What happens? How does this compare with the vinegar or orange juice test you did before? ✐

Soap solutions are mildly alkaline. If you ever had soap in your mouth, you are familiar with the bitter taste bases have. Bases usually have a slippery, soapy feel.

Bases react with acids to form salts. This process is called *neutralization*. Neutralization occurs when equal amounts of hydronium ions and hydroxide ions are in solution. These ions produce water.

$$H^+ + OH^- \rightarrow H_2O$$

Once the water is formed, the solution is no longer acidic or basic; it is neutral.

Neutralization occurs in the human body. When a person has an acid stomach, or heartburn, he or she may take an antacid tablet. The antacid tablet is a weak base that dissolves in the stomach and combines with the excess acid to neutralize it. Try the next activity to find out more about the neutralization process.

ACTIVITY

What does an antacid do?

MATERIALS

safety goggles, laboratory apron, beakers (2), dropper bottle with vinegar, blue and red litmus paper, spoon, wax paper, antacid tablet

CAUTION: Put on safety goggles and a laboratory apron, and keep them on throughout this activity.

PROCEDURE

1. Place 100 mL of water in a beaker, and add vinegar to it, one drop at a time. Test the solution with litmus paper after each drop is added. How many drops of vinegar did it take to make the solution very acidic (when the blue litmus paper turns bright red)? Record your results.

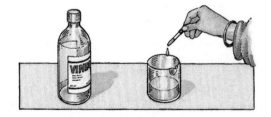

2. Place an antacid tablet on a small piece of wax paper. Crush the tablet with the back of a spoon, and dissolve it in about 100 mL of water.
3. Use litmus paper to find out whether the solution is acidic, basic, or neutral. Record your results.
4. Add vinegar, one drop at a time, to the antacid solution. Record the number of drops required to turn the litmus paper a bright pink. Compare your results with the solution that contained only vinegar and water. Compare your antacid with those of other groups.

APPLICATION

1. How does an antacid help with an acid stomach?
2. An antacid should neutralize excess acid in your stomach. Which brand worked best? Explain why you think so.

ASK YOURSELF
Describe a neutralization reaction.

Uses of Bases

Bases are as important as acids industrially. Bases are used in drain cleaners and soaps.

Sodium hydroxide is an important base. It has numerous uses that affect you. For instance, since it is used by the paper industry to remove pulp fibers from wood, it was probably used in producing this book.

Another commonly used base is ammonia. This compound is present in many cleaners because it can dissolve grease.

Figure 21–11. Sodium hydroxide is a strong base used in the paper-making process.

Many other bases are also used often. More than 16 billion kilograms of calcium hydroxide, $Ca(OH)_2$, are produced in the United States annually. This base is used to make cement, mortar, and plaster. Another base, magnesium hydroxide $Mg(OH)_2$, is used as an antacid. These are only a few examples of bases used every day.

 ASK YOURSELF

List two common bases and their uses.

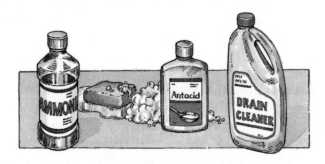

Figure 21–12. Most households use many bases.

SECTION 1 REVIEW AND APPLICATION

Reading Critically

1. What makes an acid or a base strong or weak?
2. List three main differences between an acid and a base.
3. Why do unripe fruits taste sour?

Thinking Critically

4. An acid is sometimes called a proton donor. Why is an acid such as HCl thought of as a proton donor?
5. Would you use hydrochloric acid to clean silverware? Explain.
6. Why is magnesium hydroxide used as an antacid instead of sodium hydroxide?

SKILL Designing an Experiment

▶ **MATERIALS**
- to be determined by student-designed experiment

▼ PROCEDURE

1. In earlier chapters you learned about the importance of controls and only changing one variable at a time when you perform an experiment. Suppose you wanted to find out more about the effect acid rain might have on different kinds of rocks. How would you go about designing an experiment that would provide that information?

2. Before doing an experiment, most scientists do library research to find out what is known about the subject. Check in the library for information on acid rain and the types of rocks that you might find in your area.

3. Write the hypothesis you are going to test. Have your teacher approve it before you go on. Scientists often have other researchers examine their work and offer advice.

4. Write a complete procedure for your experiment. Don't forget to include information on your variables. To complete this step in designing an experiment, you need to decide what acid you will use, what measurements you will make, and what types of rocks you want to test. Make a list of the materials you will need. Be sure to include any safety precautions you should use.

▶ **APPLICATION**

Collect your materials. Now there is only one thing left to do: conduct the experiment! Record your data as you proceed. Then draw conclusions based on your data.

✳ ***Using What You Have Learned***

After you have completed your experiment, go back and review your procedure. Would you change anything you did? How could you revise your procedure to improve the experiment?

pH and Indicators

SECTION 2

Hank is still a bit puzzled about the way Jessie's pool test kit works. He knows that different drops produce different colored solutions. How does Jessie know whether to add acid or base to the pool? Hank decides to do some research for himself on this pH stuff Jessie keeps talking about.

Red or Blue—Acid or Base

The pH of a solution is a measure of the hydronium ions in the solution. The pH scale ranges from 0 to 14. Acids, bases, and neutral solutions are separated into regions on the scale. The middle point, pH = 7, is neutral. Thus, a solution with a pH of seven is neither an acid nor a base.

Acids have a pH of less than 7. The stronger the acid, the lower the number on the pH scale. For example, the pH of lemon juice is about 2.3, while tomato juice, which is less acidic, has a pH of 4.

Bases have a pH higher than 7. The higher the number above 7 on the pH scale, the stronger the base. The chart below shows the pH values for some common substances.

Objectives

Classify *various substances based on their pHs.*

Define *the term* indicator, *and give three examples of indicators.*

Determine *whether a substance is acidic, basic, or neutral by interpreting the color of an indicator.*

Figure 21–13. This chart shows the pH of some common substances.

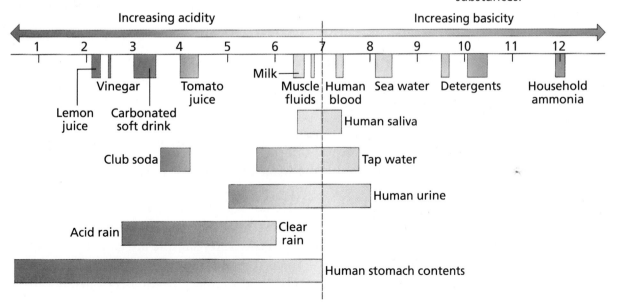

Increasing acidity — Increasing basicity

1 2 3 4 5 6 7 8 9 10 11 12

Lemon juice — Vinegar — Carbonated soft drink — Tomato juice — Milk — Muscle fluids — Human blood — Sea water — Detergents — Household ammonia

Club soda

Human saliva

Tap water

Human urine

Acid rain — Clear rain

Human stomach contents

Figure 21–14. This scientist is testing lake water for acid rain contamination. She is using a pH meter.

Many plants and animals are very sensitive to changes in pH. Lake water usually has a pH between 6 and 7. However, due to acid rain, some lakes in the northeast United States have a pH as low as 3. This drastic change kills any plants and animals that are sensitive to pH changes.

Scientists are experimenting with methods to bring the pH of these acidic lakes back to normal levels. In order to do this, large quantities of weak bases, such as ground limestone or sodium bicarbonate, are dumped from planes into the lakes. Scientists hope that the addition of bases will neutralize the excess acid. The results of these efforts, however, will be only temporary if the sources of pollution are not stopped.

Your body is also very sensitive to changes in pH. The normal pH of human blood is between 7.38 and 7.42. This range is very narrow; any change in the concentration of hydronium ions has a marked effect on the body. If the pH is above 7.8 or below 7.0, the body cannot function normally. If this change is not corrected rapidly, it can be fatal.

 ASK YOURSELF

What is the pH range for acids? For bases?

Indicators

How can we know the pH of pool water or any solution if the hydronium ions and hydroxide ions cannot be seen? This question bothered Hank until he remembered that the drops Jessie added to the water sample made the water different colors. The colors must tell Jessie what the pH is. He decides to ask Jessie about it that afternoon.

"Good thinking, Hank," says Jessie. "I know what the pH is and how much acid or base to add by the color of the water solution. The drops I add to the water sample are called indicators. **Indicators** are substances that change color in an acid or base. In fact, indicators can be used to measure the pH of just about

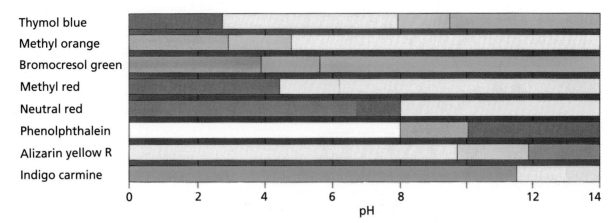

Thymol blue
Methyl orange
Bromocresol green
Methyl red
Neutral red
Phenolphthalein
Alizarin yellow R
Indigo carmine

pH

Figure 21–15.
Indicators turn different colors to show acids or bases.

anything. My grandmother has a pH test kit for her garden so she can test the soil to see whether it is acidic or basic. Vegetables grow best within certain pH ranges. Grandma can add substances to the soil to get the pH she wants."

Some indicators show only two color changes—one for acids and another for bases. Other indicators, however, show a range of color changes depending on the pH. One type of indicator paper, called *pH paper*, is dyed with universal indicator. Universal indicator is a mixture of several different indicators, so the paper gives a different color for each pH. The advantage of this type of indicator is that it shows the exact pH of a solution rather than just whether the solution is an acid or a base.

 **by** *Doing* _____

Your teacher will give you some pH paper. Use it to test some common household materials such as shampoo, detergent, and ketchup. Record your findings. 🖉

 **ASK YOURSELF**
How can scientists determine pH?

Figure 21–16. pH paper can be used to determine the pH of almost any liquid.

SECTION 2   *REVIEW AND APPLICATION*

Reading Critically
1. Define the term *pH*.
2. You have two bases, pH = 8 and pH = 12. Which is more alkaline? Why?

Thinking Critically
3. How could testing the pH of a lake be beneficial to people who fish?
4. What is the ideal pH of Jessie's pool? Why?

INVESTIGATION

Making Indicators

▶ MATERIALS

- safety goggles
- laboratory apron • beakers, 50 mL (8–10) • mortar
and pestle • flowers (4–5) • graduate, 10 mL • ethanol, 95%, 10 mL • hot
plate • stirring rod • funnel • filter paper • test tubes (5) • test-tube rack
• wax pencil • medicine dropper; • prepared solutions of the following
acids and bases, approximately 5 mL each:

HCl, hydrochloric acid (pH = 1)	NH$_4$OH, ammonium hydroxide (pH = 11)
HC$_2$H$_3$O$_2$, vinegar (pH = 3)	NaOH, sodium hydroxide (pH = 13)
Distilled water (pH = 7)	

**CAUTION: Put on safety goggles and a laboratory apron, and keep them on
throughout this investigation. Be sure your area is well ventilated.**

▼ PROCEDURE

1. Prepare a table like the one shown here.

TABLE 1: COLOR CHANGES OF FLOWER INDICATORS

Solution	Flower Name	Flower Name	Flower Name	Flower Name
HCl				
HC$_2$H$_3$O$_2$				
Distilled water				
NH$_4$OH				
NaOH				

2. Using the mortar and pestle, crush the petals of one flower. Put them in a beaker.
3. CAUTION: Ethanol is flammable. Do not heat over an open flame. Add 10 mL of ethanol to the beaker, and heat it gently in a hot-water bath or on a hot plate for 15 minutes. Stir often.
4. Allow the mixture to cool, and filter it into a clean beaker. Your indicator may not have any color.
5. Repeat steps 2–4 for each type of flower.
6. Label each test tube with the name of one of the prepared acid or base solutions. Place 5 mL of each solution into the correct test tube.
7. To each solution, add 1 or 2 drops of the indicator from one flower. Record any color changes you observe. If the color change is not clear, add more indicator—one drop at a time.
8. Clean the dropper, and repeat steps 6–7 for each indicator.

▶ ANALYSES AND CONCLUSIONS

1. Which flower made the best indicator? Give reasons for your answer.
2. Could two flowers of different colors be used to indicate the same pH? Explain your answer.

▶ APPLICATION

Find how many different pigments are in your flower indicator, by completing a paper chromatograph.

Discover More

Test other natural substances for their suitability as indicators.

*H*IGHLIGHTS

The Big Idea

Acids and bases affect natural systems as small as a swimming pool or as large as an ecosystem. Acids interact with nonliving substances such as building materials and rocks. Living organisms are also sensitive to changes in pH. In fact, most living systems are so sensitive to changes of this type that they cannot live if the pH changes drastically.

Acids and bases interact to form neutral substances in a process known as neutralization. By understanding the interactions of acids and bases with each other as well as with the living and nonliving parts of the ecosystem, you can make practical decisions about everyday occurrences. You will know what chemicals to add to your garden soil, how to counteract an acid spill, or what types of pollution would increase acid rain.

For Your Journal

At the beginning of the chapter, you wrote down your ideas about acids and bases. Look back at your ideas. How have your ideas changed? Revise your journal entries to show what you have learned.

Connecting Ideas

Below is a concept map about acids. Study it carefully, and, in your journal, make a matching concept map about bases.

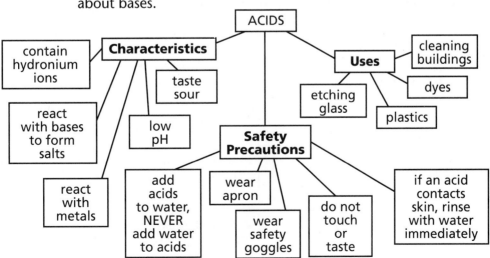

> **Understanding Vocabulary**

1. Explain how the following terms are related:
   acid (567), base (573), pH (577),
   indicator (578)

> **Understanding Concepts**

MULTIPLE CHOICE

2. An acid produces what type of ions in
   solution?
   a) oxygen
   b) hydronium
   c) hydroxide
   d) sulfur

3. One common acid is
   a) water.
   b) milk of magnesia.
   c) vinegar.
   d) bleach.

4. A base produces what type of ions
   in solution?
   a) oxygen
   b) hydronium
   c) hydroxide
   d) sulfur

5. A common base is
   a) ammonia.
   b) water.
   c) lemon juice.
   d) milk.

6. A substance with a pH of 9
   would be
   a) neutral.
   b) an acid.
   c) a base.
   d) a salt.

SHORT ANSWER

7. Explain the differences between strong and
   weak acids and strong and weak bases.

8. Describe the chemical reaction that takes
   place when an acid and a base neutralize
   each other.

9. Explain how you would use litmus paper to
   determine if a substance were an acid or a
   base.

> **Interpreting Graphics**

10. Study the graph below.
    Is the solution being
    added acidic or basic?
    Explain your answer.

11. Which litmus paper
    sample indicates an
    acid? a base?

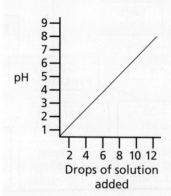

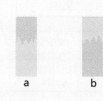

Reviewing Themes

12. *Systems and Structures*
What products would be formed if you mixed hydrochloric acid (HCl) with sodium hydroxide (NaOH)? Write an equation for the reaction.

13. *Environmental Interactions*
Chlorine is added to swimming pools as a disinfectant. It is necessary to treat the resulting chlorinated water with basic compounds. Why must these bases be added?

Thinking Critically

14. Many insects bite. The reason the bite hurts is that some toxin is introduced into the victim. Often, the toxin is some form of acid. The centipede in the photo has been killed by a marauder ant, which injected it with formic acid.

Sometimes when people get an insect bite or sting, they treat it with a product similar to the one shown here. These products contain ammonium hydroxide. Using your knowledge of acids and bases, explain how ammonium hydroxide could stop a bite from itching or hurting.

15. Explain why a base can also be described as a proton acceptor.

16. Why is it important to control the emission of sulfur compounds into the atmosphere?

17. In what way might acids and bases be considered opposites?

18. Solutions A, B, C, and D have pH values of 12, 9, 1, and 5, respectively. The solutions' concentrations are equal. Identify the strongest acid and the strongest base in this group. Explain your answer.

19. Describe the reaction below in your own words.

$$Mg + 2HBr \longrightarrow MgBr_2 + H_2$$

20. Predict the approximate pH of a solution formed when a weak base reacts with a weak acid. Explain your prediction.

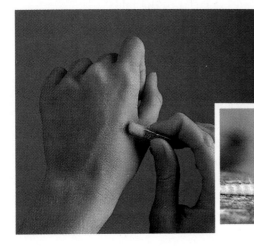

Discovery Through Reading

Thompson, Jon. "East Europe's Dark Dawn." *National Geographic* 179 (June 1991): 36–63. This extensive article focuses on the perils of acid rain and other environmental pollutants in eastern European countries.

CHAPTER 22

CARBON COMPOUNDS

Carbon is one of the most ordinary and versatile elements; it is present in countless compounds. Yet, in its pure state, carbon can excite romance, intrigue, and violence. The rarest form of this ordinary element is the diamond, a stone of matchless hardness and beauty.

The history of diamonds is filled with legend and suspense. One legend surrounds the famous Koh-i-Noor, which passed among the Mogul, Persian, and Afghan conquerors of India.

Following the Mogul invasion, the diamond passed into the hands of Sultan Baber,...founder of the Mogul Empire in India. A much treasured possession of his, Baber refers to it in his diary in 1526 as "the famous diamond" of such value that it would pay "half the expenses of the world." It remained in the ownership of his descendants for the next two centuries, thus giving some substance to the legend that "he who owns the Koh-i-Noor rules the world." In 1739, however, it was lost to the Persians who, under their ruler Nadir Shah, invaded India and sacked Delhi. There is a story that for fifty-eight days the stone could not be found because the conquered Mogul emperor Mohammed Shah had hidden it in the folds of his turban. Told the secret by a member of the ex-emperor's harem, Nadir Shah invited him to a feast and, observing an ancient Oriental custom, proposed an exchange of turbans. Mohammed was in no position to refuse. Once he had the turban, Nadir Shah ran to his tent and on seeing the great diamond among the silk of the unrolled turban, cried "Koh-i-Noor," which means Mountain of Light, thus giving it the name it has borne ever since.

The diamond's history doesn't end there. Eventually, the diamond was seized from India by the British government. Pakistan's government has asked that the diamond be returned. The British government has replied that the gem's history is so complicated that no one has a claim stronger than theirs.

Diamonds—such a simple element about which people have such complex feelings!

For Your Journal

🖉 Diamonds are one form of carbon. What other forms do you know about?

🖉 Carbon is a major part of all organisms. How can an element that forms diamonds also form living tissue?

🖉 In what ways do you think carbon affects your life?

From Soot to Computers

Objectives

Suggest a reason for the existence of such a variety of carbon compounds.

Summarize the uses for each of the elements in the carbon family.

Describe the problems associated with lead.

W hat came into your mind when you read the title of this section? "The author has lost it!" might have been a thought. What does soot have to do with computers? They don't seem to have anything in common. Yet some atoms in both belong to a prolific family of elements, the carbon family.

Soft as Soot, Hard as Diamond

Carbon is the common element that makes up soot and charcoal. What makes carbon and the carbon family so unique? Try the next activity to find out.

 **BY** Doing _____

Compare the "lead" in your pencil with the soot collected on a beaker. Both are forms of carbon. How are they alike? How are they different? ✐

The different forms, or allotropes, of carbon have different properties. Graphite, at the tip of your pencil, feels slick and oily. Because of this, graphite is used as a lubricant in machines. Look at the diagram to see how the atoms of carbon are arranged in graphite. The "sheets" of carbon molecules can glide over each other, causing graphite's slipperiness.

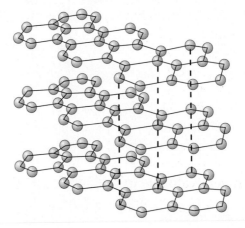

Figure 22–1. Graphite atoms are arranged in sheets that slide over each other.

Another allotrope of carbon, with very different physical properties, is diamond. In a diamond the carbon atoms are bonded in a rigid crystalline structure. This structure is responsible for the extreme hardness of diamonds.

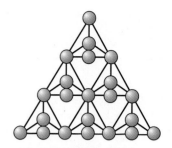

Figure 22–2. The crystalline structure of diamond is very rigid.

Scientists once thought the soot you observed in the activity was made of carbon that was not bonded in any particular pattern. Recent research, however, indicates that soot contains molecules made of 60 to 70 carbon atoms bonded into ball-shaped configurations. These molecules are called *Buckyballs* or *Buckminsterfullerenes* after Buckminster Fuller, the architect who invented geodesic domes.

The most important thing to remember about carbon is that it forms four bonds, either to other elements or to itself. This allows carbon to form long chains of carbon atoms or to combine in many different ways with other common elements, such as hydrogen, nitrogen, and oxygen.

Figure 22–3. Carbon forms four bonds when it combines with itself or other elements.

There are so many carbon compounds that there is an entire field of chemistry dedicated to them. The area of chemistry in which carbon compounds are studied is called *organic chemistry.* Organic chemists study the properties of carbon compounds. They also make important molecules that are used in medicines, insecticides, plastics, and other synthetic materials. Why do you think the study of carbon is called organic chemistry?

 ASK YOURSELF

Why does carbon form different allotropes?

Silicon: The Element That Can't Make Up Its Mind

Figure 22–4. Computer chips made of silicon

The second member of the carbon family is silicon. It combines with other elements to form a great number of compounds, but it does not make as many different types of compounds as carbon. Unlike carbon, pure silicon has a metallic shine; it is a metalloid. It may conduct electricity under certain conditions. Because of this property, silicon has been very valuable in making computer chips and semiconductors.

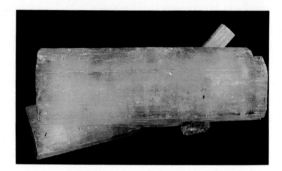

Figure 22–5. Many minerals contain silicon. The silicon-containing minerals shown here, clockwise from top left, are mica, amethyst, emerald, and beryl.

Even though silicon is a metalloid, its most common compounds are not metallic in character. It is hard to believe that the same substance that is used to make computer chips is found almost everywhere—as sand. Sand is a silicon compound called silicon dioxide (SiO_2). Other silicon-oxygen compounds are found in many minerals and gemstones.

Silicon, like carbon, can form four covalent bonds. Dr. Larry Hench of the University of Florida has used the stability of silicon and its similarity to carbon to make a unique kind of glass. This glass—made of silicon and salts of sodium, calcium, and phosphorus—can be cast and carved into the shapes of bones. The human body accepts the glass as if it were a natural bone transplant. Dr. Hench is currently studying many other applications of this glass in the body.

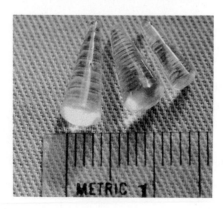

Figure 22–6. These glass cones are used as implants in the jaws of people who have lost some of their teeth. The implants reduce the shrinking of the gums.

There are many silicon compounds in your home. The caulking used to seal windows and fill cracks is made from silicon compounds called *silicones*. Silicones are long chains of silicon and oxygen atoms with different carbon chains attached. They are also used as lubricants and as water-repellent coatings on masonry and cement.

ASK YOURSELF
How are carbon and silicon similar? How are they different?

Figure 22–7. Silicones are so heat resistant that this kitten can sit on a slab of silicone and not be burned by the flame below. This property of silicones makes them useful for heat-resistant tiles on spacecraft.

Cousins—Near and Far

What other elements in the carbon family are you familiar with? Do you know how they are used? How are they similar to or different from carbon and silicon? Other members of the carbon family might not be as abundant as carbon or silicon, but they are still important.

Germanium is a metalloid. It is a semiconductor used in some electric devices such as transistors. Small amounts of germanium can be added to glass to produce filters in optical instruments and in cameras.

The other two elements in the carbon family are tin and lead. They have metallic bonds rather than the covalent bonds found in the other members of the family. Since the bonds are different, their properties are also different. As metals, both tin and lead can be easily shaped.

Figure 22–8. Germanium is used to make camera filters like these.

Tin is used to line the inside of metal cans and to make pewter. How would you like to clean your teeth with tin? What? Your teeth are not tough enough to stand a metal scouring. Nevertheless, you probably already use tin when you brush your teeth. It is found in the compound stannous fluoride. That's right, the compound that provides fluoride, which helps prevent tooth decay, contains tin.

Figure 22–9. These serving utensils are made of pewter, a metal alloy that contains tin.

Lead is a poisonous metal. At one time, most gasolines contained lead to prevent engine knock. Today, most cars use unleaded gasoline. Why the switch? Cars using leaded gasoline release small amounts of lead compounds into the air. These lead compounds can accumulate in the human body, eventually causing lead poisoning. Some of the results of lead poisoning are mental retardation and even death.

Some lead compounds are brightly colored. For this reason, they were used as pigments in house paints. The colors produced were attractive, but the lead caused problems. As the lead-containing paint peeled off walls, children were exposed to the lead. Small children are easily attracted to bright colors and some ate the paint

Figure 22–10. Compounds that contain lead occur in many bright colors. For this reason, they were once used to color paints.

because many lead compounds taste sweet. This gave them lead poisoning. Because of this, leaded paint is no longer sold. Some artists' paints still do contain lead, however, as does some colored clay pottery. You can check for lead by using an indicator stick that will change color if rubbed against a substance that contains lead compounds.

Some historians believe that the downfall of the Roman Empire might have been related to the use of lead. The early Romans used lead to line their aqueducts and seal their wine containers. Traces of lead have been found in human remains of that period. Imagine an element having such an impact on the course of history!

Figure 22–11. This Roman aqueduct contained pipes that were lined with lead.

 ASK YOURSELF

Describe one use each for germanium, tin, and lead.

SECTION 1   *REVIEW AND APPLICATION*

Reading Critically

1. What is the difference between silicon and silicones?

2. Name three things in your home that are made from silicon, germanium, tin, or lead.

Thinking Critically

3. How is the chemistry of carbon different from the chemistry of the other elements in its family?

4. Sugar, which is a typical organic compound, does not conduct electricity when dissolved in water. What can you conclude about the type of bonding in organic compounds from this information?

Organic Compounds and Their Uses

Objectives

Sketch *structural formulas for simple carbon compounds.*

Explain *what structural isomers are, and give two examples.*

Define *the term* polymer, *and give an example of a polymer used in your home.*

Have you ever heard of swamp gas? If you have ever paddled a canoe through a marsh or bayou, you might have seen bubbles of gas rising to the surface. This swamp gas is called methane and is formed by decomposing organic matter at the bottom of the swamp.

Dot, Dot, Dash

Methane, the principal component in natural gas, can be used to illustrate the properties of many organic compounds. Methane has the molecular formula CH_4. How could you draw a molecule of methane? Look at the diagram. The dashes that connect the atoms represent covalent bonds. This type of formula is called a **structural formula**.

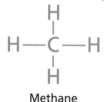
Methane

Figure 22–12. Structural formula for methane

Problem Solving

Carbon can form only four bonds. Draw the structural formula for the molecule C_2H_6. ✐

The molecule you just drew is known as *ethane*. There are many compounds that are made of only carbon and hydrogen. These compounds are called **hydrocarbons.** When hydrocarbons form, long chains of carbons can bond to each other. You can find out more about how this happens by doing the next activity.

DISCOVER BY Doing

Using a molecular model kit, make models of methane and ethane. Then try to make two more hydrocarbons that have more than two carbon atoms. Use as many hydrogens as you need. In your journal, draw your compound and write the correct formula for it. Compare your models with those of other students. Are they all the same? Explain. ✎

Hydrocarbons are very common. You probably use some every day. One of their properties is that they are very flammable. Because of this, they are used as fuels. Methane, ethane, and propane are used to heat homes and to cook food.

The length of the carbon chain in a hydrocarbon determines the physical state of the compound. Hydrocarbons with less than five carbon atoms are gases at room temperature. Those with five to 20 carbon atoms are usually liquids. Gasoline is an example. Hydrocarbons containing more than 20 carbon atoms tend to be solids at room temperature. They are usually waxy to the touch.

Many organic compounds contain other elements besides carbon and hydrogen. Oxygen is a common element found in organic compounds. Find out more about these compounds by trying the next activity.

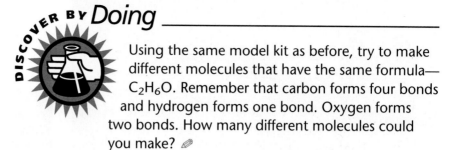

Using the same model kit as before, try to make different molecules that have the same formula—C_2H_6O. Remember that carbon forms four bonds and hydrogen forms one bond. Oxygen forms two bonds. How many different molecules could you make?

There are two compounds that have the formula you worked with in the activity. Did you figure them both out? You can see them drawn below. Ethanol, also known as ethyl alcohol or grain alcohol, is shown on the left, and dimethyl ether is on the right.

Ethanol   Dimethyl ether

Figure 22–14.
Structural formulas for ethanol and dimethyl ether

Think about the properties of both. Ethanol can be drunk; it is the alcohol found in alcoholic beverages. Dimethyl ether, on the other hand, can be used as an anesthetic and is poisonous if drunk. Even though these two compounds have the same molecular formula, they have very different properties. Why do you think this is so? You guessed it—their structures are different.

These two compounds are known as structural isomers. An **isomer** is one of two or more compounds that have the same chemical formula but different structures. The structure refers to the arrangement of atoms. The number of isomers that a compound has increases with the number of carbon atoms in the molecule; more atoms available allow for more variations in how they are arranged.

 **ASK YOURSELF**
Diagram structural formulas for two isomers of C_4H_{10}.

A Train of Carbon

The Riverfront Mall has a small shop that attracts a crowd. In it you can see a man dressed in a candy-striped jumpsuit making taffy. He kneads the taffy and pulls it into long strands almost a meter in length. The taffy seems to stretch forever.

Look at the beaker in the Figure 22–15. How is the substance inside the beaker similar to taffy?

The photograph shows two liquids in the beaker, one above the other. A long strand of nylon is being pulled out of the center. Yes, this stuff is the same material that is used to make tents and windbreakers!

Look around you, and make a list of all the things you see that are made of plastic or synthetic fibers. Did you know that many car parts are plastic? Your desk or counter top may be made from synthetic materials. Maybe the clothes you are wearing are synthetic. The pen you are using is probably made of plastic. Your world is filled with materials that did not exist 50 years ago. These synthetic materials are made of giant molecules called *polymers*. **Polymers** are large molecules made by "hooking" many identical small molecules together, much like railroad cars are linked to one another. The units in a single polymer might number in the thousands. The small units that make up the polymer are like the boxcars that make up a train. They are called **monomers.**

Figure 22–15. On the right, Dr. Hill, a co-inventor of nylon, is pulling a strand of nylon from the beaker. On the left is a close-up. How long do you think the strand can get?

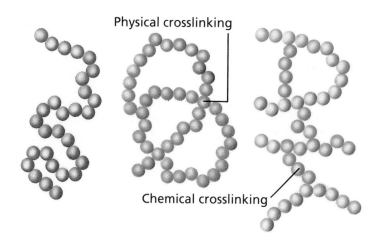

Figure 22–16. Polymers form as long chains of molecules. Those polymers with no cross-bridges are flexible. Polymers with physical bridges or chemical crosslinks are more rigid. The greater the number of bridges or crosslinks, the harder the polymer.

Physical crosslinking

Chemical crosslinking

Why are some polymers elastic like rubber bands, while others are hard or brittle? In order to understand this, picture some cooked spaghetti (no sauce) on a plate. If you jiggle the plate, the strands shift every which way. Certain polymers are like that; they tend to be elastic.

In contrast, other polymers are hard. The seat you are sitting doesn't squirm like spaghetti. How can scientists change the properties of polymers so much? They hook the strands of polymer together with short bridges. This process is known as crosslinking. Crosslinked polymers have lost some degree of movement. These bridges make the polymers stiff so that they keep their shape. They tend to be less elastic and harder. You can make your own polymer in the next activity.

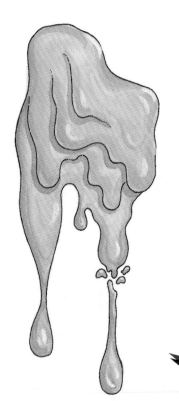

DISCOVER BY Doing

Your teacher will give you 20 mL of polyvinyl alcohol in a paper cup. Describe the liquid. Polyvinyl alcohol is a polymer. Add 2 to 3 drops of food coloring and stir. While you stir, your teacher will add a small amount of a second solution. Keep stirring until you can tell that some crosslinking has occurred. How would you know this? Describe what happens. **CAUTION: You can play with your slime, but don't eat it.** ✍

Polyethylene is an example of one type of polymer that can be either elastic or rigid. You are probably familiar with many polyethylenes. Study Table 22–1 to find polymers you recognize.

ASK YOURSELF

What is the difference between a polymer and a monomer?

Table 22-1

Polyethylene and Its Thermoplastic Derivatives

Monomer	Polymer	Uses
$H_2C = CH_2$ ethylene	$\{H_2C{-}CH_2\}_n$ polyethylene	packaging, bottles, toys
$H_2C = C \overset{H}{\underset{CH_3}{}}$ propylene	$\begin{bmatrix} H_2C{-}CH \\ \quad\ \ CH_3 \end{bmatrix}_n$ polypropylene	fibers, films, carpets, laboratory equipment, kitchenware
$H_2C = C{-}H$ (benzene ring) styrene	$\begin{bmatrix} H_2C{-}CH \\ \text{(benzene ring)} \end{bmatrix}_n$ polystyrene	Styrofoam insulation, packaging and packing material, household articles, toys
$H_2C = C \overset{H}{\underset{Cl}{}}$ vinyl chloride	$\begin{bmatrix} H_2C{-}CH \\ \quad\ \ Cl \end{bmatrix}_n$ polyvinyl chloride	floor coverings, garden hoses, phonograph records, packaging
$H_2C = C \overset{H}{\underset{O-C-CH_3}{}}$, $\underset{O}{\overset{\ }{\parallel}}$ vinyl acetate	$\begin{bmatrix} H_2C{-}CH \\ \quad\ \ O \\ O = C{-}CH_3 \end{bmatrix}_n$ polyvinyl acetate	latex paints
$H_2C = C \overset{H}{\underset{C{\equiv}N}{}}$ acrylonitrile	$\begin{bmatrix} H_2C{-}CH \\ \quad\ \ CN \end{bmatrix}_n$ polyacrylonitrile (Orlon or Acrilan)	textile fibers
$H_2C = CCl_2$ vinylidene chloride	$\{H_2C{-}CCl_2\}_n$ polyvinylidene chloride (Saran)	self-adhering food wrap
$H_2C = C \overset{CH_3}{\underset{O = C-OCH_3}{}}$ methylmethacrylate	$\begin{bmatrix} \quad\ O \\ \quad\ \parallel \\ CH_3{-}C{-}C{-}O{-}CH_3 \\ \quad\ CH_2 \end{bmatrix}_n$ polymethylmethacrylate (acrylic, Lucite, Plexiglas)	unbreakable glass for windows, windshields; water-based latex paints
$F_2C = CF_2$ tetrafluoroethane	$\{F_2C{-}CF_2\}_n$ polytetrafluoroethene (Teflon)	chemically inert items, gasket materials, cookware coatings

Carbon Keeps You Clean?

Look at the ingredients listed on a bar of soap. There are probably many that you do not recognize. Today's soaps are made in laboratories using some synthetic substances. However, from Roman times to the late 1900s, soap was made by a different method. Large kettles containing animal fat, ashes, and water were heated and stirred over open fires. As the soap particles formed they were skimmed off the top of the liquid. Then the lumps were pressed together to form cakes.

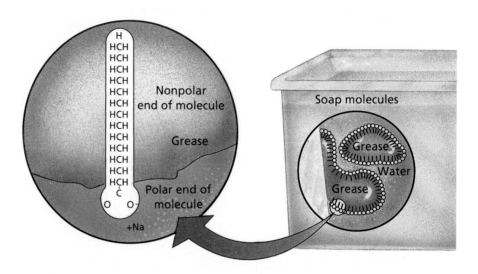

Figure 22–17. One end of a soap molecule bonds to dirt, and the other bonds to the water in which the items, such as your hands or dirty dishes, are being washed. These bonds pull the dirt off the items being washed.

Soap molecules have two ends with different properties. One end of the molecule is a hydrocarbon; the other end is a salt. The hydrocarbon part bonds to fats (grease and dirt), while the salt end sticks to water. The grease is broken up into small droplets surrounded by soap molecules. These droplets can then be washed away with water.

 ASK YOURSELF

Why can soap remove oil stains that cannot be removed by water alone?

Fuel for Thought

You have probably realized how important organic chemistry is to you. Did you know that without hydrocarbons you would not be able to ride the school bus or drive to the mall? Vehicle fuels are hydrocarbons. Where do these fuels come from? The sources of many hydrocarbons in use today are natural gas and petroleum. Petroleum is found trapped in pockets under the earth's surface. Natural gas deposits are usually found in the same places. Natural gas is a mixture of several hydrocarbons with very low boiling points.

You can learn more about different types of hydrocarbons by studying Table 22–2. It shows the different hydrocarbons that are present in petroleum.

Table 22-2 — Hydrocarbons Separated from Petroleum

Fraction	Number of carbons	Boiling point ($°C$)	Common uses
Gas	1–5	Less than 30	Fuel
Petroleum ether	5–7	30–90	Solvent
Gasoline	5–12	40–200	Motor fuel
Kerosene	12–16	175–275	Fuel
Fuel oil Diesel oil	15–18	250–400	Fuel for furnaces and diesel engines, raw materials for petrochemicals
Lubricating oils Greases Petroleum jelly	More than 16	Higher than 350	Lubricants
Paraffin (wax)	More than 20	Melts at 50–55	Candles, matches, waterproofing

Figure 22–18. Wax, like that on the outside of these apples, is a long-chain hydrocarbon.

Figure 22–19.
Petroleum is separated using fractional distillation. The process takes place in fractionation towers such as those shown here.

Petroleum can be compared to a mixture of many different colored paints. Imagine trying to separate such a mess in order to see each color individually. It would be quite a job. Petroleum can be separated using a process called *fractional distillation,* which separates the hydrocarbons using the differences in their boiling points. The petroleum is heated at the bottom of a distillation column, and the compounds are turned into gases. Those with higher boiling points condense at the lowest part of the column. The substances with very low boiling points condense at the highest part.

Besides petroleum and natural gas, many other important organic compounds are found in nature. Compounds produced by living organisms are called *natural products.* These compounds are used in medicines, pesticides, perfumes, and food additives, among other things. Organic chemists try to isolate these natural products and to investigate their properties.

In the early 1990s, there was much medical interest in a substance found in a tree called a Pacific yew. The substance,

Figure 22–20. The Pacific yew may have been saved from extinction now that the chemicals it contains can be made in the laboratory.

called *taxol,* interested scientists because it can be used to fight certain types of cancer. The problem is that the Pacific yew is not very common, and thousands of trees must be cut in order to make just a few grams of taxol. When an important substance is found in too small a quantity in nature, scientists may try to synthesize it, or make it in the laboratory. By 1993, organic chemists had synthesized taxol inexpensively in the laboratory. These chemists made taxol using similar chemicals found in European yews. These trees are very common and in no danger of extinction.

What other useful things can organic chemists make? What about those wonderful scents that come from certain flowers and fruits? You might think that laboratories "stink," but scientists can prepare wonderful scents also. Many sweet, flowery scents are produced by organic compounds called *esters*. As you might expect, esters are very valuable to the perfume industry. What may surprise you is that esters are also used in prepared foods to enhance their aroma and flavor. You can experience some of these odors by trying the next activity.

Figure 22–21. This French scientist is testing the odors of various perfumes.

ACTIVITY

How can you identify the odors of different natural oils?

MATERIALS

filter paper strips with different natural oils (5), strips of filter paper with unknown oils (3)

PROCEDURE

1. Make a table like the one shown.

TABLE 1: IDENTIFYING ODORS		
Strip	Odor	Matches unknown
1		
2		
3		
4		
5		

2. Your teacher will give you five pieces of filter paper with a different natural oil on each one. Do not let the strips touch each other. Smell and become familiar with the odor of each strip.

Do not smell any of the strips for too long because it may tire your sense of smell. Describe the odors on your data table.

3. Get one to three unknowns on strips of filter paper. Each unknown may have from one to four of the different oils you have already smelled. Again, don't tire your sense of smell. Try to identify the odors of the unknowns. Refer to the known samples if necessary.

APPLICATION

1. What were the oils in each of the unknowns? Refer to them by number if you do not know their names.

2. Find out how some of the natural oils you smelled are produced.

3. Insects can detect odors about 10^4 times better than humans. Why do you think a sense of smell is important to insects?

Figure 22–22.
Kongfrontation at Universal Studios in Orlando, Florida

Have you ever been to a theme park? Universal Studios has a park in Orlando, Florida, in which one ride uses esters! Does that surprise you? It often surprises the people on the ride, too. The ride is called *Kongfrontation*. During one part of the ride, King Kong comes up to the cablecar you are riding in and breathes on you. His breath smells just like bananas. Now you know the odor is the result of synthetic esters!

ASK YOURSELF

What is a natural product?

SECTION 2   *REVIEW AND APPLICATION*

Reading Critically

1. Bottled gas (propane) contains hydrocarbons. How many carbon atoms does each molecule of this hydrocarbon have?

2. What is a polymer?

Thinking Critically

3. Is petroleum a mixture or a complex compound? How can you tell? Be specific.

4. Orlon, a polymer used in synthetic fibers, has the structure shown. What is the monomer that makes up Orlon?

$$-CH_2-CH-CH_2-CH-CH_2-CH-CH_2-$$
$$CN \qquad CN \qquad CN$$

SKILL Using Models

▶ **MATERIALS**
- pencil ● paper ● mirror

▼ **PROCEDURE**

1. Print TOT on a sheet of paper. Place it in front of a mirror. On another sheet of paper draw the image you see. Is the mirror image the same as the original? How can you tell?

2. Now print POP on a sheet of paper, and repeat the process. Are the images the same? How are they different from the images in step 1?

3. Make a model of a molecule by taking a plastic-foam ball and inserting four different-colored straws in it. Place your model in front of the mirror, and make another model that is like the image in the mirror. Are the two models the same? Explain why or why not.

▶ **APPLICATION**

Your hands are mirror images of each other. You can place them facing each other and they match. However, if you put one on top of the other, they do not match. Some molecules are like your hands. They are alike but are mirror images of each other. These molecules are called *enantiomers* (ihn AN tee uh muhrz).

1. Which of the words you looked at in the mirror is an enantiomer?

2. Was the model you made an enantiomer?

3. Are the compounds shown here mirror images? Are they enantiomers? Explain.

4. Are three-dimensional models necessary in order to understand enantiomers? Why or why not?

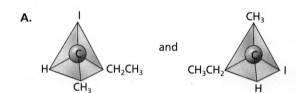

A.

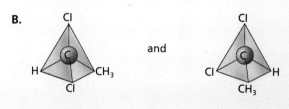

B.

※ **Using What You Have Learned**
How did using a model help you understand the structure of enantiomers? Describe another situation in which using a model could help you understand a new idea.

Molecules of Life

Compare and **contrast** carbohydrates, proteins, and fats.

Differentiate between saturated and unsaturated fats.

So far in this chapter, we have discussed the importance of carbon in many manufactured goods. Your life is filled with carbon compounds; your life is also made up of them. You would not move, breathe, or exist if it were not for carbon compounds. The study of carbon compounds in living organisms is called *biochemistry*.

Fuel for Your Body

Jim is getting ready for a marathon in the morning. He must be in peak condition to win the race. Tonight he is meeting with other runners for a spaghetti feast. Jim does this before every marathon. He knows that spaghetti will provide the energy he needs to do well tomorrow. It is the fuel his body will burn as he runs. Bread, pasta, and rice are foods that contain carbohydrates. **Carbohydrates** are compounds made of carbon, hydrogen, and oxygen in a ratio of CH_2O. Figure 22–23 shows some common carbohydrates.

Figure 22–23. Carbohydrates are found in all the foods shown here.

Sugars, starches, and cellulose are examples of carbohydrates. Glucose ($C_6H_{12}O_6$) is a simple sugar that makes up more complicated carbohydrates. All living cells use glucose as their "fuel." There are other simple sugars besides glucose. Fructose, for example, is commonly found in fruit. Simple sugars such as these are used as monomers in making carbohydrate polymers.

The sugar that you put in your lemonade is called *sucrose* ($C_{12}H_{22}O_{11}$). It is a compound formed by joining two simple sugars, glucose and fructose. Look closely at Figure 22–24. Find where the two molecules have joined and what is formed in the reaction. Notice that as the larger carbohydrate is formed, a molecule of water is released. This is how plants synthesize complex sugars and starches. In your body, the reverse reaction occurs. Water is used to break down the bigger molecules into simple sugars that can be used by your body to produce energy.

Figure 22–24. A dehydration reaction produces a large molecule from two smaller molecules. The reaction shown here is the formation of sucrose (table sugar) from glucose and fructose.

Glucose	Fructose	Sucrose	Water
$C_6H_{12}O_6$	$C_6H_{12}O_6$	$C_{12}H_{22}O_{11}$	

 ASK YOURSELF

Why are carbohydrates useful?

Muscle-Building Molecules

Other types of molecules important to your body are proteins. What kinds of food are rich in protein?

Like polymers, **proteins** are huge molecules made up of smaller ones. The building blocks of proteins are called *amino*

Figure 22–25. These foods are rich in protein.

Figure 22–26. Human hair is formed from complex proteins. A magnified human hair is shown above. A computer-generated image of a complex protein is shown on the right.

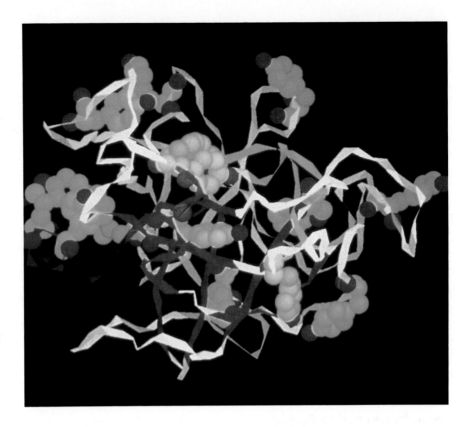

acids. There are two chemical groups by which you can recognize an amino acid. The -NH_2 part is an amino group, and the -COOH group is an organic acid. Put them together, and you have an amino acid. The other parts of this molecule can vary. These variable parts are called *R groups.* Table 22–3 shows the characteristic groups for some of the more common amino acids.

Figure 22–27. Amino acids are the building blocks of proteins.

$$\begin{array}{c} H \\ | \\ R-C-COOH \\ | \\ NH_2 \end{array}$$

Table 22-3

Some Common Amino Acids

Amino acid	Characteristic R group
Glycine	-H
Alanine	-CH_3
Serine	-CH_2OH
Valine	CH_3—CH—CH_3
Leucine	CH_3—CH—CH_3 CH_2

There are 20 important amino acids. The number of different proteins, however, may be in the millions. If you had 20 different colored blocks and had to arrange them in a straight line, how many different combinations could you make? To give you an idea of how many combinations there could be, try the next activity.

Discover By Doing

Get six different-colored blocks from your teacher. Arrange them in as many straight-line patterns as you can. How many combinations did you find? ✎

Amino acids link together just like boxcars in a train to make proteins. Your muscles, hair, tendons, and even parts of your blood are all made of proteins. Other proteins actually control all the chemical reactions that go on inside a cell. These proteins are called *enzymes*. These are of critical importance to living cells. Just one misplaced amino-acid boxcar in the protein train can sometimes mean the difference between life and death of a cell or even an entire organism.

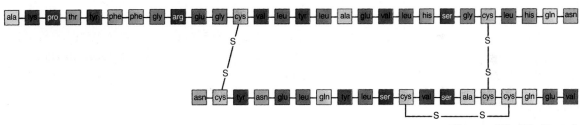

Figure 22–28. Proteins, like this one, are formed from long chains of amino acids.

In a protein, each boxcar represents one amino acid. Scientists use the first three letters of each amino acid to write a protein sequence. For example, glycine would be written as GLY and alanine as ALA. The diagram shown (Figure 22–28) illustrates an amino acid sequence in a protein. The -S-S- structure represents a sulfur bridge. These bridges increase the rigidity of the molecule. To what molecules do the sulfur bridges attach?

 ASK YOURSELF

What are amino acids, and what function do they perform in the human body?

Figure 22–29. When food supplies are low, some animals, such as these bears, hibernate and live off the fat stored in their bodies.

Slick Globs

What is the difference between vegetable shortening and vegetable oil? Why is one solid and the other liquid? These materials are all lipids. **Lipids** contain atoms of carbon, hydrogen, and oxygen—just as carbohydrates do. Lipids that are solids are called *fats*. Liquid lipids are called *oils*.

Most living organisms produce lipids as a storage fuel. Bears can hibernate through the winter without starving because they draw energy from the layer of fat they have built up over the summer. After your body burns all its sugar for energy, it will draw fat from its reserves.

Figure 22–30. Fats (orange) are stored in the bodies of plants and animals. The reaction (inset) shows how fats are formed.

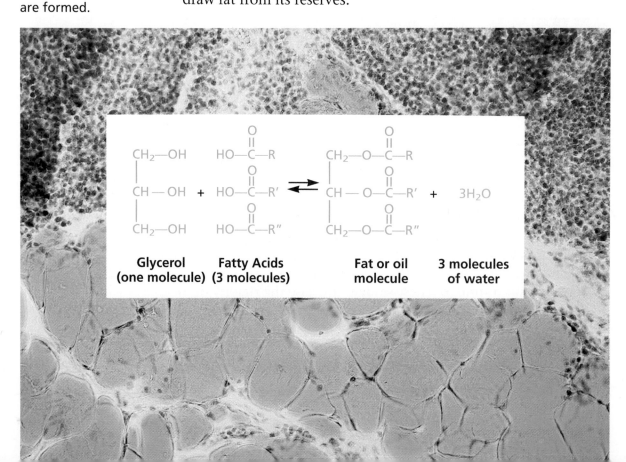

$$
\begin{array}{l}
\text{CH}_2\text{—OH} \quad \text{HO—}\overset{\displaystyle O}{\overset{\|}{\text{C}}}\text{—R} \\
\text{CH—OH} \ + \ \text{HO—}\overset{\displaystyle O}{\overset{\|}{\text{C}}}\text{—R}' \ \rightleftharpoons \ \text{CH—O—}\overset{\displaystyle O}{\overset{\|}{\text{C}}}\text{—R}' \ + \ 3\text{H}_2\text{O} \\
\text{CH}_2\text{—OH} \quad \text{HO—}\overset{\displaystyle O}{\overset{\|}{\text{C}}}\text{—R}''
\end{array}
$$

| Glycerol (one molecule) | Fatty Acids (3 molecules) | Fat or oil molecule | 3 molecules of water |

You have probably heard commercials claiming that cooking with vegetable oil is better than using animal fat. The commercial might even have used the term *polyunsaturated fats*. What are polyunsaturated fats?

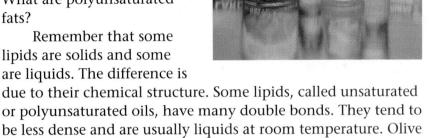

Figure 22–31. Saturated and unsaturated fats, such as those shown here, are available for use in cooking.

Remember that some lipids are solids and some are liquids. The difference is due to their chemical structure. Some lipids, called unsaturated or polyunsaturated oils, have many double bonds. They tend to be less dense and are usually liquids at room temperature. Olive oil, corn oil, safflower oil, and peanut oil are examples of unsaturated lipids.

Saturated lipids do not have double bonds. They have only single bonds and are, therefore, denser than unsaturated ones. Saturated lipids tend to be solids at room temperature. Research has shown that saturated fats contribute to heart disease. The American Heart Association suggests that eating a diet low in saturated fats is better for your health. This means a reduction of most animal fats in your diet.

 ASK YOURSELF

What is the difference between saturated and unsaturated fats?

SECTION 3   *REVIEW AND APPLICATION*

Reading Critically

1. What characteristics do carbohydrates, lipids, and proteins have in common?

2. What molecule is formed when sugars combine to form complex carbohydrates? When fats or oils are formed?

3. Carbohydrates and lipids both contain carbon, hydrogen, and oxygen. How are these molecules different?

Thinking Critically

4. Saturated fats can be responsible for clogging arteries. Based on what you know about the properties of saturated and unsaturated fats, why do you think this is true?

5. Different proteins may contain the same amino acids arranged in a different order. How many different combinations are possible with one unit of each of the following three amino acids: glycine, alanine, and serine? Write the different amino acid sequences.

INVESTIGATION

Testing for Nutrients

▶ MATERIALS

- safety goggles
- laboratory apron • wax pencil • test tubes (12)
- test tube rack • egg white solution • green onion shoots • saltine crackers • graduate, 10 mL • stirring rod • glucose test strips (4)
- iodine solution • sodium hydroxide, 10% • stoppers (4) • copper sulfate, 0.5%

▼ PROCEDURE

CAUTION: Put on safety goggles and a laboratory apron, and leave them on throughout this Investigation.

1. Copy the table shown, and record your observations on it.

TABLE 1: NUTRIENTS IN FOOD			
Substance Tested	**Test results**		
	Glucose test strip	**Iodine**	**NaOH/CuSO₄**
Egg white			
Green onion			
Cracker			
Control			

2. Using a wax pencil, label four test tubes 1–4. Put 5 mL of water in each. Place a small amount of egg white solution in #1, green onion shoots in #2, and a piece of cracker in #3. Leave only water in #4; this is your control.

3. **Test for Sugar** Add 4 mL of water to each test tube, and stir. Use a different glucose test strip for each test tube. Moisten the test strip in the liquid, remove it from the test tube, and note the color after 10 seconds. A violet to purple color indicates the presence of glucose. The darker the color, the more glucose in the mixture.

4. **Test for Starch** Repeat step 2, and test the mixtures with a few drops of iodine solution instead of the glucose strip.

5. **Test for Protein** Repeat step 2, and test the mixtures with 5 mL of sodium hydroxide solution. Stopper each solution, and shake well. Add several drops of copper sulfate solution, shaking well after each drop. Watch for a change of color.

▶ ANALYSES AND CONCLUSIONS

1. Explain the results of the glucose test strip test. Which material(s) tested contained glucose?

2. Iodine solution turns blue-black in the presence of starch. Which substance(s) gave a positive test for starch?

3. What was a positive test for protein? Which material(s) test positive for protein?

▶ APPLICATION

Glucose test strips are used to test urine for sugar. Many diabetics check themselves this way. What advantage would this method give a diabetic?

Discover More

Choose other foods to test for sugar, starch, and protein. Make a chart to show your results. What conclusions can you draw about certain types of foods and their nutrients?

The Big Idea

Differences in structure often mean differences in properties. Structural isomers such as ethyl alcohol and dimethyl ether have very different properties. Lipids can be harmful to your health depending on how many double bonds they contain. Combining amino acids in different orders produce different proteins. Crosslinking in polymers produces a stronger material.

The structure of a molecule affects the function of the molecule. Changing the shape of the molecule, even when the components remain the same, may change the function of the molecule drastically.

For Your Journal

You read about diamonds at the begining of the chapter. Now you have learned more about compounds of carbon. Revise your journal entry to include new ideas about these questions: How can an element that forms diamonds also form living tissue?

Connecting Ideas

Below are some photographs of compounds containing carbon. Look at the photographs, and write a paragraph in your journal that explains how each compound affects your lifestyle.

a. Coal

b. Wood chips

c. Nylon running clothes

REVIEW

Understanding Vocabulary

1. Explain how the terms in each set are related.
   a) isomer (594), polymer (595), monomer (595), structural formula (592)
   b) hydrogen (593), carbon (593), hydrocarbons (593)

2. Explain the similarities and differences among carbohydrates (604), proteins (605), and lipids (608).

Understanding Concepts

MULTIPLE CHOICE

3. All of the following are members of the carbon family except
   a) arsenic.
   b) germanium.
   c) lead.
   d) tin.

4. The maximum number of covalent bonds a carbon atom can form is
   a) one.　　　　b) two.
   c) three.　　　d) four.

5. The focus of most hydrocarbon reactions involves changes in bonds between
   a) hydrogen atoms.
   b) carbon atoms.
   c) oxygen atoms.
   d) nitrogen atoms.

6. The study of carbon compounds in living organisms is called
   a) physics.
   b) organic chemistry.
   c) inorganic chemistry.
   d) biochemistry.

7. Materials that have hydrocarbons and salts at opposite ends of their molecules are called
   a) esters.
   b) natural gas.
   c) petroleum products.
   d) soaps.

8. Which of the following characteristic ratios of elements represents a carbohydrate?
   a) CHO
   b) CHN
   c) CH_2O
   d) CNH_2

9. Silicones are long chains of what types of atoms with different carbon chains attached?
   a) hydrogen and oxygen atoms
   b) silicon and hydrogen atoms
   c) silicon and oxygen atoms
   d) silicon and nitrogen atoms

SHORT ANSWER

10. Explain the difference between the chemical formula and the structural formula of a hydrocarbon.

11. Why does the number of isomers of a carbon compound increase with the number of carbon atoms in the compound?

12. Every gram of fat provides 38 kJ (kilojoules) of energy, as compared to carbohydrates and proteins, which each provide 17 kJ of energy per gram. How many kJ of energy can be supplied by 15 grams of carbohydrates? 15 grams of protein? Why do you think eating too much fat would make you gain weight?

Interpreting Graphics

Classify each pair below as structural isomers or the same substance.

13. CH₃—CH₂
          |
          CH₂—CH3

CH₃—CH₂—CH₂—CH₃

14. CH₃—CH₂—CH₂—CH₂—CH₃

CH₃
    \
     CH—CH₂—CH₃
    /
CH₃

15. CH₃—CH₂—CH = CH₂

CH₃—CH = CH—CH₃

16. CH₃—CH—CH₃
              |
              CH₂
              |
              CH₂
              |
              CH
             / \
          CH₃   CH₃

CH₃   CH₃
    \   /
     CH—CH₂—CH₂—CH
    /   \
CH₃   CH₃

Reviewing Themes

17. *Systems and Structures*
Glucose exists as two enantiomers. (See the Skill Activity on page 603.) One enantiomer can be used to supply energy by your body, but the other cannot. Why do you think this so?

18. *Technology*
If the polyethylene band that holds a six-pack of soda is stretched, it doesn't return to its original shape, as would a rubber band. However, if the polyethylene band is warmed in the sun, it begins to return to its original shape. Explain why this happens.

Thinking Critically

19. In this chapter you have studied organic chemistry. Another area of study is inorganic chemistry. Based on what you know about organic chemistry, what do you think an inorganic chemist studies?

20. One of the most interesting uses of polymers is in artificial body parts. Most artificial body parts, such as artificial knees or hips, contain polyurethane. What properties must a polymer have in order to function as a body part?

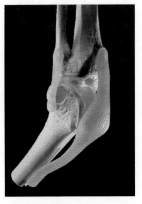

Discovery Through Reading

Edelson, E. "Buckyball: The Magic Molecule." *Popular Science* 239 (August 1991): 52–57+. The discovery, properties, and possible uses of buckminsterfullerenes, a new class of hydrocarbons, are discussed in the article.

Science PARADE

Chemical Communication in Nature

Chemists are constantly trying to discover practical uses for chemicals, such as new medicines and insecticides. Where do chemists get their ideas? Sometimes they just look at nature.

Silent Communications

Animals have developed many complicated ways to communicate using chemicals. These special chemicals are called pheromones (FEER uh mohnz).

Honeybees

Swarm of monarch butterflies

Pheromones can alert animals to danger or let them know where others of their species are located. Another important use of pheromones is to attract a mate. Some pheromones serve as a defense against predators, while other pheromones attract prey.

The pheromones used by honeybees have been widely studied. The queen bee produces a chemical that tells the worker bees there is a queen in the hive and another one is not needed at that time. It also tells the workers to feed substances to the eggs so that the eggs will produce only other workers.

Bees release a substance called *isoamyl acetate* when they sting. Isoamyl acetate is the same substance that gives bananas their characteristic odor. It produces an aggressive behavior in other bees. This explains why, when a person is stung by several bees, the bees tend to sting in the same area. The isoamyl acetate directs the bees to the spot.

Caribbean fruit flies attracted to artificial pheromones

Flies use their pheromones to attract mates. The female gives off the pheromone and the male fly comes to her.

Once the pheromone can be reproduced, it can be used to trap the insects. Sticky traps, sprayed with the pheromone and placed in groves, could attract and hold the flies without the use of insecticides.

Another insect that uses pheromones is the ant. Ants leave chemical trails so that other ants can follow them. If one ant finds a food source, its trail will lead other ants to the food. This is one way in which ants communicate.

Fly Away

Pheromones are combinations of different chemicals in very specific amounts. If one chemical is missing, or if the amounts are not right, the pheromone does not work. Today there is much interest in insect pheromones. Researchers hope to use these pheromones to control insect growth and reproduction without pesticides. The Caribbean fruit fly, for example, is a pest that attacks orange trees. Chemists are trying to make the fly's pheromone in the laboratory.

Ants following pheromone-marked trail

Defend Yourself

Many animals use chemicals other than pheromones for defense. The bombardier beetle, for example, fires a chemical spray that quickly repels its enemies. Certain

Snail defending itself using a pheromone foam

types of millipedes store a compound in their bodies that can quickly produce hydrogen cyanide, a deadly poison. The compound, called *mandelonitrile,* is converted to hydrogen cyanide when the millipede is in danger. The hydrogen cyanide cannot be stored in the millipede's body because the chemical would kill the millipede.

Some animals do not make their own defensive chemicals but can obtain them from other sources. For example, the monarch butterfly eats milkweed plants. A chemical in the plant is stored in the body of the butterfly. When a bird attacks and eats a monarch butterfly, the chemical makes the bird sick. Even though the individual butterfly did not survive the attack, the bird may remember its bad experience. In the future, the bird may avoid eating another monarch butterfly. This

defense has worked so well that most birds avoid any type of butterfly that has the same wing colors as the monarch. This color copying, or mimicry, protects butterflies with similar coloration.

Nature still holds many secrets about chemistry and chemicals. Examples like these can help you understand how nature and chemistry are related. ◆

Bombardier beetle squirting a deadly pheromone at a predator

Steroids Build Muscles and Destroy Health

by A. T. McPhee from *CURRENT SCIENCE*

Eight runners settled into the starting blocks. The runners gazed toward the finish line, 100 meters (110 yards) away, and wondered who would cross the line first. The winner would bring home an Olympic gold medal.

The crowd fell silent. A referee pointed a starting pistol into the air, paused, and fired.

The men grunted. Bodies lurched forward. Feet pounded the track in a blur of blazing speed. A few blinks of an eye later, Canadian track star Ben Johnson raised his right hand in a winner's salute. Mr. Johnson had set a world's record, beaten an arch rival, and won an Olympic gold medal—all in just 9.79 seconds.

Yet, 18 hours later, Mr. Johnson lost all three prizes. Tests done after the race revealed that Mr. Johnson had been taking an anabolic steroid, a drug that might have improved his performance. Olympic rules forbid athletes to take such drugs. So that night, officials disqualified Mr. Johnson

How much extra power did steroids give Ben Johnson, shown here running in the Olympics? Maybe none. Many doctors say steroids don't increase power so much as they drive the athlete to train harder.

617

from further competition, canceled his world's record, and gave the gold medal to his rival, Carl Lewis.

Mr. Johnson's disqualification marked what many experts say is a major problem in sports: steroid abuse. Sports organizations such as the International Olympic Committee are playing tough with athletes who take steroids. These organizations say athletes will pay a price for steroid abuse—a price that for these athletes will mean losing their own good health.

Building Big Muscles, Hot Tempers

Why do athletes risk victory and good health by taking steroids? "Clearly," says Dr. Charles Yesalis, professor of health policy at Pennsylvania State University in University Park, "steroids give the athlete an advantage over people who aren't taking [the steroids]." That advantage may stem in part from the ability of anabolic steroids to increase muscle size and promote aggression.

Anabolic steroids are synthetic substances, usually taken in pill form, that resemble the natural male hormone testosterone. Testosterone promotes facial hair growth, deepens the voice, and increases muscle size.

Most athletes who take anabolic steroids do so for the drugs' effect on muscle size. These athletes insist that bigger muscles make them stronger.

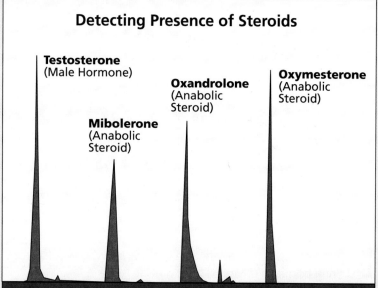

Detecting Presence of Steroids

Testosterone (Male Hormone)

Mibolerone (Anabolic Steroid)

Oxandrolone (Anabolic Steroid)

Oxymesterone (Anabolic Steroid)

This tracing from a urine test shows the presence of three anabolic steroids. A urine test taken from someone who has not used an anabolic steroid would show no such peaks. Ben Johnson's urine test showed he had taken a powerful steroid many times.

Many doctors doubt such claims. Steroids do make muscles larger, the doctors say, but not necessarily stronger.

Studies show that anabolic steroids, taken in moderate doses, increase strength only in women, children, and men with a low level of testosterone. Dr. Glenn Braunstein, professor of medicine at the University of California at Los Angeles, says, "The gains in muscle strength [from steroids] are minimal at best and do not occur in all individuals."

Even if steroids *do* increase strength somewhat, as some athletes and doctors claim, the drugs cause a far greater increase in aggressive behaviors. Dr. Robert Voy, chief medical officer for the U.S. Olympic Committee, says, "People don't realize that athletes on steroids become mean." The problem, say doctors, is that athletes can't leave their aggression on the field.

One teen "strung out" on steroids had a friend videotape him while he slammed his car into a tree. Another steroid-pumping teen threw a wild temper tantrum at a friend's house. The teen smashed his friend's stereo, tossed a TV through a window, ripped a refrigerator door off its hinges, and was finally arrested when he beat up and severely injured a former teammate. Such aggression may help win football games, say doctors, but it wreaks havoc in the athletes' personal lives.

Steroids Halt Growth, Hurt Liver

Perhaps the greatest harm from anabolic steroids comes from their effects on the bones and the liver. Studies show that

Why People Use Steroids...

An athlete uses anabolic steroids in order to gain muscle mass to help him or her excel in a sport. A nonathlete uses anabolic steroids to look good either at the beach or in the gym. Steroids produce many harmful side effects in both males and females. Males may experience scarring acne, hair loss, and "roid rage"—an increase in anger and aggression. Females may notice a deepening of their voice, hair loss, and significant growth of facial hair. Anabolic steroids may also cause liver and heart damage, which can result in death.

long-term steroid use can stop a teen from growing as tall as he or she would grow without steroids.

Bones in the arms and legs grow by producing new bone tissue at areas near the ends of the bones. These areas are called *growth plates*. Growth plates normally stop producing bone tissue during the late teenage years. After the plates stop producing new bone, they close, or *fuse*. This fusion stops further growth at the plates.

Anabolic steroids can fuse growth plates much earlier than normal. Dr. Yesalis explains, "A kid who might grow to be 6 feet 2 inches tall [without steroids] could end up 5 feet 8 inches tall [with steroids]."

Besides stopping growth early, steroids also can cause extensive damage to the liver. Doctors say this damage may lead to liver cancer. "Teenagers don't understand [the dangers of steroid use]," says Armond Colombo, football coach at Brockton (Mass.) High School. "Liver cancer really doesn't frighten them."

Nor do teens seem frightened by other possible side effects of steroid use—abnormal hair loss, kidney disease, high blood pressure, and an inability to have children. Some kids seem bent on taking steroids just to "look good at the beach."

Avoiding Steroid Abuse

World famous Olympic track star Edwin Moses says more frequent testing would help lower the number of athletes now using steroids. New lab tests done at some athletic events can determine whether an athlete took a certain drug days or even weeks before being tested.

Most experts agree that more frequent testing would help curb steroid abuse. They add that adults should be honest with kids. "You can't tell kids that steroids don't work," says Pat Croce, a physical therapist in Philadelphia, Penn., "because they know they do."

Mr. Croce believes kids should take a healthier route to bigger muscles. "If you want to get big," Mr. Croce tells teens, "you have to eat right and sweat hard."

Former professional wrestling champion Ken Ventura understands the value of Mr. Croce's advice. Mr. Ventura took steroids until 1981, when he stopped because "I didn't want [the drugs] to tear my body apart."

Today, Mr. Ventura's message—and the message of experts the world over—is simple: "Don't pump trouble. Stay away from steroids." ◆

This teenager built up his muscles lifting weights. Some other teens take anabolic steroids to build muscles. One recent survey says that nearly one in ten high school seniors has used steroids.

Gladys Emerson (1903-1984)

When Gladys Emerson was growing up in Caldwell, Kansas, she enjoyed listening to and playing music. As she grew older, she found it difficult to choose a career, because, in addition to enjoying the arts, she also liked science. When she went to college, she decided to take classes in both fields. Upon graduation from Oklahoma College for Women, she received two degrees—one in physics and chemistry and the other in English and history. She attended graduate school at Stanford University, where she completed a master's degree in history.

When she was in her twenties, Emerson taught nutrition at the University of California at Berkeley. That position allowed her to earn her Ph.D. in nutrition and biochemistry. It also started her on a road that would take her to the Garvan Medal, which is awarded for distinguished work in chemistry.

Among her many contributions were her work in helping to isolate vitamin E and in studying the chemistry and effects of vitamin B complex. She also studied the effect of diet on tumors. Gladys Emerson helped groups such as the World Health Organization and UNICEF work toward the goal of achieving good nutrition for all citizens. ◆

Patrick Montoya (1962-)

As a young adult, Patrick Montoya wanted to build and design automobiles. Once in college, however, he decided to combine his interest in engineering with classes in biology and chemistry. Now, Montoya's work as a biomedical engineer offers both challenges and rewards. It involves observing the processes of the human body, using computer models to study the way our lungs work, and designing artificial machines to temporarily take the place of organs.

Montoya designs machines that can perform vital functions. Sometimes, especially during an operation, a patient's lungs are not able to transfer oxygen to the blood. When this happens, physicians may use an oxygenator, better known as an artificial lung.

This oxygenator removes blood from the body, exchanges the carbon dioxide in the blood for oxygen, and returns the oxygen-rich blood to the body. Montoya's research has involved designing the pump that drives the blood through the oxygenator. This pump is a type of artificial heart, and together these machines form a heart-lung machine. He uses a computer model to understand better how the transport systems of the human body work and what improvements he should make to his designs. Although Montoya remains interested in automobile design, he has found fulfillment in using his engineering background to help people live longer and more healthful lives. ◆

Annie Johnson, Dietitian

Annie Johnson knows about the nutritional value of school lunches. She is a dietitian, someone who chooses menus and supervises meals, often in hospitals, schools, and other institutions. In the following interview, she talks about what it means to be a dietitian.

What made you choose diet and nutrition as a career? I always enjoyed baking, and when I went to college, I decided to study food and nutrition. When I went to college, we studied everything about nutrition. Now students specialize in a particular field, such as therapeutic nutrition for hospitals. I graduated as a registered dietitian and started working in a hospital.

What does a dietitian do in a hospital? When I worked in the hospital, I helped teach patients about nutrition. I also helped individualize patients' diets according to their physicians' instructions. In the hospital, I always had the feeling I was helping someone.

What is your work like now? Now I supervise school lunch programs in an urban school district in Florida. Every day I'm out in the schools at lunch time. I check all the menus to see that they're balanced nutritionally. I allow cafeteria managers to set their own menus, so there is room for creativity.

How do school cafeteria programs teach students about nutrition? Cafeteria managers go into the classroom to help teach students about nutrition. They let students "taste test" unfamiliar foods. In the cafeteria, we encourage students to try new foods that they may never have at home. Some foods, like beets, aren't favorites, even among adults. I see the school lunch as a learning situation.

What has changed about school lunches since your career began? Years ago school cafeteria food was higher in fat. For example, all vegetables were cooked in butter. Now, vegetables are steamed, and meats are broiled instead of fried in fat. Instead of using fruit canned in heavy syrup, we buy fresh fruit. Now many school cafeterias offer salad bars and low-fat milk.◆

Discover More

For more information about careers for dietitians, write to the

American Dietetic Association
216 W. Jackson Blvd, Suite 800
Chicago, IL 60606

Artificial Body Parts

The human body is a biological machine whose parts may fail or simply wear out from use. Scientists and physicians are learning how to replace the body's parts with artificial organs that extend and improve human life.

Artificial Organs and Joints

People with cataracts have regained their sight through the implantation of lenses in the eye. Artificial knees and hips have enabled arthritis patients to walk. The hearing impaired may soon have their hearing restored by an electronic middle ear. Researchers continue to work on artificial replacements for the body's most complex organs, such as the heart and liver.

Even with artificial joints, the most common implant, scientists want a better match between the body and the artificial part. Artificial joints are usually made of space-age metals and plastics. However, the body's bone and muscle tissues do not easily bond to these foreign materials. Researchers have learned to coat the metal joint with "bioactive" ceramics (the material that clay pots are made of) that interact with body tissues. The ceramics actually stimulate bone growth and help cement the artificial joint in place.

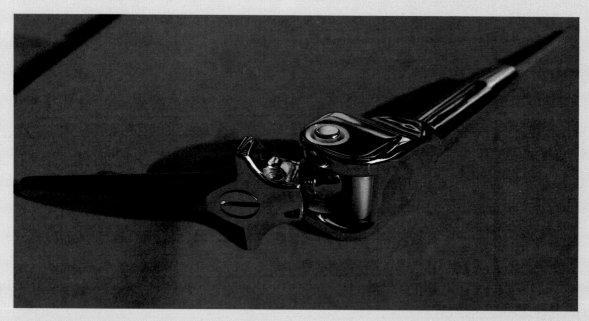

Knee implants can now be custom designed for patients.

Researchers are also developing bioactive polymers (long chains of molecules) that are friendly to the body's tissues. Fibers coated with bioactive polymers serve as artificial ligaments to which muscle tissue will attach. Scientists are experimenting with a gel-like polymer that can be made to contract and relax like muscle.

Saving Lives with Artificial Organs

Newspaper headlines often tell of young patients waiting for kidney or liver transplants. Physicians hope they soon can routinely replace vital organs with artificial substitutes. Researchers continue to test several designs for an artificial heart. The most famous is the Jarvik-7, which has kept patients alive for several weeks.

Researchers are developing artificial cells that mimic the chemistry of the liver and kidneys. The artificial cells contain enzymes that convert the body's toxins into usable amino acids. Diabetics may soon benefit from an artificial pancreas. Scientists take cells that can manufacture insulin from a dog or calf and encase them in an artificial sac. The sac protects the animal cells from the body's rejection but allows glucose to reach the new cells and insulin to flow into the bloodstream.

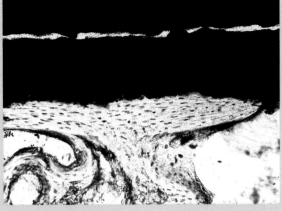

This micrograph shows a bioceramic coating (brown) that encourages bone cells (purple) to grow close to the surface of an artificial hip.

Many challenges remain in creating artificial body parts that effectively mimic the body's complex functions. Scientists are hampered especially by the difficulties of testing on humans. However, some researchers predict that technology may someday make replacement organs that are better than the originals. ◆

Many types of artificial joints have been developed. At far right, a technician checks an artificial hand; center and below respectively, artificial knee and hip joints.

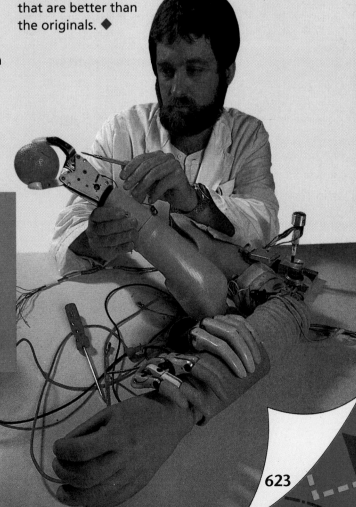

REFERENCE SECTION

SAFETY GUIDELINES

Participating in laboratory investigations should be an enjoyable learning experience. You can ensure both learning and enjoyment from the experience by making the laboratory a safe place in which to work. Carelessness, lack of attention, and showing off are the major causes of laboratory accidents. It is, therefore, important that you follow safety guidelines at all times. If an accident should occur, you should know exactly where to locate emergency equipment. Practicing good safety procedures means being responsible for your classmates' safety as well as your own.

You will be expected to practice the following safety guidelines whenever you are in the laboratory.

1. **Preparation** Study your laboratory assignment in advance. Before beginning your investigation, ask your teacher to explain any procedures you do not understand.
2. **Neatness** Keep work areas clean. Tie back long, loose hair and button or roll up long sleeves when working with chemicals or near an open flame.
3. **Eye Safety** Wear goggles when handling liquid chemicals, using an open flame, or performing any activity that could harm the eyes. If a solution is splashed into the eyes, wash the eyes with plenty of water and notify your teacher at once. Never use reflected sunlight to illuminate a microscope. This practice is dangerous to the eyes.

4. **Chemicals and Other Dangerous Substances** Some chemicals can be dangerous if they are handled carelessly. If any solution is spilled on a work surface, wash the solution off at once with plenty of water.
   - Never taste chemicals or place them near your eyes. Never eat in the laboratory. Counters and glassware may contain substances that can contaminate food. Handle toxic substances in a well-ventilated area or under a ventilation hood.
   - Never pour water into a strong acid or base. The mixture produces heat. Sometimes the heat causes splattering. To keep the mixture cool, pour the acid or base slowly into the water.
   - When noting the odor of chemical substances, wave the fumes

toward your nose with your hand rather than putting your nose close to the source of the odor.

• Do not use flammable substances near a flame.

5. Safety Equipment Know the location of all safety equipment, including fire extinguishers, fire blankets, first-aid kits, eyewash fountains, and emergency showers. Report all accidents and emergencies to your teacher immediately.

6. Heat Whenever possible, use an electric hot plate instead of an open flame. If you must use an open flame, shield the flame with a wire screen that has a ceramic center. When heating chemicals in a test tube, do not point the test tube toward anyone.

7. Electricity Be cautious around electrical wiring. Do not let cords hang loose over a table edge in a way that permits equipment to fall if the cord is tugged. Do not use equipment with frayed cords.

8. Knives Use knives, razor blades, and other sharp instruments with extreme care. Do not use double-edged razor blades in the laboratory.

9. Glassware Examine all glassware before heating. Glass containers for heating should be made of borosilicate glass or some other heat-resistant material. Never use cracked or chipped glassware.

• Never force glass tubing into rubber stoppers.

• Broken glassware should be swept up immediately, never picked up with the fingers. Broken glassware should be discarded in a special container, never into a sink.

10. Unauthorized Experiments Do not perform any experiment that has not been assigned or approved by your teacher. Never work alone in the laboratory.

11. Cleanup Wash your hands immediately after any laboratory activity. Before leaving the laboratory, clean up all work areas. Put away all equipment and supplies. Make sure water, gas, burners, and electric hot plates are turned off.

Remember at all times that a laboratory is a safe place only if you regard laboratory work as serious work.

The instructions for your laboratory investigations will include cautionary statements when necessary. In addition, you will find that the following safety symbols appear whenever a procedure requires extra caution:

 Wear safety goggles

 Biohazard/disease-causing organisms

 Electrical hazard

 Wear laboratory apron

 Flame/heat

 Gloves

 Sharp/pointed object

 Dangerous chemical/poison

 Radioactive material

LABORATORY PROCEDURES

READING A METRIC RULER

1. Examine your metric ruler. The numbers on it represent lengths in centimeters. The usual metric ruler is about 30 cm long. There are 10 marked spaces within each centimeter, which represent tenths of centimeters (0.1 cm).

2. To measure the width of a piece of paper, place the ruler on the paper. The zero end of the ruler must line up exactly with one edge of the paper. Look at the other edge of the paper to see which of the marks on the ruler is closest to that edge. In Figure A, for example, the edge of the paper is nearest to the second line beyond the 7. Therefore, the width of the paper is 7.2 cm.

3. The edge of the paper might fall exactly on one of the centimeter marks. In Figure B, the edge is just on the 5-cm mark. The width of this paper is 5.0 cm. You must write in the .0 to indicate that the measurement is accurate to the nearest tenth of a centimeter; that is, it is more than 4.9 cm and less than 5.1 cm.

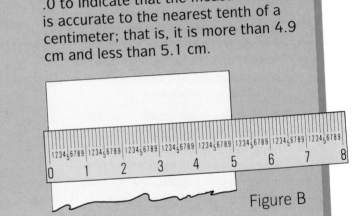

Figure B

4. Sometimes you may want to make a reading with more accuracy. It is possible to estimate readings to the nearest hundredth of a centimeter, but you must be very careful. Look at Figure A again. You can guess the number of tenths in the distance between the marks. The edge of the paper is about 3 tenths of the space between 7.2 and 7.3. The best estimate, then, is that the width of the paper is 7.23 cm.

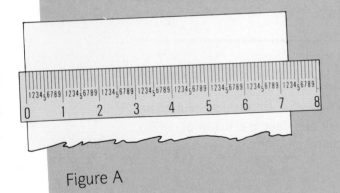

Figure A

5. In Figure C, the edge of the paper falls exactly on the 8.6 mark. If you are taking careful readings, accurate to the nearest hundredth of a centimeter, you must record the width as 8.60 cm.

6. Note the general rule: You can estimate scale readings to the nearest tenth of a scale division. If the scale is marked in tenths, you can estimate the hundredths place but never more than that.

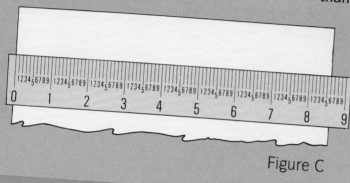

Figure C

CONVERTING SI UNITS

In SI, it is easy to convert from unit to unit. To convert from a larger unit to a smaller unit, move the decimal to the right. To convert from a smaller unit to a larger unit, move the decimal to the left. Figure D shows you how to move the decimals to convert in SI.

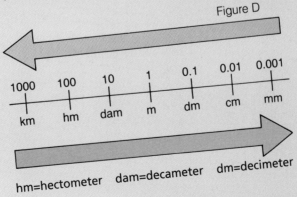

Figure D

SI Conversion Table

SI Units		Converting SI to Customary		Converting Customary to SI	
Length		1 km	= 0.62 mile	1 mile	= 1.609 km
kilometer (km)	= 1000 m	1 m	= 1.09 yards	1 yard	= 0.914 m
meter (m)	= 100 cm		= 3.28 feet	1 foot	= 0.305 m
		1 cm	= 0.394 inch	1 foot	= 30.5 cm
centimeter (cm)	= 0.01 m	1 mm	= 0.039 inch	1 inch	= 2.54 cm
millimeter (mm)	= 0.001 m				
micrometer (μm)	= 0.000 001 m				
nanometer (nm)	= 0.000 000 001 m				
Area		1 km²	= 0.3861 square mile	1 square mile	= 2.590 km²
square kilometer (km²)	= 100 hectares	1 ha	= 2.471 acres	1 acre	= 0.4047 ha
hectare (ha)	= 10 000 m²	1 m²	= 1.1960 square yards	1 square yard	= 0.8361 m²
square meter (m²)	= 10 000 cm²			1 square foot	= 0.0929 m²
		1 cm²	= 0.155 square inch	1 square inch	= 6.4516 cm²
square centimeter (cm²)	= 100 mm²				
Mass		1 kg	= 2.205 pounds	1 pound	= 0.4536 kg
kilogram (kg)	= 1000 g	1 g	= 0.0353 ounce	1 ounce	= 28.35 g
gram (g)	= 1000 mg				
milligram (mg)	= 0.001 g				
microgram (μg)	= 0.000 001 g				
Volume of Solids		1 m³	= 1.3080 cubic yards	1 cubic yard	= 0.7646 m³
1 cubic meter (m³)	= 1 000 000 cm³		= 35.315 cubic feet	1 cubic foot	= 0.0283 m³
		1 cm³	= 0.0610 cubic inch	1 cubic inch	= 16.387 cm³
1 cubic centimeter (cm³)	= 1000 mm³				
Volume of Liquids		1 kL	= 264.17 gallons	1 gallon	= 3.785 L
kiloliter (kL)	= 1000 L	1 L	= 1.06 quarts	1 quart	= 0.94 L
liter (L)	= 1000 mL	1 mL	= 0.034 fluid ounce	1 pint	= 0.47 L
milliliter (mL)	= 0.001 L			1 fluid ounce	= 29.57 mL
microliter (μL)	= 0.000 001 L				

READING A GRADUATE

1. Examine the graduate and note how the scale is marked. The units are milliliters (mL). A milliliter is a thousandth of a liter and is equal to a cubic centimeter. Note carefully how many milliliters are represented by each scale division on the graduate.

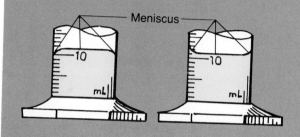

2. Pour some liquid into the cylinder and set the cylinder on a level surface. Notice that the upper surface of the liquid is flat in the center and curved at the edges. This curve is called the *meniscus* and may be either upward or downward. In reading the volume, you must ignore the curvature and read the scale at the flat part of the surface.

3. Bring your eye to the level of the surface and read the scale at the level of the flat surface of the liquid.

USING A LABORATORY BALANCE

1. Make sure the balance is on a level surface. Use the leveling screws at the bottom of the balance to make any necessary adjustments.

2. Place all the countermasses at zero. The pointer should be at zero. If it is not, adjust the balancing knob until the pointer rests at zero.

3. Place the object you wish to mass on the pan. **CAUTION: Do not place hot objects or chemicals directly on the balance pan, because they can damage its surface.**

4. Move the largest countermass along the beam to the right until it is at the last notch that does not tip the balance. Follow the same procedure with the next largest countermass. Then move the smallest countermass until the pointer rests at zero.

5. Determine the readings on all beams and add them together to determine the mass of the object.

6. When massing crystals or powders, use a piece of filter paper. First, mass the paper; then add the crystals or powders and remass. The actual mass is the total minus the mass of the paper. When massing liquids, first mass the empty container, then mass the liquid and container. Finally, subtract the mass of the container from the mass of the liquid and the container to get the mass of the liquid.

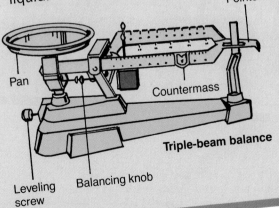

Triple-beam balance

LABORATORY PROCEDURES

USING A BUNSEN BURNER

1. Before lighting the burner, observe the locations of fire extinguishers, fire blankets, and sand buckets. Wear safety goggles and an apron. Tie back long hair and roll up long sleeves.
2. Close the air ports of the burner and turn the gas full on by using the valve at the laboratory outlet.

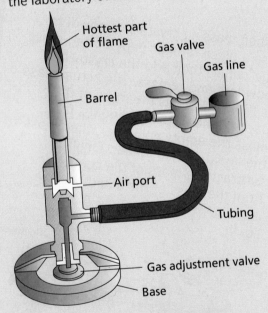

Hottest part of flame
Gas valve
Gas line
Barrel
Air port
Tubing
Gas adjustment valve
Base

3. Hold the striker in such a position that the spark will be just above the rim of the burner. Strike a spark.
4. Open the air ports until you can see a blue cone inside the flame. If you hear a roaring sound, the ports are open too wide.
5. **CAUTION: If the burner is not operating properly, the flame may burn inside the base of the barrel. Carbon monoxide, an odorless gas, is released from this type of flame. Should this situation occur, immediately turn off the gas at the laboratory gas valve. Do not touch the barrel of the burner.** After the barrel has cooled, partially close the air ports before relighting the burner.
6. Adjust the gas-flow valve and the air ports on the burner until you get a flame of the desired size with a blue cone. The hottest part of the flame is just above the tip of the blue cone.

FILTERING TECHNIQUES

1. To separate a precipitate from a solution, pass the mixture through filter paper. To do this, first obtain a glass funnel and a piece of filter paper.

2. Fold the filter paper in fourths as shown. Then open up one fourth of the folded paper. Put the paper, pointed end down, into the funnel.

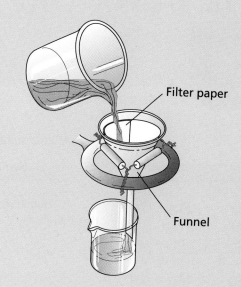

Filter paper

Funnel

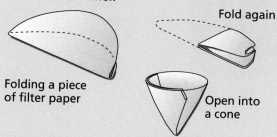

Folding a piece of filter paper

Fold again

Open into a cone

3. Support the funnel, and insert its stem into a beaker.

4. Stir the mixture to be filtered and pour it quickly into the filter paper within the funnel. Wait until all the liquid has flowed through.

5. You may wish to wash the solid that is left in the filter paper. If this is the case, pour some distilled water into the filter paper. Your teacher will tell you how much water to use.

USING REAGENTS

1. For safety reasons, it is important to learn how to pour a reagent from a bottle into a flask or a beaker. Begin with the reagent bottle on the table.

2. While holding the bottle steady with your left hand, grasp the stopper of the bottle between the first and second fingers of your right hand. Remove the stopper from the bottle. **DO NOT** put the stopper down on the table top.

3. While still holding the stopper, lift the bottle with your right hand and pour the reagent into your container.

4. Replace the bottle on the table top and replace the stopper.

If you are left handed, reverse the instructions.

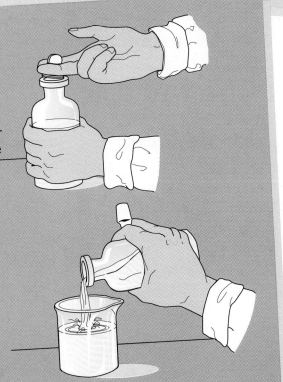

THE PERIODIC TABLE

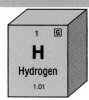

1	G
H	
Hydrogen	
1.01	

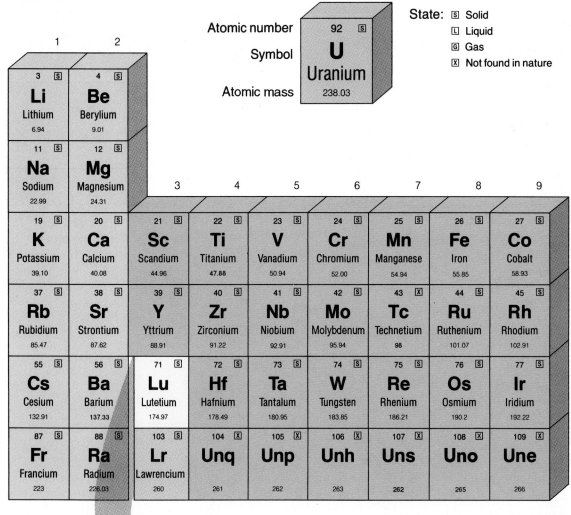

Atomic number

Symbol

Atomic mass

92	S
U	
Uranium	
238.03	

State:
- ⒮ Solid
- ⒧ Liquid
- ⒢ Gas
- ⓧ Not found in nature

	1		2		3		4		5		6		7		8		9	

3 S	4 S
Li	**Be**
Lithium	Berylium
6.94	9.01

11 S	12 S
Na	**Mg**
Sodium	Magnesium
22.99	24.31

19 S	20 S	21 S	22 S	23 S	24 S	25 S	26 S	27 S
K	**Ca**	**Sc**	**Ti**	**V**	**Cr**	**Mn**	**Fe**	**Co**
Potassium	Calcium	Scandium	Titanium	Vanadium	Chromium	Manganese	Iron	Cobalt
39.10	40.08	44.96	47.88	50.94	52.00	54.94	55.85	58.93

37 S	38 S	39 S	40 S	41 S	42 S	43 X	44 S	45 S
Rb	**Sr**	**Y**	**Zr**	**Nb**	**Mo**	**Tc**	**Ru**	**Rh**
Rubidium	Strontium	Yttrium	Zirconium	Niobium	Molybdenum	Technetium	Ruthenium	Rhodium
85.47	87.62	88.91	91.22	92.91	95.94	98	101.07	102.91

55 S	56 S	71 S	72 S	73 S	74 S	75 S	76 S	77 S
Cs	**Ba**	**Lu**	**Hf**	**Ta**	**W**	**Re**	**Os**	**Ir**
Cesium	Barium	Lutetium	Hafnium	Tantalum	Tungsten	Rhenium	Osmium	Iridium
132.91	137.33	174.97	178.49	180.95	183.85	186.21	190.2	192.22

87 S	88 S	103 S	104 X	105 X	106 X	107 X	108 X	109 X
Fr	**Ra**	**Lr**	**Unq**	**Unp**	**Unh**	**Uns**	**Uno**	**Une**
Francium	Radium	Lawrencium						
223	226.03	260	261	262	263	262	265	266

57 S	58 S	59 S	60 S	61 X	62 S
La	**Ce**	**Pr**	**Nd**	**Pm**	**Sm**
Lanthanum	Cerium	Praseodymium	Neodymium	Promethium	Samarium
138.91	140.12	140.91	144.24	145	150.4

89 S	90 S	91 S	92 S	93 X	94 X
Ac	**Th**	**Pa**	**U**	**Np**	**Pu**
Actinium	Thorium	Protactinium	Uranium	Neptunium	Plutonium
227.03	232.04	231.04	238,03	237.05	244

Metals

Transition Metals

Nonmetals

Noble gases

Lanthanide series

Actinide series

18					
					2 G **He** Helium 4.00
13	14	15	16	17	

13	14	15	16	17	18
5 S **B** Boron 10.81	6 S **C** Carbon 12.01	7 G **N** Nitrogen 14.01	8 G **O** Oxygen 16.00	9 G **F** Fluorine 19.00	10 G **Ne** Neon 20.18
13 S **Al** Aluminum 26.98	14 S **Si** Silicon 28.09	15 S **P** Phosphorus 30.97	16 S **S** Sulfur 32.07	17 G **Cl** Chlorine 35.45	18 G **Ar** Argon 39.95

10	11	12					
28 S **Ni** Nickel 58.69	29 S **Cu** Copper 63.55	30 S **Zn** Zinc 65.39	31 S **Ga** Gallium 69.72	32 S **Ge** Germanium 72.61	33 S **As** Arsenic 74.92	34 S **Se** Selenium 78.96	35 L **Br** Bromine 79.90	36 G **Kr** Krypton 83.80
46 S **Pd** Palladium 106.42	47 S **Ag** Silver 107.87	48 S **Cd** Cadmium 112.41	49 S **In** Indium 114.82	50 S **Sn** Tin 118.71	51 S **Sb** Antimony 121.75	52 S **Te** Tellurium 127.60	53 S **I** Iodine 126.90	54 G **Xe** Xenon 131.29
78 S **Pt** Platinum 195.08	79 S **Au** Gold 196.97	80 L **Hg** Mercury 200.59	81 S **Tl** Thallium 204.38	82 S **Pb** Lead 207.2	83 S **Bi** Bismuth 208.98	84 S **Po** Polonium 209	85 S **At** Astatine 210	86 G **Rn** Radon 222

63 S **Eu** Europium 151.96	64 S **Gd** Gadolinium 157.25	65 S **Tb** Terbium 158.93	66 S **Dy** Dysprosium 162.50	67 S **Ho** Holmium 164.93	68 S **Er** Erbium 167.26	69 S **Tm** Thulium 168.93	70 S **Yb** Ytterbium 173.04
95 X **Am** Americium 243	96 X **Cm** Curium 247	97 X **Bk** Berkelium 247	98 X **Cf** Californium 251	99 X **Es** Einsteinium 252	100 X **Fm** Fermium 257	101 X **Md** Mendelevium 258	102 X **No** Nobelium 259

ELECTRIC CIRCUIT SYMBOLS

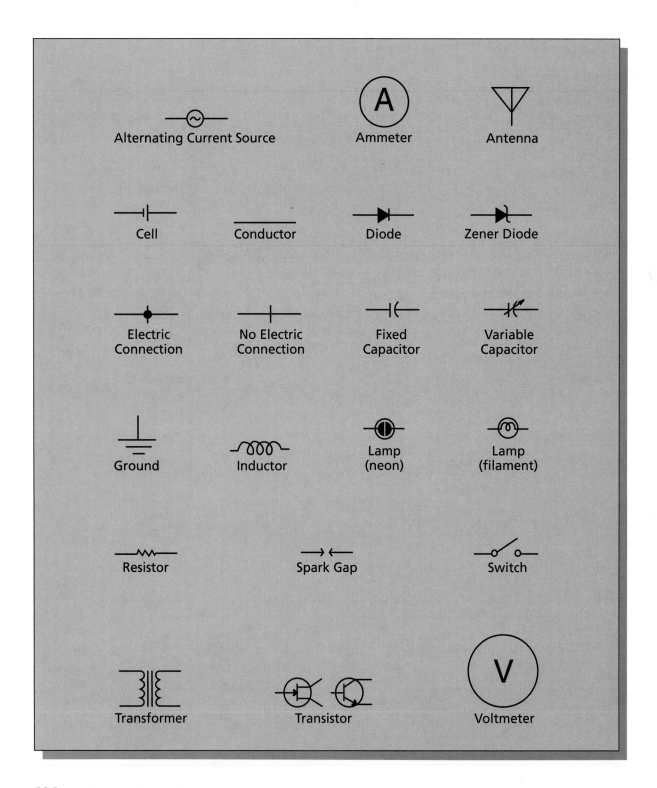

Alternating Current Source

Ammeter

Antenna

Cell

Conductor

Diode

Zener Diode

Electric Connection

No Electric Connection

Fixed Capacitor

Variable Capacitor

Ground

Inductor

Lamp (neon)

Lamp (filament)

Resistor

Spark Gap

Switch

Transformer

Transistor

Voltmeter

PHYSICAL SCIENCE EQUATIONS

$$\text{density} = \frac{\text{mass}}{\text{volume}}$$

$$d = \frac{m}{V}$$

$$\text{mechanical advantage} = \frac{\text{force output}}{\text{force input}}$$

$$\text{MA} = \frac{F_o}{F_i}$$

$$\text{acceleration} = \frac{\text{distance}}{\text{time}^2}$$

$$a = \frac{d}{t^2}$$

$$\text{kinetic energy} = \frac{\text{mass} \times \text{speed}^2}{2}$$

$$\text{KE} = \frac{1}{2}mv^2$$

$$\text{acceleration} = \frac{\text{force}}{\text{mass}}$$

$$a = \frac{F}{m}$$

$$\text{average speed} = \frac{\text{total distance}}{\text{total time}}$$

$$v = \frac{d}{t}$$

$$\text{acceleration} = \frac{\text{final velocity} - \text{initial velocity}}{\text{time}}$$

$$a = \frac{v_f - v_i}{t}$$

$$\text{power} = \frac{\text{work}}{\text{time to do work}}$$

$$P = \frac{W}{t}$$

$$\text{work} = \text{force} \times \text{distance}$$
$$W = Fd$$

$$\text{power} = \text{current} \times \text{potential difference}$$
$$P = IV$$

$$\text{wave speed} = \text{wavelength} \times \text{frequency}$$
$$v = \lambda f$$

GLOSSARY

Acceleration

A

acceleration the rate at which velocity changes **(75)**

acid a compound that produces hydrogen ions (H⁺) in solution **(567)**

acid rain rain formed when air pollutants (sulfur oxides and nitrogen oxides) combine with oxygen and water vapor in the atmosphere to form weak sulfuric and nitric acids **(182)**

acoustics the branch of physics that deals with the transmission of sound **(252)**

active solar heating complex solar heating systems; method of heating that uses mechanical devices such as fans to move heat from one place to another **(161)**

allotropes one of two or more molecular forms of the same element; graphite and diamond are allotropes of carbon **(472)**

alloy mixture of two or more metals, or a mixture of metals and nonmetals **(459)**

alpha decay type of nuclear decay in which an atom emits an alpha particle (a helium nucleus), which is made up of two protons and two neutrons **(485)**

alternating current (AC) electric current in which the electrons repeatedly change direction rather than flowing in a single direction **(350)**

amplitude the height of a wave, or the distance a wave moves from a rest position **(220)**

analog something that represents something else; an audio signal is a series of electric impulses and is an analog of a sound wave **(408)**

atomic mass the sum of the relative masses of the protons and neutrons in an atom **(439)**

atomic number the number of protons in the nucleus of an element **(440)**

Allotropes

B

base a compound that produces hydroxide ions (OH⁻) in solution **(573)**

beta decay type of nuclear decay in which an atom emits a beta particle, which is formed when a neutron is converted into a proton and an electron **(485)**

Base

C

carbohydrates compounds made of carbon, hydrogen, and oxygen in a ratio of CH_2O **(604)**

chain reaction a reaction that occurs over and over as the result of a single beginning action **(497)**

characteristic property a property that always stays the same and is characteristic of a particular kind of matter **(41)**

chemical bond the force that holds the atoms in a compound together **(542)**

chemical equation a list of the compounds or elements involved in a reaction and the new compounds or elements that form as a result **(553)**

Circuit breaker

circuit breaker a device that protects circuits from overloads **(405)**

coherent light the type of light created when all the light waves in a beam are in step **(313)**

colloid a mixture in which the particles are so small that they stay in solution **(533)**

compound a pure substance that has molecules made of two or more different kinds of elements that have been combined chemically **(35; 520)**

compound machines two or more simple machines combined to do work **(115)**

compressions the areas where the particles in a longitudinal wave are close together **(219)**

concave lens a lens that is thinner in the middle than it is at the edges; it causes light rays to diverge **(268)**

conclusion a judgment based on the analysis of data gathered from experiments or field studies **(8)**

conduction the transfer of heat energy from one substance to another by direct contact **(142)**

conductor a material that readily allows the transfer of heat or the flow of electric current **(143; 355)**

conservation the careful and efficient use of resources **(188)**

convection the transfer of heat in liquids and gases as groups of molecules move in currents **(144)**

convex lens a lens that is thicker in the middle than it is at the edges; it causes light rays to converge **(268)**

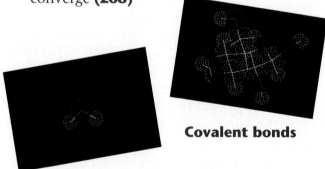

Covalent bonds

covalent bond a bond formed between two or more atoms in which the valence electrons are shared **(546)**

crests the high points of a transverse wave **(217)**

D

decibels (dB) unit used to measure the volume of sound **(246)**

decomposition reaction a reaction in which a compound is broken down into two or more substances **(556)**

density the ratio of the mass of an object to its volume **(24)**

diatomic molecules molecules composed of two atoms **(470)**

digital signals a series of electric pulses that stand for two digits, 0 and 1 **(409)**

direct current (DC) electric current in which the electrons flow steadily in one direction **(350)**

Doppler effect a change in the frequency of waves caused by a moving wave source or a moving observer **(252)**

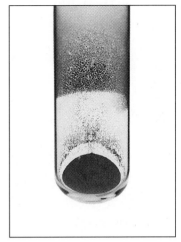

Decomposition reaction

E

electric circuit the path electricity follows **(355)**

electric current flow of electrons through a conductor **(348)**

electric motor a device that converts electrical energy into mechanical energy **(394)**

electromagnet a strong magnet created when a solenoid with an iron core has electric current passed through it **(381)**

electromagnetic induction the process by which a moving magnetic field produces an electric current in a conductor **(383)**

electromagnetic spectrum all types of electromagnetic waves **(285)**

electromagnetic waves transverse waves that carry both electric and magnetic energy **(284)**

electron the negatively charged particle found in atoms **(432)**

element a pure substance that contains only one kind of atom **(34)**

energy levels orbits in which electrons circle the nucleus **(435)**

excited state the state of an atom in which it contains stored energy **(314)**

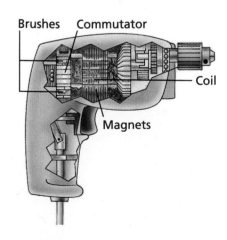

Brushes Commutator

Coil

Magnets

Electric motor

F

focal length the distance from the lens to the focus **(268)**

force a push or a pull **(78)**

fossil fuels fuels formed from plant and animal material that was buried in sediment millions of years ago; coal and petroleum **(153)**

frequency the number of vibrations a wave produces each second **(221)**

friction the force that opposes motion between surfaces that touch **(81)**

fuse an electric safety device containing a short strip of metal with a low melting point **(404)**

G

gamma decay type of nuclear decay in which an atom emits high-energy radiation similar to X-rays **(486)**

Geiger counter a hand-held instrument that measures the level of radiation **(490)**

geothermal energy natural heat within the earth **(166)**

greenhouse effect the increase in the temperature of Earth due to the increase of carbon dioxide in the atmosphere **(179)**

ground state the state an atom returns to as it gives off the energy gained to reach an excited state **(314)**

groups sets of elements arranged in vertical columns on the periodic table **(449)**

Geothermal Energy

H

heat the measure of the total kinetic energy of the random motion of the atoms and molecules of a substance **(139)**

hologram an interference pattern created by a laser on a piece of film **(321)**

hydrocarbons compounds that are made of only carbon and hydrogen **(593)**

hypothesis an educated guess about what might happen **(5)**

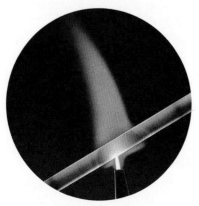

Hydrocarbon

inclined plane   a simple machine consisting of a flat, sloping surface; a ramp **(114)**

indicator   a substance that changes color in an acid or a base **(578)**

inertia   the resistance any object has to a change in velocity **(88)**

insulator   a material that is a poor conductor of heat or electric current **(143; 355)**

ion   an atom or group of atoms that has an electric charge as the result of losing or gaining electrons **(459)**

ionic bond   a chemical bond resulting from the attraction between a positive and a negative ion **(544)**

isomer   one of two or more compounds that have the same chemical formula but different structures **(594)**

isotopes   atoms of the same element that have different numbers of neutrons **(441)**

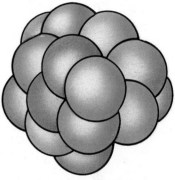

Isotope, C–14

joule (J)   the amount of work equal to the force of 1 N moving an object a distance of 1 m **(104)**

kinetic energy   the energy an object has due to its mass and its motion **(127)**

L

larynx the organ that produces the human voice; the small box made of cartilage located in the front of your neck **(254)**

law a summary of many experimental results and observations **(9)**

law of conservation of momentum the law that states that momentum can be transferred from one object to another but cannot change in total amount **(92)**

lever a simple machine; a bar used for prying or dislodging something **(110)**

light pipe a clear plastic rod that can be bent into a curved shape to carry light into otherwise unreachable places **(322)**

lipid solid fat containing atoms of carbon, hydrogen, and oxygen **(608)**

longitudinal wave a wave in which the particles move parallel to the path of the wave **(219)**

lux a unit of measure for the brightness of light **(263)**

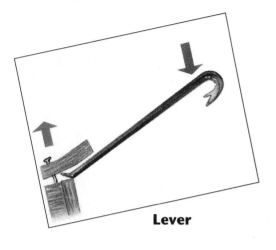

Lever

M

machine a device that helps to do work by changing the size or direction of a force **(101)**

magnetic field any region in which magnetic forces are present **(375)**

magnetism a force of repulsion or attraction between like or unlike poles **(368)**

mass the measure of the amount of matter in an object **(21)**

mass number the whole number nearest the atomic mass of an atom; its total protons + neutrons **(439)**

mechanical advantage (MA) the number of times a machine multiplies force; the ratio of the force that comes out of a machine to the force that is put into the same machine **(106)**

metalloids substances that exhibit some, but not all, of the properties of metals **(462)**

mixture a combination of two or more different substances that are not combined chemically **(36)**

momentum an object's mass multiplied by its velocity **(77)**

monomers the small chemical units that make up a polymer **(595)**

N

neutron a particle found in the nucleus of an atom that has no charge **(434)**

newton the force required to move 1 kg of mass a distance of one meter per second each second **(79)**

noise pollution noise that is harmful to human health **(246)**

nonrenewable resources resources that cannot be replaced after they are used **(153)**

nuclear reactors complex devices that are used to convert nuclear energy into heat energy **(170)**

nuclear waste spent fuel and other radioactive products produced by nuclear reactions **(185)**

Nuclear reactor

O

optical fiber a long, thin light pipe that uses total internal reflection to carry light **(323)**

P

parallel circuit an electric circuit that has two or more separate paths through which electricity can flow **(359)**

passive solar heating simple solar heating systems; heating as a result of the simple absorption of solar energy **(160)**

period a row of elements in the periodic table **(445)**

periodic table the organization of elements by atomic number into a table; similar elements are grouped in rows (periods) and columns (groups) **(445)**

phosphors the chemicals that are used to coat the inside of TV screens; when struck by electrons they glow to produce a picture **(411)**

physical change a change in matter that does not produce a new kind of molecule **(43)**

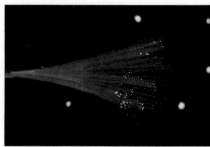

Optical fibers

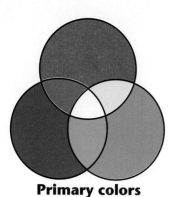
Primary colors

Protein

physical property   a property of matter that can be observed or measured without changing the composition of the matter **(42)**

pitch   the highness or lowness of a sound; determined by frequency **(245)**

plasma   a phase of matter consisting of a low-density, super-heated gas in which the electrons and nuclei are free to move at random **(499)**

polymers   large molecules that are made by hooking many identical small molecules together **(595)**

potential difference   a value expressed in volts related to electrical potential energy that is measured by a voltmeter **(352)**

potential energy   the energy stored in an object due to its position **(125)**

power   the measure of the amount of work done in a certain period of time **(102)**

power grid   a network of electric transmission lines that provide electricity to homes and other buildings **(391)**

primary colors   red, green, and blue; when combined these three colors can produce any other color **(302)**

protein   huge molecules made up of amino acids; needed for tissue growth and repair **(605)**

proton   the positive particle found in the nucleus of an atom **(434)**

pulley   a simple machine that consists of a wheel that is free to spin on an axle **(112)**

pumping   the process of raising atoms to an excited state during the production of a laser beam **(314)**

pure substance   matter that contains only one kind of molecule **(34)**

R

radiation   the transfer of energy by electro-magnetic waves **(145)**

radioactive   the condition in which the nuclei of unstable atoms spontaneously break apart, giving off particles or high-energy radiation **(483)**

rarefactions   the areas where the particles in a longitudinal wave are far apart **(219)**

real image   the image created when a collection of points of light is focused by a lens **(271)**

refinery a large industrial plant that separates crude oil into products such as gasoline, diesel fuel, heating fuel, petroleum jelly, and asphalt **(155)**

reflection the bouncing back of a wave after it strikes a barrier **(226)**

refraction the change of direction when a wave enters a different medium **(228)**

replacement reaction a reaction in which one atom or group of atoms in a compound is replaced with another atom or group of atoms **(557)**

resistance a measure of how much a substance opposes the flow of electricity; the ratio of potential difference to electric current **(356)**

reverberation the combination of many small echoes occurring very close together; caused when reflected sound waves meet at a single point, one right after another **(252)**

Refinery

saturated solution a solution in which the maximum amount of solute is dissolved in a given amount of solvent **(526)**

screw a simple machine composed of an inclined plane wrapped around a cylinder **(115)**

secondary colors yellow, cyan, and magenta; colors formed by combining two primary colors **(302)**

series circuit an electric circuit in which there is only one path for the current to follow **(357)**

short circuit any accidental connection that allows electric current to take an unintended path instead of passing through an appliance **(402)**

simple machines machines that have only one or two parts; the lever, the wheel and axle, the pulley, the inclined plane, the wedge, and the screw **(109)**

solar energy energy from the sun **(159)**

solenoid a wire coil through which electric current can flow **(381)**

solubility the maximum amount of solute that can be dissolved in 100 g of solvent at a specific temperature and pressure **(527)**

solute the substance in a solution that is dissolved in the solvent **(523)**

solution a homogeneous mixture **(523)**

Solar energy

Static electricity

solvent the material in a solution in which the solute is dissolved **(523)**

sonar a navigation system that uses echoes of ultrasonic waves to find the depth of water; an acronym for **so**und **na**vigation **r**anging **(247)**

speed the measure of how fast an object moves over a certain distance in a specific time **(72)**

standard units units that are agreed upon and used by large numbers of people **(15)**

standing waves waves that form a pattern where portions of the waves do not move and other portions move with increased amplitude **(232)**

static electricity a buildup of an electric charge on an object **(343)**

stimulated emission the process of forcing identical atoms into step **(315)**

structural formula a simplified drawing that shows the pattern in which atoms are bonded in a molecule **(592)**

sublimation a phase change in which a solid changes directly into a gas **(45)**

suspension a heterogeneous mixture in which one of the parts is a liquid **(531)**

synthesis reaction a chemical reaction in which two or more substances combine to form one new substance **(555)**

T

telecommunications the science and technology of sending messages over long distances **(297)**

temperature the measure of the average kinetic energy of the moving atoms and molecules of a substance **(140)**

theory an explanation of why things work the way they do **(9)**

thermal expansion a physical change that occurs when the volume of a substance increases as the temperature increases **(49)**

total internal reflection in an optical fiber, the complete reflection of light from the inside surface; in this way, light waves travel long distances through the fibers **(322)**

Telecommunications

transformer a device that uses electromagnetic induction to change the electric potential (voltage) of alternating current **(397)**

transverse wave the wave created when the particles within the wave move perpendicular to the path of the wave **(218)**

troughs depressions in a transverse wave **(217)**

velocity the speed of an object in a particular direction **(74)**

virtual image an image that forms only because the light seems to diverge from it **(272)**

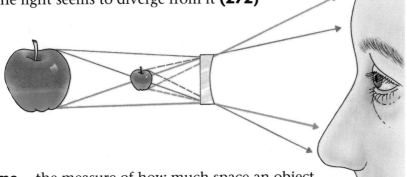

volume the measure of how much space an object takes up; the softness or loudness of a sound **(20; 244)**

W

wave a disturbance that travels through matter or space **(216)**

wavelength the distance between one point on one wave and the identical point on the next wave **(221)**

wedge a simple machine consisting of an inclined plane; used to pry objects apart **(114)**

weight the force of gravity pulling one object toward the center of another object **(85)**

wheel and axle a simple machine consisting of a lever connected to a shaft **(111)**

wind turbines machines that convert the energy of wind into electricity **(163)**

work the measure of the force required to make something move **(101)**

Wind turbines

INDEX

Boldface numbers refer to an illustration on that page.

CREDITS

Art Library/SuperStock; 369(bl), Joseph Palmieri/Uniphoto; 369(br), Science Museum, London/Bridgeman Art Library/SuperStock; 370(br), E. R. Degginger; 371, William E. Ferguson; 372(bl), Dan McCoy/Rainbow; 372(bc), William E. Ferguson; 375(br), Culver Pictures, Inc.; 376, William E. Ferguson; 377(t), Richard Megna/Fundamental Photographs; 377(b), E. R. Degginger; 381(tr), E. R. Degginger; 381(bc), Westinghouse Electric Corporation Research & Development Center; 381(bl), Westinghouse Electric Corporation Research & Development Center; 382, Mike Small/Photri; 383, William E. Ferguson; 385, HRW Photo by Michelle Bridwell; 387, Cara Moore/The Image Bank; 388, Bob Gomel/LIFE Magazine © 1965 Time Warner, Inc .; 390, NOAA; 391, Joe Bator/The Stock Market; 392, E. R. Degginger; 393, Aldo Mastrocola/Lightwave; 394, SuperStock; 397(l), James Kamp/Black Star; 397(r), William E. Ferguson; 398, Tony Freeman/PhotoEdit; 399, The Bettmann Archive; 401(r), Claude Charlier/Black Star; 401(l), James Kamp/Black Star; 402, William E. Ferguson; 404, Aldo Mastrocola/Lightwave; 405, Tom Stack & Associates; 409, William E. Ferguson; 410(t), Tony Freeman/PhotoEdit; 410(b), Frank P. Rossotto/The Stock Market; 412(l), HBC Photo by Jerry White; 412(inset), HBC Photo by Oscar Burtrago; 413, Bob Gomel/LIFE Magazine © 1965, Time Warner, Inc.; 415(l), J. P. Laffont/Sygma; 415(r), J. P. Laffont/Sygma; 416, Peter Angelo Simon/Phototake; 419(all), Peter Menzel; 421(t), U.S. Department of the Interior, National Park Service, Edison National Historic Site; 421(b), Steve Hansen/TIME Magazine; 422(both), Jonathan A. Meyers; 423(tr), Reginald C. Fortier/Francis Bitter National Magnet Laboratory/MIT; 423(c), Bill Pierce/Rainbow; 423(bl), Kyodo News Service; 424-425, Chuck Davis; 426, photo courtesy of the Collection of the J. Paul Getty Museum, Malibu, California; 440, CNRI/Science Photo Library/Photo Researchers; 445, Brown Brothers; 448(l), HBC Photo; 448(r), Library of Congress Collection, HRW Photo by Eric Beggs; 451, photo courtesy of the Colllection of the J. Paul Getty Museum, Malibu, California; 454, 455, Used by permission of DC Comics; 456(l), Comstock; 460(tr)(bc)(bl), Used by permission of DC Comics; 460(tl), David Falconer/After Image/Tony Stone Images; 461, E. R. Degginger; 463(tl), M. Serraillier/Photo Researchers; 463(bl), Weinberg/Clark/

The Image Bank; 463(tc), Dailloux/Rapho/Photo Researchers; 463(br), Bill Nation/Sygma; 463(tr), Frank Wing/The Image Bank; 467(both), E. R. Degginger; 468(b), Brown Brothers; 468(t), J. Imber/H. Armstrong Roberts; 469, Ginger Chih/Peter Arnold, Inc.; 471(br), E. R. Degginger; 471(t), Dr. Leland C. Clark, Jr.; 472(tr), Rich Treptow/Visuals Unlimited; 472(tl), E. R. Degginger; 473, NASA/Science Source/Photo Researchers, Inc.; 474, Park Street; 475(t), E. R. Degginger/Bruce Coleman, Inc.; 475(b), Leonard Lee Rue III/Bruce Coleman, Inc.; 475(c), E. R. Degginger/Bruce Coleman, Inc.; 477(t), Used by permission of DC Comics; 477(cl), Yoav Levy/Phototake; 477(cr), Rich Treptow/Photo Researchers; 477(br), Yoav Levy/Phototake; 477(bl), Rich Treptow/Photo Researchers; 479 (r), Anthony Mercieca/Photo Researchers; 479(l), Brenda L. Lewison/The Stock Market; 480, Los Alamos National Laboratory; 481(tr), Guy Fleming/courtesy of Lansing Lamont, author of DAY OF TRINITY; 482(c), E. R. Degginger; 482(r), E. R. Degginger; 482(l), HRW Photo by Lisa Davis; 483(t), Culver Pictures, Inc.; 483(b), David Parker/Science Photo Library/Custom Medical Stock Photos; 487, TASS/Sovfoto/Eastfoto; 489(t), E. R. Degginger; 489(b), David Parker/Science Photo Library/Photo Researchers; 490(l), E. R. Degginger/Bruce Coleman, Inc.; 490(r), SIU/Peter Arnold, Inc.; 491, Dan McCoy/Rainbow; 492(c), Simon Fraser/Medical Physics, RVI, Newcastle-Upon-Tyne/Science Photo Library/Photo Researchers; 492(l), Peticolas/Megna/Fundamental Photographs; 492(r), Simon Fraser/Medical Physics, RVI, Newcastle-Upon-Tyne/Science Photo Library/Photo Researchers; 493, Debra Page, courtesy of CURRENT SCIENCE Magazine; 495, NASA; 498, SEUL/Science Source/Photo Researchers; 499, NASA/Science Source/Photo Researchers; 500, Argonne National Laboratory; 501, Dr. Robert Gale/Sygma; 503, Los Alamos National Laboratory; 506, Dylan Kibler/courtesy Mel Fisher Maritime Heritage Society; 507(cr), Ralph Oberlander, Jr./Stock, Boston; 507 (bl, br), courtesy Mel Fisher Maritime Heritage Society; 507(bc), William Curtsinger/Photo Researchers, Inc.; 508, Photri; 509(t), Richard Nelridge/Phototake; 509(b), Richard A. Nelridge/Phototake; 510(t), Photoworld/FPG; 510(b), UPI/Bettmann Newsphotos; 511(l), Gary C. Bublitz/DOW Chemical Co.; 511(br), printout courtesy of Dr. Leaphart, DOW Chemical Co.; 512,

Will McIntyre/Photo Researchers, Inc., 513, Science Photo Library/Photo Researchers, Inc.; 514-515, Whit Bronaugh; 515, Rivera Collection SuperStock; 517, HRW Photo by John Langford; 519(tr), Ronnie Kaufman/The Stock Market; 520(all), Yoav Levy/Phototake; 521(l), Michael Keller/FPG; 521(r), Park Street; 523, Todd Merritt Haiman/The Stock Market; 525, courtesy of Jack Brown Cleaners of Austin; HRW Photo by Michelle Bridwell; 526(br), Mark E. Gibson/The Stock Market; 526(tl), HBC Photo by Rodney Jones; 526(bl), E. R. Degginger; 528(both), E. R. Degginger; 529, E. R. Degginger; 530, Simon Fraser/Science Photo Library/Photo Researchers, Inc.; 532(r,l), Yoav Levy/Phototake; 532(bl), O.A. Battista Research Institute; 533(both), HBC Photo by Richard Haynes; 534(both), HBC Photo by Richard Haynes; 535(b), Tim Thompson/AllStock; 535(t), Kip & Pat Peticolas/Fundamental Photographs; 537, Rivera Collection/SuperStock; 539, Van Bucher/Photo Researchers, Inc.; 540(t), Ken Eward/Science Source/Photo Researchers, Inc.; 540(b), Chemical Design/Science Photo Library/Photo Researchers; 542, Bob LeRoy/ProFiles West; 546(r), Loctite Corporation; 546(tl,bl), Ken Eward/Science Source/Photo Researchers, Inc.; 548, E. R. Degginger; 550(t), Tony Freeman/PhotoEdit; 550(c), J. Robbins/The Stockhouse; 550(b), Park Street; 551, David Young-Wolff/PhotoEdit; 552(tl), Burlington/Visuals Unlimited; 552(bl,br,tr), Park Street; 553, Chip Clark; 555, E. R. Degginger; 556(t), Chip Clark; 556(c), AP/Wide World Photos; 556(b), Tom Tracy/The Stock Shop; 557(t), E. R. Degginger; 557(b), Chip Clark; 558, Michael Melford/The Image Bank; 562, Ken Eward/Science Source/Photo Researchers; 563(r), Sea Studios, Inc./Peter Arnold, Inc.; 563(l), Chip Clark; 564(bkgrd), Craig Aurness/Westlight; 564, Laurence Pringle; 564(inset), David R. Frazier Photolibrary; 565, Bill Weedmark/Panographics; 567, HBC Photo by Richard Haynes; 569(tl), HBC Photo by Richard Haynes; 569(tr,c), E. R. Degginger; 570(r), Cary Wolinsky/Stock, Boston; 570(tl), Ned Haines/Photo Researchers, Inc.; 570(l), Martin Rogers/Stock, Boston; 571(tc), Eric Anderson/Stock, Boston; 571(br), The Corning Museum of Glass, Corning, NY; 571(tl), Ted Horowitz/The Stock Market; 571(tr), Brownie Harris/The Stock Market; 572(tl), Artstreet; 572(bl,tr), E. R. Degginger; 573, E. R. Degginger; 575, Gabe Palmer/Kane/After Image/Tony Stone Images; 578, Grapes/Michaud/

ILLUSTRATIONS